W9-DAG-330

About Island Press

Island Press is the only nonprofit organization in the United States whose principal purpose is the publication of books on environmental issues and natural resource management. We provide solutions-oriented information to professionals, public officials, business and community leaders, and concerned citizens who are shaping responses to environmental problems.

In 2001, Island Press celebrates its seventeenth anniversary as the leading provider of timely and practical books that take a multidisciplinary approach to critical environmental concerns. Our growing list of titles reflects our commitment to bringing the best of an expanding body of literature to the environmental community throughout North America and the world.

Support for Island Press is provided by The Bullitt Foundation, The Mary Flagler Cary Charitable Trust, The Nathan Cummings Foundation, Geraldine R. Dodge Foundation, Doris Duke Charitable Foundation, The Charles Engelhard Foundation, The Ford Foundation, The George Gund Foundation, The Vira I. Heinz Endowment, The William and Flora Hewlett Foundation, W. Alton Jones Foundation, The John D. and Catherine T. MacArthur Foundation, The Andrew W. Mellon Foundation, The Charles Stewart Mott Foundation, The Curtis and Edith Munson Foundation, National Fish and Wildlife Foundation, The New-Land Foundation, Oak Foundation, The Overbrook Foundation, The David and Lucile Packard Foundation, The Pew Charitable Trusts, Rockefeller Brothers Fund, The Winslow Foundation, and other generous donors.

Large Mammal Restoration

Large Mammal Restoration

Ecological and Sociological Challenges in the 21st Century

Edited by
David S. Maehr
Reed F. Noss
Jeffery L. Larkin

ISLAND PRESS
Washington ◆ Covelo ◆ London

Library of Congress Cataloging-in-Publication Data
Large mammal restoration : ecological and sociological considerations on the 21st century / edited by David S. Maehr, Reed F. Noss, and Jeffery L. Larkin ; foreword by John F. Eisenberg.
     p.   cm.
Includes bibliographical references.
   ISBN 1–55963–816–8 (hardcover : alk. paper) — ISBN 1–55963–817–6 (pbk. : alk. paper)
   1. Mammals—Reintroduction—United States. I. Maehr, David S., 1955–
II. Noss, Reed F. III. Larkin, Jeffery L.
   QL717 .L37   2001
   333.95'4153'0973—dc21   2001004424

British Cataloguing-in-Publication Data available.

Contents

Foreword

In the late sixties I had the privilege of working in Sri Lanka. Since the Pleistocene extinctions, this island nation had been able to maintain its fauna and flora in a relatively intact state, in spite of various foreign occupations. The conservation of the elephant on this island, the size of West Virginia, had been a concern since the high cultures of A.D. 1153 and perhaps before. Preserves with nonexploited cores surrounded by zones of varying intensities of human use, wildlife corridors, and translocations of elephant herds were and are topics of concern for wildlife officials on Sri Lanka. I was impressed with what had been accomplished and what has continued by the efforts of biologists during the last 30 years all in spite of a destructive civil war and a doubling of the human population (Jayewardene 1994). Public education, a respect for life, and dedicated wildlife professionals have combined to preserve the island's unique ecosystems. It should be an inspiration to all of us.

During the last 50 years in North America there have been noticeable shifts in human land-use patterns. The decline in the numbers of small family farms over much of the United States, but especially in the eastern regions, New England, and Appalachia, has led to an extension of forested tracts through natural succession with or without human assistance. The shift away from dry land farming in the northern Great Plains has been accelerated as the cost of irrigation has risen with conflicting demand for water use. Recent shifts in continental rainfall patterns and the concerns of global warming have prompted a reevaluation of human land-use patterns, especially agriculture. The net result has been the increase in available land for occupancy by displaced plants and animals. Some of this colonization has come about without human intervention.

Restocking ungulates has had a long history. Poland has worked with ungulates of the Bialowieza Forest for more than 200 years. South Africa began restocking programs more than 80 years ago (Harper 1945). It is only in recent years, however, that large predator reintroductions have been actively pursued and the goals of restoring a functional ecosystem have been approached. Restoring the integrity of an ecosystem is obviously a major task, but it is a worthy one. It is also a task where the predator component becomes pivotal.

One can encourage reoccupation of suitable range with the use of habitat corridors from a source, or one can release translocated animals. The latter course of action is fraught with potential problems, as eastern elk researchers have learned. The inadvertent introduction of pathogens is dealt with in two chapters of this volume. Capture and release is one thing, but release of captive-bred stock presents special problems, especially in the case of predators, and in this volume the case of the Mexican wolf is fairly treated as a test program.

The predator component cannot be introduced until the natural prey base is restored. This volume discusses the problems and prospects of the planning, execution, and restoration of functional predator-prey systems. The focus is on North America and the larger herbivores and carnivores that once widely occupied the continent. The authors are drawn from a wide spectrum of disciplines and present a balanced review of the problems and prospects associated with the undertaking. The preservation of biological diversity is a worthy aim. The expertise, public approval, and opportunities now exist to restore many of North America's large mammals, especially in light of the successes in much more challenging arenas such as Sri Lanka.

This book derives from a symposium held at the Wildlife Society annual meeting in 1999—the content reflects the enthusiasm and expertise of the participants. There is much new material deriving from ongoing research that is now made available through this publication to a wider audience of students, professionals, and the interested public. This is an important and timely contribution.

JOHN F. EISENBERG
Eminent Scholar, Emeritus
University of Florida, Gainesville

Literature Cited

Harper, F. 1945. *Extinct and Vanishing Mammals of the Old World.* Special Publication 12. New York: American Committee for International Wildlife Protection, New York Zoological Park.

Jayewardene, J. 1994. *The Elephant in Sri Lanka.* Colombo, Sri Lanka: Mortlake Press.

Acknowledgments

This book is an outgrowth of a daylong symposium held at The Wildlife Society (TWS) annual meeting in 1999. What began as a simple effort to bring together leaders in large mammal conservation grew into this diverse collection of viewpoints. Without the support of TWS in putting on the symposium—and the incredible attendance—we would likely have ended our efforts in Austin. As it turned out, we enjoyed SRO audiences for most of the day. Even Dr. Harris's stimulating lecture, which extended into the dinner hour, maintained everyone's rapt attention.

We thank the University of Kentucky, College of Agriculture, for encouraging us to pursue this timely book. Early drafts of chapters were reviewed by P. Barrows, D. Forrester, T. Hoctor, P. Jackson, D. Shindle, D. Smith, W. Tzilkowski, and M. Vaughan—whose comments were quite helpful. Our editors, B. Dean and B. Youngblood, helped us muddle through the production gauntlet and, as always, assured us that we were no worse than most of their distracted academic contributors.

Introduction: Why Restore Large Mammals?

REED F. NOSS

Imagine a world where all the animals are smaller than a breadbox. No more sounds of crashing through the underbrush as you round the bend in the trail. No more steaming mounds of scat. No more rush of adrenaline as you confront a beast far bigger, stronger, and single-minded than you. I have never come face to face with a large carnivore, but a bull moose (*Alces alces*) in rut standing in the trail before me on Isle Royale a few years ago was humbling enough. Losing large mammals, however, is more than a matter of diminished human experience. Without large beasts, many ecosystems would be quite different from what they are today. This is not idle speculation. We know more than a few places that have lost most or all of their megafauna, and these places are changing in insidious ways.

Are we really losing large mammals? Two decades ago, Michael Soulé suggested that the evolution of large organisms is coming to an end (Soulé 1980; Frankel and Soulé 1981). The argument goes like this: Nature is becoming an archipelago of small, isolated natural areas in a sea of habitat modified and degraded by humans. Although evolutionary biologists and paleontologists have shown that isolation promotes speciation, this is not always the fate of populations confined to such areas. Large organisms require considerable space. In tiny islands of habitat, their populations are too small to have much chance of persisting long enough to speciate. Transporting organisms from reserve to reserve will not help, Soulé argues: it will only counteract local adaptation and divergence. Over the years, Soulé has refined his argument to concentrate on large mammals. The diversity of large mammals today, he suggests, may be the greatest we will see for many millennia.

I do not accept Soulé's scenario completely. If we consider extinctions of mammalian species since 1500—virtually all of them human-caused—it turns out that small-bodied species dominate the list (MacPhee and Flemming 1999). This finding stands in striking contrast to the extinction spasm at the close of the Pleistocene, also caused by humans, when megafauna such as mammoths (*Mammuthus* spp.), ground sloths (*Glossotherium harlani*, for example), and dire wolves (*Canis dirus*) were disproportionately affected. Only 4.5 percent of the mammalian extinctions since 1500 were of ungulates and 2.3 percent of carnivores—whereas 52.3 percent of the recently extinct species are rodents, 12.5 percent are insectivores, and 10.2 percent are bats (MacPhee and Flemming 1999). Small mammals, especially on islands, have fared much worse than large ones. Moreover, mammals in general seem to be doing better today than many other groups of animals. According to data from The Nature Conservancy, 16 percent of the mammal species in the United States are currently at risk of extinction. This is an appallingly high number, but it pales when compared to the 69 percent of freshwater mussels or 51 percent of crayfishes considered at risk (Master et al. 2000).

The evolution of large mammals is not entirely at an end. Many species in the deer family (*Cervidae*), for example, do all too well in humanized landscapes so long as human hunting is regulated. Our first conservation actions on this continent were laws protecting deer from overharvest (Noss and Cooperrider 1994). Similarly, as reviewed in case 1 and in chapter 11, the first wildlife translocations of deer and elk in the late nineteenth century were designed to restore populations depleted by hunting. Today, however, deer populations in many areas could stand considerably more hunting pressure—or natural predation—than they presently receive. Across North America, but best documented in eastern deciduous forests, the elevated density of deer (especially white-tailed deer, *Odocoileus virginianus*) in the absence of large carnivores has resulted in severe impacts from browsing and grazing on plant regeneration, abundance, and distribution, threatening some rare plant species. (See Hough 1965; Frelich and Lorimer 1985; Alverson et al. 1988; Strole and Anderson 1992; Miller et al. 1992; Augustine and Frelich 1998.) Changes in flora and habitat structure caused by intensive browsing by deer affect other animals in the community, too—for example, shrub-nesting birds and butterflies that forage on understory herbs (Diamond 1992; McShea and Rappole 1997). High densities of deer also are responsible for increased incidence of zoonotic diseases affecting a number of species, including humans (Wilson and Childs 1997).

Nevertheless, we should not be too optimistic about the fate of large mammals. The world never has been so dominated by human activity, never so fragmented as today. Many large mammals, especially carnivores, do not thrive like deer in human-dominated landscapes. In myriad places the elimination of large carnivores appears to have triggered the irruption of herbivores and the cascade of ecological impacts I just described. Several large herbivores, too—bison (*Bison bison*), elk (*Cervus elaphus*), moose, caribou (*Rangifer tarandus*), pronghorn (*Antilocapra americana*), and bighorn (*Ovis canadensis*)—have been extirpated from much of their previous distributions in North America. And in some cases we know that the ecosystems they inhabited have changed as a consequence of their loss.

It is not just the total loss of certain taxa that should concern us. Drastic declines in ecologically pivotal species have repercussions across an ecosystem long before the species become fully extinct. Ecological extinction precedes demographic extinction. Restoring populations of large mammals to ecologically functional levels is a good idea, therefore, wherever and whenever it can be accomplished with ecological, social, and ethical good sense. Beyond this, as discussed in chapter 16, the recovery of large animals should be pursued within the broader context of restoring entire regional biotas and functioning ecosystems. This introductory chapter explores some of the reasons for restoring large mammals—as well as some of the complications. Other chapters present specific case studies of such efforts. I believe we have an ethical obligation to restore native species we have eliminated. And this obligation is strongest for ecologically pivotal species such as many large mammals.

Ecological Reasons

What are the ecological reasons for restoring large mammals? I used the term "ecologically pivotal species." By this I mean something similar to keystone species—defined as species whose impacts on community structure are disproportionately large for their abundance (Power et al. 1996). Many ecologists exclude dominant species, which in some ecosystems include large herbivores such as bison or caribou, from the keystone category because their impacts, although large, are not necessarily disproportionately so. The concept of ecologically pivotal species includes all species that have a determining effect on the structure, function, or composition of an ecosystem. Removal, or even decline, of such species will have a pronounced effect on ecosystem parameters.

Among the large herbivores in North America, the bison is especially well documented as a significant ecological force. Many studies (summarized by Knapp et al. 1999) have found that bison increase the spatial heterogeneity of grassland ecosystems. By grazing predominantly on grasses, bison allow a greater diversity of forbs to coexist. Essentially bison select species-poor sites dominated by grasses and convert them to species-rich sites (Hartnett et al. 1996; Collins et al. 1998). The wallowing behavior of bison promotes additional heterogeneity—for example, by creating depressions that collect rainwater in the spring and support plant species adapted to ephemeral wetlands. The nitrogen-rich urine of bison produces a fertilization effect. By grazing preferentially in urine-treated patches of grass, bison contribute further to the patchiness of the landscape. Retention of nitrogen in grasslands is enhanced by bison activity, counteracting fire, which is the major pathway of nitrogen loss. Importantly, bison generally concentrate in recently burned watersheds, which helps bring nitrogen back into these systems. Their grazing also reduces nitrogen losses by lowering levels of aboveground detritus and increasing the patchiness of fires. Finally, bison carcasses produce localized nutrient pulses that exceed all others in North American grasslands (Knapp et al. 1999).

Imagine, then, the ecological effects of reducing the bison herds on the Great Plains from an estimated 30–60 million individuals in the eighteenth century to a few thousand by 1880. Although bison populations have rebounded somewhat, today they are localized. The bison as a significant ecological agent at the scale of a biome is a thing of the past. Or so it would seem. Perhaps we can bring the bison back and help restore the biodiversity of the Great Plains and tallgrass prairie—ecosystems truly unique to North America. The Buffalo Commons idea (Callenbach 2000) is not radical: it makes economic as well as ecological sense. The only thing standing in the way of such proposals is a multitude of small minds. For other species, restoration will be more difficult or even impossible. We have lost forever our largest herbivores, whose prodigious consumption of woody vegetation may have played a critical role in the development of the North American grassland biome at the close of the Pleistocene. The expansion of bison populations during the Holocene may have depended on the prior actions of the mega-browsers (Axelrod 1985; Hartnett et al. 1997).

The Pleistocene megaherbivores of North America, which included mammoths, mastodonts (*Mammut americanum*), glyptodonts (*Glyptotherium floridanum*), camels (*Camelops hesternus*), horses (*Equus* spp.), ground sloths, and other extraordinary beasts, were extinguished at least in part at the hand of humans (Martin and Klein 1984). These creatures and their predators— saber-toothed cats (*Smilodon fatalis*), American lions (*Panthera leo atrox*), American cheetahs (*Miracinonyx trumani*), giant short-faced bears (*Arctodus simus*), dire wolves, and others—cannot be dreamed back. But the premier predator, *Homo sapiens*, is still here. And some members of our species are now wishing we had not been so ruthless 13,000 years ago. Paul Martin and David Burney (1999) have proposed, in perfect seriousness, that African and Asian elephants (*Loxodonta africana* and *Elephas maximus*, respectively) be introduced to western North America and allowed to evolve here to fill the ecological roles presently vacant. They write (p. 58) that

> the gomphotheres, a family of Neotropical elephants that prospered in the Americas for well over ten million years . . . vanished at the end of the Pleistocene around 13,000 years ago, along with mammoths and mastodons. All deserve consideration as a natural part of Wild America. With such a rich fossil record and such a late American extinction, it is natural to consider restarting New World evolution of the Proboscidea with whatever taxa of elephants are left.

Martin and Burney (1999) and Janzen and Martin (1982) have pointed out that such trees as mesquite (*Prosopis*), honey locust (*Gleditsia*), and monkey ear (*Enterolobium*) developed their sweet pods in coevolution with large herbivores that served as seed dispersers. In most cases today, these trees have no reliable dispersers. Studies of megaherbivores on continents where they still exist show strong keystone roles of these species (Owen-Smith 1988). In fact, Owen-Smith (1989) concludes that the elimination of megaherbivores from the Americas, northern Eurasia, and Australia caused a major reduction in habitat mosaic diversity and forage quality—in turn precipitating the extinction of lesser large-mammal species. "Beyond Pleistocene parks," Martin and Burney (1999:64) write, "we need Pleistocene proving grounds, places to fathom as well as to celebrate our lost wildness. Above all the time has come to consider restarting elephant evolution by enabling elephants to reinvent their ecology on the continent that once constituted an important part of their global range." Their proposal is almost convincing.

Large mammalian herbivores in many other parts of the world play ecologically pivotal roles. In Africa elephants are well known for their importance in dispersing and favoring the germination of seeds and as regulators of succession on a landscape scale (Lieberman et al. 1987; Owen-Smith 1988, 1989). Hippopotamus (*Hippopotamus amphibius*) strongly modify the physical environment and hydraulic conditions of river floodplains in southern Africa, especially through their daytime wallowing in pools and their nocturnal movements to and from foraging areas. Their movements create a maze of trails and canals that, among other functions, serve as corridors for the movements of other animals (Naiman and Rogers 1997). Wallowing by Cape buffalo (*Synceros caffer*), elephant, and other large mammals increases the size of natural depressions in savanna landscapes. Larger depressions retain larger volumes of water during a prolonged dry season, hence benefiting other species (Naiman and Rogers 1997).

Taking a step up the trophic ladder, let us consider the mammalian carnivores. Popular wisdom has it that big predators control the numbers and distribution of their prey. Ecologists are not always so sure. The argument over top-down versus bottom-up regulation of ecological communities is one of the longest-running controversies in academic ecology. Hairston et al. (1960) have proposed the general hypothesis that "the world is green" because predators keep herbivores in check, thus allowing green biomass to accumulate. Other ecologists have seen the role of predators, especially large ones, differently. Because of their lower abundance and their scale-dependent physiological processes, such ecologists argue, these animals are less significant within biological communities than smaller organisms (Peters 1983). The regulatory role of predators remains controversial among ecologists. In his introductory ecology textbook, Colinvaux (1973) disputes the idea that predators are the "main controlling agents" of prey populations and has branded Leopold's (1943) Kaibab Plateau story a "fiction; a history of embroidered data." Empirical evidence of predators regulating prey populations and exerting top-down control of community structure has been sketchy. The best evidence of such effects comes from marine ecosystems (Paine 1966, 1980; Estes et al. 1978, 1989) and freshwater ecosystems (Carpenter et al. 1985; Carpenter and Kitchell 1988; Power 1990). Demonstrations of top-down effects in these ecosystems have been followed by demonstrations of bottom-up effects. (See Hunter and Price 1992 for an overview.) The controversy between top-down and bottom-up regulation has been most intense with respect to terrestrial ecosystems. (See, for example, Erlinge et al. 1984, 1988; Terborgh 1988; Hunter and Price 1992; Power 1992; Strong 1992; Wright et al. 1994; Estes 1995, 1996; Polis and Strong 1996; Kay 1998; Pace et al. 1999; Terborgh et al. 1999.)

Scientific support for the idea that large mammalian predators are capable, at least at times, of controlling the distribution or abundance of ungulates and other prey—and by so doing enhance the overall biodiversity or ecological integrity of the landscape—is mounting. (See the review by Terborgh et al. 1999.) It is well established, for example, that mainland populations of barren-ground caribou across the range of the species in the Old and New Worlds are limited primarily by wolf (*Canis lupus*) predation. The rate of increase (r) of caribou introduced to predator-free environments averages 0.27 ± 0.18, which is close to the theoretical maximum for the species. In contrast, lightly hunted herds coexisting with wolves and other natural predators have an average rate of increase of −0.009 (Bergerud 1988). On Isle Royale, Michigan, growth rates of balsam fir (*Abies balsamea*) are regulated by moose density, which in turn is controlled by wolf predation. This top-down regulation, however, can shift suddenly (if only temporarily) to bottom-up regulation when stand-replacing disturbances occur at times of low moose density; at such times, vegetation growth escapes control (McLaren and Peterson 1994). In a similar ecosystem at Algonquin Park, Ontario, wolves do not appear to control herbivore numbers at any time; hence, bottom-up control predominates (Forbes and Theberge 1994, 1996). In the Neotropics, jaguars (*Panthera onca*) appear to exert top-down control by taking a wide range of prey in proportion to their abundance, thus providing some balance in abundance among prey species (Terborgh 1988; Miller and Rabinowitz in press).

Less direct but nonetheless compelling evidence of the importance of mammalian predators in maintaining ecological integrity comes from the documented superabundance of ungulates and medium-sized mammals (including several "opportunistic mesopredators") in landscapes from which large carnivores have been eliminated (Wilcove 1999). Particularly convincing are comparisons between predator-inhabited and predator-free sites. For example, although small mammals (which are not prey of large predators) are about equally abundant on Barro Colorado Island in Panama (isolated by flooding associated with construction of the Panama Canal) and comparable mainland sites, such medium-sized mammals as agouti (*Dasyprocta* spp.), paca (*Agouti paca*), armadillo (*Cabassous centralis, Dasypus novemcinctus*), and coati (*Nasua narica*) are more than an order of magnitude more abundant on the island, which lacks large predators. The effects of these differences in the mammal community are apparent throughout the food web. (See Terborgh and Winter 1980; Terborgh 1988, 1992; Glanz 1990; Janson and Emmons 1990; Wright et al. 1994; Terborgh et al. 1999.) In Spain, the Iberian lynx (*Felis pardina*), although it preys to some extent on rabbits (*Oryctolagus cuniculus*), controls populations of such smaller predators as Egyptian mongoose (*Herpestes ichneumon*) and by so doing promotes higher densities of rabbits (Palomares et al. 1995). In southern California, absence of coyotes (*Canis latrans*) from fragmented patches of scrub has resulted in proliferation of domestic cats (*Felis sylvestris catus*), opossum (*Didelphis marsupialis*), raccoon (*Procyon lotor*), and gray fox (*Urocyon cinereoargenteus*) in these patches, in turn resulting in declines and local extinctions of scrub-breeding birds (Soulé et al. 1988; Crooks and Soulé 1999). Wilcove (1985) found that predation rates on songbird nests were higher in woodlots surrounded by suburbs than in woodlots in agricultural land, apparently because of the higher densities of mesopredators in suburbs.

A recent study by Joel Berger and colleagues (2001) presents empirical evidence of top-down regulation and sheds more light on the question of secondary effects of predator removal. Berger et al. show that the local extinction of the wolf and grizzly bear (*Ursus arctos*) in the southern Greater Yellowstone Ecosystem triggered a cascade of events—chiefly an irruption of moose, a substantial alteration of riparian vegetation by their herbivory, and a reduction in populations of songbirds that nest in riparian willows. These effects were not observed in nearby, carefully paired areas open to human hunting, where moose densities were depressed. Berger et al. pose three restoration options: do nothing and let biodiversity continue to erode in riparian areas; replace carnivore predation by human predation throughout the ecosystem; or allow natural recovery of the ecosystem by encouraging the dispersal of grizzly bears and wolves into previously occupied habitat. These options are politically charged, of course, and may be mutually exclusive. For example, hunting on national forest land—and the road access that allows it—likely will have a negative effect on recovery of carnivores. Direct killing by humans is often the greatest source of mortality for grizzly bears and wolves in the Rocky Mountains and elsewhere. (See Mattson et al. 1996; Peek and Carnes 1996; chapters 1, 6, 8, and 10 in this volume.) These carnivores will have the best chance of recovery in areas with low road and trail density where hunting is prohibited.

In their review of the role of large carnivores in regulating terrestrial ecosystems, Terborgh et al. (1999:58) concluded:

> Our current knowledge about the natural processes that maintain biodiversity suggests a crucial and irreplaceable regulatory role of top predators. The absence of top predators appears to lead inexorably to ecosystem simplification accompanied by a rush of extinctions. Therefore, efforts to conserve North American biodiversity in interconnected mega-reserves will have to place a high priority on reestablishing top predators wherever they have been locally extirpated.

Evolutionary Reasons

Although I will not dwell on this issue here, predation as a selective force on prey populations is a critical long-term concern. Without large carnivores to prey on them, are prey species turning into wimps? In the presence of predators, herbivores typically behave as time-minimizers: they are secretive, spend most of their time in secure places, and minimize the time spent feeding and exposed to predators. In the absence of predators, these same herbivore species may become energy-maximizers: they spend much more time feeding than hiding and, as a result, increase their fecundity (Terborgh et al. 1999). Over a long period of time in the absence of predators, evolutionary theory predicts that prey species will lose their defensive behavior. Observations of prey species on islands lacking predators confirm this prediction: the animals appear "tame." It is not unreasonable to suggest that the wholesale removal of large predators as a selective force across much of this continent is altering the course of evolution.

When reintroduction of predator or prey species to formerly occupied regions is proposed, evolutionary, genetic, and selective issues again become important. Translocations of wild-born animals are more successful than those involving captive-reared individuals. (See Griffith et al. 1989 and chapters 6 and 8 in this volume.) Perhaps the most obvious recommendation for genetic management is that released animals must show some genetic diversity among them. Furthermore, assuming that natural selection allows animals to adapt to local environments, animals selected for reintroduction should be similar genetically to those that originally inhabited the release site (Miller et al. 1999). Because present-day subspecies frameworks often do not reflect the distribution of genetic variation within species (Ryder 1986; Avise 1989), molecular genetic data should be used to determine appropriate genetic subdivisions. Nevertheless, it appears that large carnivores, many of which have amazing dispersal capacities, often show low levels of differentiation, even among widely separated populations, whereas small carnivores show genetic differences that reflect dispersal barriers (Mercure et al. 1993). Not all large carnivores are good dispersers, however, so each species has to be considered individually (Miller et al. 1999).

Beyond the details of translocation planning, many of which are discussed in the chapters ahead, it is clear that the presently limited and fragmented distribution of many large herbivores and most large carnivores has already resulted in diminished genetic variation and evolutionary potential. Although I doubt we will see the end to evolution for all these species, we have certainly limited their possibilities. Restoring populations of these species brings some of those possibilities back.

Human Reasons

As I suggested at the beginning of this essay, there is an element of excitement in a landscape where large mammals still roam. Aesthetic and spiritual reasons figure prominently in arguments to restore populations of wild creatures. (See, for example, chapter 4 in this volume.) When I am hiking in grizzly bear country, for instance, my adrenaline levels are somewhat higher than when I'm sitting in front of my computer putting together a budget for a grant proposal. I notice everything around me—including every tree I might scamper up should Old Scarface pop up from behind a boulder. This state of mind is different from fear. It is more like the attitude one achieves in Zen meditation: aware, yet calm and unhurried. Breathing comes easily. There is no hint of fatigue and very little distraction. You hold out your hand and it's as steady as the mountain in the distance. Could this be the original harmony with nature that modern humanity has lost?

Knowing that we are not all-powerful, that there are animals around us bigger and meaner than we are, and that how we conduct ourselves is of some consequence is an enlightening experience. It is the kind of experience that would do people in our overcivilized society a world of good. Of course, one can be humbled by other phenomena in nature: thunderstorms, fires, tornadoes, hurricanes, volcanic eruptions, and earthquakes—not to mention diseases. All teach the receptive observer that we are not entirely in control of our destiny. Nevertheless, something about a potential encounter with a large, hairy, and possibly violent creature really catches our attention. I believe that the feeling of humility one acquires in country inhabited by wild beasts carries over to other contexts. It makes us better persons. It also makes us acutely aware of the emptiness of landscapes that now lack their native megafauna.

Biologists should not be too quick to dismiss the aesthetic, emotional, spiritual, and ethical arguments for protecting and restoring populations of large mammals—or indeed for protecting nature in general. Scientific facts and technical literature do not move people to change their behavior. Virtually everyone, including scientists, is motivated by emotion and sensory experience. Emotion is complementary to science, not opposed to it. Our aesthetic experiences of nature spur us to care about wild creatures, which in turn makes us want to learn all we can about them and to conserve them. And when these creatures have been extinguished from a landscape, our emotions correctly tell us that bringing them back is the *right* thing to do.

Challenges

Beyond their trophic role in promoting the integrity of natural ecosystems, large mammals have a place in conservation strategy that few other species could duplicate. This is especially true for the large carnivores. Aldo Leopold, among others, believed that carnivores present the critical test of a society's commitment to conservation (Meine 1988). They are among the most challenging of all species to conserve. Wide-ranging and dependent on adequate populations of prey, many large carnivores require immense areas of mostly undeveloped land—land that many humans would like to use for economic purposes. For the most part, large carnivores do not intrinsically "need" wilderness; many are quite tolerant of humans. Humans, however, often do not tolerate them. (See Noss et al. 1996 and many chapters in this volume.) These animals prey on many of the same species sought by human hunters and occasionally take livestock and pets. Rarely, some of them attack people. For these reasons humans have persecuted carnivores relentlessly. With a few exceptions, only in relatively remote wildlands do these animals find sufficient security from people to thrive.

Species with very low resilience to disturbance, such as grizzly bear and wolverine (*Gulo gulo*), are particularly in need of large refugia from humans (Weaver et al. 1996). Some large-carnivore populations, however, display remarkable adaptability to human-altered landscapes. David Mladenoff (pers. comm.) points out: "In the Lake States, the wolf has recovered in a highly altered ecosystem . . . in part because of high human [habitat] alteration, producing too many deer, and education of the human population to stop killing wolves." Wolves in western North America, however, appear to be ecologically less plastic and may be genetically distinct, perhaps a separate species, from wolves in the east (P. Paquet, pers. comm.). Moreover, an analysis of ten species of large carnivores from four continents shows that conflict with people on reserve borders is the major cause of mortality—in fact, border areas represent population sinks (Woodroffe and Ginsberg 1998).

Public opinion about carnivores is changing. Many people now see carnivores as beautiful, inspiring creatures. They want them to persist in the wildlands around their homes and to be returned to areas where they once roamed. Rather than following the path of convenience, conservationists should rise to meet the challenge of carnivore conservation (Noss 1996).

Despite all the arguments in favor, restoring populations of large mammals to regions where they have been extirpated will seldom be easy. As pointed out by Larkin and coauthors (chapter 5 in this volume), Murphy's Law operates vigorously in translocation programs. Innumerable practical and ethical challenges confront us (Bekoff 1999). Many elements of our society—most notably farmers and ranchers—generally oppose reintroduction of large herbivores or, especially, carnivores. This is partly an economic issue, and on a local scale the concerns are sometimes legitimate. In other cases the concerns seem to reflect little else but ignorance and prejudice. One would think that conservationists would have more unity of opinion for restoring extirpated animals to the places they call home. Yet in some regions—for example, the Pacific Northwest and California—even conservation activists are split over the issue of reintroducing large carnivores. The "G-word," in particular, scares many of them to death—they find hiking considerably more serene without grizzly bears. As discussed by Maehr and coauthors (chapter 15 in this volume), local opposition to reintroduction of Florida panthers (*Puma concolor coryi*) to northern Florida has overwhelmed broad public support and caused state and federal agencies to shift from a holistic, landscape-oriented recovery program to an approach emphasizing intensive, symptom-oriented genetic management.

Conservation biologists accept that if a species was once native to an area, it should be reintroduced if habitat conditions remain favorable (or can be restored) and reintroduction is technically feasible. The issue of what is native can be problematic, however. Most biologists agree that if a species got to a region under its own powers of dispersal, it is native. If it was introduced by humans, it is nonnative. But what if the species was native in the Pleistocene and subsequently went extinct? Does it matter whether the cause of extinction was human hunting or something else? Earlier I noted the case of the Pleistocene megafauna and Martin and Burney's (1999) proposal to bring back elephants to North America. This is a temporally broader concept of native than probably most biologists would accept. What about wild horses and burros? Horses were native to North America in the Pleistocene and earlier, but they have been missing for the past 13,000 years. In the dispute over the elimination of burros (*Equus asinus*) from Grand Canyon National Park, some scientists have argued that burros should be considered the ecological equivalent of the native equids that existed there in the Pleistocene. Other scientists disagree: North American equids were related to African progenitors of the burro only at the subgeneric level, they point out, and plant communities in the Grand Canyon today differ markedly from those in the Pleistocene (Houston et al. 1994).

Similar arguments have erupted over management of apparently exotic species in other national parks. I recently chaired a panel of scientists evaluating the ecological impact and management of mountain goats (*Oreamnos americanus*) in Olympic National Park. One of the questions put forth by opponents of mountain goat control—who in fact have sued the National Park Service—is whether the mountain goat might in fact be native to the Olympic peninsula. Virtually everyone agrees that the present population of goats is derived from individuals introduced from British Columbia and Alaska in the 1920s. Some paleontologists, however, have suggested that mountain goats may have been native to the Olympics during the late Pleistocene and early Holocene (Lyman 1998). No evidence—neither fossils nor archaeological artifacts—exists to support this claim. Moreover, the mountain goat is missing from the oral history of the Indians native to the peninsula. Nevertheless, opponents of mountain goat control cite the problem of "negative evidence" and put the burden of proof on the National Park Service. Our review (Noss et al. 2000) concluded that the hypothetical presence of mountain goats on the Olympic peninsula thousands of years ago is irrelevant to the National Park Service's exotics policy. Given current policy, only if the goats were driven to extinction by Europeans would a strong case exist for their reintroduction.

Finally, the way in which we reintroduce a population can determine the success of the reintroduction, both biologically and socially. Griffith et al. (1989) reviewed the status of translocations as a conservation tool and cite several factors that contribute to success. The chapters in this book elaborate on these factors. Griffith and colleagues concluded that native game species are more likely to be reintroduced successfully than are threatened and endangered species. Translocations into core habitat of the species' historical range are more successful than those into peripheral habitat. Herbivores fare better than carnivores or omnivores. Translocations are less successful when potential competitors are present. And, finally, large founder populations are more successful up to a point—the relationship is asymptotic, with little increase in success rate above 80 to 120 individuals for birds and only 20 to 40 individuals for game mammals (Griffith et al. 1989). All observers agree that translocations must be based on sound research. As Griffith et al. (1989:479) point out: "Without high habitat quality, translocations have low chances of success regardless of how many organisms are released or how well they are prepared for the release. Active management is required. Limiting factors must be identified and controlled and assurances of maintenance of habitat quality obtained prior to translocation."

As illustrated by the introduction of plains bison to Wood Buffalo National Park, discussed by Carbyn and Watson (chapter 9 in this volume), translocations often run afoul when political concerns are allowed to override the advice of biologists. The plains bison translocation took place back in the 1920s, but politically driven mistakes still happen today. A classic case of how not to conduct a translocation is provided by the recent and ongoing reintroduction of lynx (*Lynx canadensis*) to southwestern Colorado. In this instance a hasty and poorly planned translocation was prompted by the pending listing of the lynx under the federal Endangered Species Act. As reasoned by John Seidel, who directed the reintroduction program for the Colorado Division of Wildlife (CDOW) in its early stages, reintroduction might preempt listing under the act and thus prevent restrictions on land use (Bekoff 1999).

A few native lynx may persist in Colorado. The last confirmed sighting was in 1973 near an area in which Vail Associates plans to construct four ski lifts, 12.3 miles of roads, and a 20,000-square-foot restaurant (Hansen 1999a; Morson 1999). Vail Associates gave $200,000 to CDOW for its lynx reintroduction program—a generous gift that carried with it a request for a quid pro quo in which the state would release Vail Associates from all further obligations for preserving and protecting lynx habitat in the Vail ski area expansion zone (Hansen 1999b). Although CDOW denies agreeing to these terms, they accepted the gift and the ski area expansion proceeded (Hansen 1999a, 1999b).

The story gets more sordid still. The U.S. Fish and Wildlife Service (USFWS), at both the regional and D.C. offices, was involved in this scheme as well. The Vail expansion zone was classified by the service as "critical habitat" for the lynx (Hansen 1999a). Biologists with the service's Colorado office, however, reported that superiors ordered that lynx habitat not be classified as "in jeopardy" under any circumstances (Hansen 1999a; Morson 1999). Meanwhile the reintroduction into southwestern Colorado was rushing ahead without sufficient background research or feasibility studies such as habitat modeling or intensive censuses of snowshoe hare (*Lepus americanus*), the lynx's primary prey. Biologists Rich Reading and Brian Miller, both with the Denver Zoo and experts on carnivore reintroduction, were asked by CDOW to review the reintroduction proposal. According to Miller (pers. comm., 13 May 1999): "For the record, they took very few of our suggestions. DOW pushed the process way too fast. They should not have released animals this year. They needed a careful feasibility study examining all the biological and social variables, but they chose shortcuts." Miller and other biologists also suspect that hare densities were overestimated—a suspicion borne out by the starvation deaths of 17 of the 41 released lynx in 1999 (eight other animals are missing) (Bekoff 2000).

Besides creating a public relations fiasco for CDOW and USFWS, the lynx story in Colorado generated considerable political opposition to reintroduction programs across the country. Although high mortality of released animals is not unusual in translocations of carnivores (Griffith et al. 1989; Miller et al. 1999), in this case the loss of individuals might well have been reduced by thorough feasibility studies, including intensive hare surveys, prior to releases. Perhaps the southern Rocky Mountains is not the best place to attempt to restore a lynx population. The reintroduction site in the San Juan Mountains is at the southern end of the species' historical range and is isolated from other lynx populations. Experience confirms that translocations to peripheral areas are more likely to fail than those in central portions of a species' range (Griffith et al. 1989). Such translocations are all the more risky considering the high probability that global warming will make habitat conditions less favorable in the near future (Scott et al. 1999).

Marc Bekoff, a noted animal behaviorist at the University of Colorado, has been outspokenly critical of the lynx reintroduction—largely out of concern for the suffering of the starving animals (Bekoff 1999, 2000). In response to Bekoff's criticism, John Seidel wrote a letter to the president of the University of Colorado threatening to withhold CDOW grant money and personal funds from the university unless Bekoff was silenced (Hansen 1999b). Such behavior on the part of bureaucrats in charge of reintroduction programs does little to inspire public support for such projects. Flippant comments made by state biologists with regard to animal suffering are not helpful either. As biologists interested in populations, species, and ecosystems, we sometimes become callous to the suffering and loss of individuals and are quick to dismiss the concerns of the animal rights crowd. I believe we do so at our peril. Living things at all levels of organization are valuable, and as human beings as well as scientists, we have a responsibility to reduce unnecessary suffering.

The lynx story may yet have a happy ending. Although it is too early to conclude much, mortality rates of released animals did decline as the reintroduction project progressed. According to reports by CDOW and USFWS in July 2000 (Environmental News Service, 4 July 2000, unpublished), the program is now proceeding better than expected and is ahead of schedule. Steve Buskirk of the University of Wyoming, an authority on forest carnivores, concludes: "The preparation and learning over the past year have really been a great investment that has really paid off. This is clearly starting out as a successful reintroduction. . . . It's highly encouraging" (Environmental News Service, 4 July 2000, unpublished). I hope Buskirk is right and the reintroduction becomes a success—even in the face of global warming. But let us learn from the mistakes made in the early phases of this project so that we do not repeat them elsewhere.

Many good reasons exist for restoring populations of large mammals to areas they once inhabited. By so doing, we help native ecosystems regain their diversity and wholeness. We provide the public, including future generations, a more complete natural heritage. We have learned much about the consequences of removing large mammals from ecosystems and quite a bit about the factors that influence reintroduction success. The chapters in this book offer many valuable lessons for translocation programs elsewhere. If we can only learn to keep politics from subverting biology, we will have come a long way indeed.

Literature Cited

Alverson, W. S., D. M. Waller, and S. L. Solheim. 1988. Forests too deer: Edge effects in northern Wisconsin. *Conservation Biology* 2:348–358.

Augustine, D. J., and L. E. Frelich. 1998. Effects of white-tailed deer on populations of an understory forb in fragmented deciduous forests. *Conservation Biology* 12:995–1004.

Avise, J. C. 1989. The role for molecular genetics in the recognition and conservation of endangered species. *Trends in Ecology and Evolution* 4:279–281.

Axelrod, D. I. 1985. Rise of the grassland biome, central North America. *Botanical Review* 51:163–201.

Bekoff, M. 1999. Jinxed lynx? Some very difficult questions with few simple answers. *Endangered Species Update* 16:26–27.

———. 2000. Redecorating nature: Reflections on science, holism, community, humility, reconciliation, spirit, compassion, and love. *Human Ecology Review* 7:59–67.

Berger, J., P. B. Stacey, L. Bellis, and M. P. Johnson. 2001. A mammalian predator-prey disequilibrium: How the extinction of grizzly bears and wolves affects the biodiversity of avian neotropical migrants. *Ecological Applications* 11.

Bergerud, A. T. 1988. Caribou, wolves and man. *Trends in Ecology and Evolution* 3:68–72.

Callenbach, E. 2000. *Bring Back the Buffalo! A Sustainable Future for America's Great Plains.* Berkeley: University of California Press.

Carpenter, S. R., and J. F. Kitchell. 1988. Consumer control of lake productivity. *BioScience* 38:764–769.

Carpenter, S. R., J. F. Kitchell, and J. R. Hodgson. 1985. Cascading trophic interactions and lake productivity. *BioScience* 35:634–639.

Colinvaux, P. A. 1973. *Introduction to Ecology.* New York: Wiley.

Collins, S. L., A. K. Knapp, J. M. Briggs, J. M. Blair, and E. M. Steinauer. 1998. Modulation of diversity by grazing and mowing in native tallgrass prairie. *Science* 280:745–747.

Crooks, K. R., and M. E. Soulé. 1999. Mesopredator release and avifaunal extinctions in a fragmented system. *Nature* 400:563–566.

Diamond, J. 1992. Must we shoot deer to save nature? *Natural History* 8(92):2–8.

Erlinge, S., G. Göransson, G. Högstedt, G. Jansson, O. Liberg, J. Loman, I. N. Nilsson, T. von Schantz, and M. Sylvén. 1984. Can vertebrate predators regulate their prey? *American Naturalist* 123:125–133.

———. 1988. More thoughts on vertebrate predator regulation of prey. *American Naturalist* 132:148–154.

Estes, J. A. 1995. Top-level carnivores and ecosystem effects: Questions and approaches. In C. G. Jones and J. H. Lawton, eds., *Linking Species and Ecosystems.* New York: Chapman & Hall.

———. 1996. Predators and ecosystem management. *Wildlife Society Bulletin* 24:390–396.

Estes, J. A., N. S. Smith, and J. F. Palmisano. 1978. Sea otter predation and community organization in the western Aleutian Islands, Alaska. *Ecology* 59:822–833.

Estes, J. A., D. O. Duggins, and G. B. Rathbun. 1989. The ecology of extinctions in kelp forest communities. *Conservation Biology* 3:252–264.

Forbes, G. J., and J. B. Theberge. 1994. Multiple landscape scales and winter distribution of moose (*Alces alces*) in a forest ecotone. *Canadian Field Naturalist* 107:201–207.

———. 1996. Cross-boundary management of Algonquin Park wolves. *Conservation Biology* 10:1091–1097.

Frankel, O. H., and M. E. Soulé. 1981. *Conservation and Evolution.* Cambridge: Cambridge University Press.

Frelich, L. E., and C. G. Lorimer. 1985. Current and predicted long-term effects of deer browsing in hemlock forests in Michigan, USA. *Biological Conservation* 34:99–120.

Glanz, W. E. 1990. Neotropical mammal densities: How unusual is the community of Barro Colorado Island, Panama? In A. H. Gentry, ed., *Four Neotropical Forests.* New Haven: Yale University Press.

Griffith, B., J. M. Scott, J. W. Carpenter, and C. Reed. 1989. Translocation as a species conservation tool: Status and strategy. *Science* 477–480.

Hairston, N. G., F. E. Smith, and L. B. Slobodkin. 1960. Community structure, population control, and competition. *American Naturalist* 94:421–424.

Hansen, B. 1999a. Man arrested in protest: Activists promise to put selves on line in summer of protests against Vail expansion. *Colorado Daily*, 15 June 1999.

———. 1999b. Science under fire: DOW official pressured CU to put limits on outspoken prof. *Colorado Daily*, 16 July 1999.

Hartnett, D. C., K. R. Hickman, and L. E. Fischer-Walter. 1996. Effects of bison grazing, fire, and topography on floristic diversity in tallgrass prairie. *Journal of Range Management* 49:413–420.

Hartnett, D. C., A. A. Steuter, and K. R. Hickman. 1997. Comparative ecology of native versus introduced ungulates. In F. Knopf and F. Samson, eds., *Ecology and Conservation of Great Plains Vertebrates.* New York: Springer-Verlag.

Hough, A. F. 1965. A twenty-year record of understory vegetation change in a virgin Pennsylvania forest. *Ecology* 46: 370–373.

Houston, D. B., E. G. Schreiner, and B. B. Moorhead. 1994. Mountain goats in Olympic National Park: Biology and management of an introduced species. USDI

National Park Service Scientific Monograph NPS/NROLYM/NRSM-94/25. Natural Resources Publication Office, Denver.

Hunter, M. D., and P. W. Price. 1992. Playing chutes and ladders: Heterogeneity and the relative roles of bottom-up and top-down forces in natural communities. *Ecology* 73:724–732.

Janson, C. H., and L. H. Emmons. 1990. Ecological structure of the nonflying mammal community at Cocha Cashu Biological Station, Manu National Park, Peru. In A. H. Gentry, ed., *Four Neotropical Forests*. New Haven: Yale University Press.

Janzen, D. H., and P. S. Martin. 1982. Neotropical anachronisms: The fruits the gomphotheres ate. *Science* 215:19–27.

Kay, C. E. 1998. Are ecosystems structured from the top-down or bottom-up: A new look at an old debate. *Wildlife Society Bulletin* 26:484–498.

Knapp, A. K., J. M. Blair, J. M. Briggs, S. L. Collins, D. C. Hartnett, L. C. Johnson, and E. G. Towne. 1999. The keystone role of bison in North American tallgrass prairie. *BioScience* 49:39–50.

Leopold, A. 1943. *Deer Irruptions*. Wisconsin Conservation Bulletin, August. Reprinted in Wisconsin Conservation Department Publication 321:1–11.

Lieberman, D., M. Lieberman, and C. Martin. 1987. Notes on seeds in elephant dung from Bio National Park, Ghana. *Biotropica* 19:365–369.

Lyman, R. L. 1998. *White Goats, White Lies: The Abuse of Science in Olympic National Park*. Salt Lake City: University of Utah Press.

MacPhee, R. D. E., and C. Flemming. 1999. Requiem Æternam: The last five hundred years of mammalian species extinctions. In R. D. E. MacPhee, ed., *Extinctions in Near Time: Causes, Contexts, and Consequences*. New York: Kluwer Academic/Plenum.

Martin, P. S., and R. Klein, eds. 1984. *Quaternary Extinctions: A Prehistoric Revolution*. Tucson: University of Arizona Press.

Martin, P. S., and D. A. Burney. 1999. Bring back the elephants! *Wild Earth* 9(1):57–64.

Master, L. L., B. A. Stein, L. S. Kutner, and G. A. Hammerson. 2000. Vanishing assets: Conservation status of U.S. species. In B. A. Stein, L. S. Kutner, and J. S. Adams, eds., *Precious Heritage: The Status of Biodiversity in the United States*. New York: Oxford University Press.

Mattson, D. J., S. Herrero, R. G. Wright, and C. M. Pease. 1996. Designing and managing protected areas for grizzly bears: How much is enough? In R. G. Wright, ed., *National Parks and Protected Areas: Their Role in Environmental Protection*. Cambridge, Mass.: Blackwell Science.

McLaren, B. E., and R. O. Peterson. 1994. Wolves, moose, and tree rings on Isle Royale. *Science* 266:1555–1558.

McShea, W. J., and J. H. Rappole. 1997. Herbivores and the ecology of forest understory birds. In W. J. McShea, H. B. Underwood, and J. H. Rappole, eds., *The Science of Overabundance*. Washington, D.C.: Smithsonian Institution Press.

Meine, C. 1988. *Aldo Leopold: His Life and Work*. Madison: University of Wisconsin Press.

Mercure, A., K. Rallas, K. P. Koepfli, and R. K. Wayne. 1993. Genetic subdivisions among small canids: Mitochondrial DNA differentiation of swift, kit, and arctic foxes. *Evolution* 47:1313–1328.

Miller, B., K. Ralls, R. P. Reading, J. M. Scott, and J. Estes. 1999. Biological and technical considerations of carnivore translocations: A review. *Animal Conservation* 2:59–68.

Miller, B., and A. Rabinowitz. In press. Why conserve jaguars? In R. A. Medellin, C. Chetkiewicz, A. Rabinowitz, K. H. Redford, J. G. Robinson, E. Sanderson, and A. Taber, eds., *El jaguar en el nuevo milenio: Una evaluación de su estado, detección deprioridades y recomendaciónes para la conservación de los jaguares en America*. Mexico City: Universidad Nacional Autonoma de México/Wildlife Conservation Society.

Miller, S. G., S. P. Bratton, and J. Hadidian. 1992. Impacts of white-tailed deer on endangered and threatened vascular plants. *Natural Areas Journal* 12: 67–74.

Morson, B. 1999. U.S. biologists say agency shows bias against lynx. *Denver Rocky Mountain News*, 19 May 1999.

Naiman, R. J., and K. H. Rogers. 1997. Large animals and system-level characteristics in river corridors. *BioScience* 47:521–529.

Noss, R. F. 1996. Conservation or convenience? *Conservation Biology* 10:921–922.

Noss, R. F., and A. Y. Cooperrider. 1994. *Saving Nature's Legacy*. Washington, D.C.: Island Press.

Noss, R. F., H. B. Quigley, M. G. Hornocker, T. Merrill, and P. C. Paquet. 1996. Conservation biology and carnivore conservation in the Rocky Mountains. *Conservation Biology* 10:949–963.

Noss, R. F., R. Graham, D. R. McCullough, F. L. Ramsey, J. Seavey, C. Whitlock, and M. P. Williams. 2000. Review of scientific material relevant to the occurrence, ecosystem role, and tested management options for mountain goats in Olympic National Park. Fulfillment of Contract 14-01-0001-99-C-05, U.S. Department of Interior, Washington, D.C.

Owen-Smith, R. 1988. *Megaherbivores: The Influence of Very Large Body Size on Ecology*. Cambridge: Cambridge University Press.

———. 1989. Megafaunal extinctions: The conservation message from 11,000 years B.P. *Conservation Biology* 3:405–412.

Pace, M. L., J. J. Cole, S. R. Carpenter, and J. F. Kitchell. 1999. Trophic cascades revealed in diverse ecosystems. *Trends in Ecology and Evolution* 14:483–488.

Paine, R. T. 1966. Food web complexity and species diversity. *American Naturalist* 100:65–75.

————. 1980. Food webs: Linkage, interaction strength and community infrastructure. *Journal of Animal Ecology* 49:667–685.

Palomares, F., P. Gaona, P. Ferreras, and M. Delibes. 1995. Positive effects on game species of top predators by controlling smaller predator populations: An example with lynx, mongoose, and rabbits. *Conservation Biology* 9:295–305.

Peek, J. M., and J. C. Carnes. 1996. Wolf restoration in the northern Rocky Mountains. In R. G. Wright, ed., *National Parks and Protected Areas: Their Role in Environmental Protection.* Cambridge, Mass.: Blackwell Science.

Peters, R. H. 1983. *The Ecological Implications of Body Size.* Cambridge: Cambridge University Press.

Polis, G. A., and D. R. Strong. 1996. Food web complexity and community dynamics. *American Naturalist* 147:813–846.

Power, M. E. 1990. Effects of fish in river food webs. *Science* 250:811–814.

————. 1992. Top-down and bottom-up forces in food webs: Do plants have primacy? *Ecology* 73:733–746.

Power, M. E., D. Tilman, J. A. Estes, B. A. Menge, W. J. Bond, L. S. Mills, G. Daily, J. C. Castilla, J. Lubchenco, and R. T. Paine. 1996. Challenges in the quest for keystones. *BioScience* 46:609–620.

Ryder, O. A. 1986. Species conservation and systematics: The dilemma of subspecies. *Trends in Ecology and Evolution* 1:9–10.

Scott, J. M., D. Murray, and B. Griffith. 1999. Lynx reintroduction. *Science* 286:49–50.

Soulé, M. E. 1980. Thresholds for survival: Maintaining fitness and evolutionary potential. In M. E. Soulé and B. A. Wilcox, eds., *Conservation Biology: An Evolutionary-Ecological Perspective.* Sunderland, Mass.: Sinauer.

Soulé, M. E., D. T. Bolger, A. C. Alberts, R. Sauvajot, J. Wright, M. Sorice, and S. Hill. 1988. Reconstructed dynamics of rapid extinctions of chaparral-requiring birds in urban habitat islands. *Conservation Biology* 2:75–92.

Strole, T. A., and R. C. Anderson. 1992. White-tailed deer browsing: Species preferences and implications for central Illinois forests. *Natural Areas Journal* 12:139–144.

Strong, D. R. 1992. Are trophic cascades all wet? Differentiation and donor-control in speciose ecosystems. *Ecology* 73:747–754.

Terborgh, J. 1988. The big things that run the world—a sequel to E.O. Wilson. *Conservation Biology* 2:402–403.

————. 1992. Maintenance of diversity in tropical forests. *Biotropica* 24:283–292.

Terborgh, J., and B. Winter. 1980. Some causes of extinction. In M. E. Soulé and B. A. Wilcox, eds., *Conservation Biology: An Evolutionary-Ecological Perspective.* Sunderland, Mass.: Sinauer.

Terborgh, J., J. A. Estes, P. Paquet, K. Ralls, K. Boyd-Heger, B. J. Miller, and R. F.

Noss. 1999. The role of top carnivores in regulating terrestrial ecosystems. In M. E. Soulé and J. Terborgh, eds., *Continental Conservation: Scientific Foundations of Regional Reserve Networks.* Washington, D.C.: Island Press.

Weaver, J. L., P. C. Paquet, and L. F. Ruggiero. 1996. Resilience and conservation of large carnivores in the Rocky Mountains. *Conservation Biology* 10:964–976.

Wilcove, D. S. 1985. Nest predation in forest tracts and the decline of migratory songbirds. *Ecology* 66:1211–1214.

———. 1999. *The Condor's Shadow: The Loss and Recovery of Wildlife in America.* New York: Freeman.

Wilson, M. L., and J. E. Childs. 1997. Vertebrate abundance and the epidemiology of zoonotic diseases. In W. J. McShea, H. B. Underwood, and J. H. Rappole, eds., *The Science of Overabundance.* Washington, D.C.: Smithsonian.

Woodroffe, R., and J. R. Ginsberg. 1998. Edge effects and the extinction of populations inside protected areas. *Science* 280:2126–2128.

Wright, S. J., M. E. Gompper, and B. de Leon. 1994. Are large predators keystone species in neotropical forests? The evidence from Barro Colorado Island. *Oikos* 71:279–294.

PART I

Feasibility

To speak correctly, it is too late in the day to write about the "conservation" of American big game. The word conservation itself is no longer a happy choice as applied to wild life. It suited well enough while we still had a sufficient stock of large mammals for our needs, recreational or commercial. But now the term conservation seems more suitable when associated with the economy of soil, minerals and oil, the non-renewable resources. It has much too narrow a meaning to convey to the public a proper appreciation of what ought to be done if we are to restore useful or beautiful species, control overpopulation of game and epidemics of disease, and preserve wilderness areas where the ordinary traveller may see wild game in an unspoiled environment or one as little spoiled as possible. It is so obvious that game cannot be restored or even maintained nowadays by merely passive, restrictive measures, such as the word conservation implies, that we now use the terms "game management" or "restoration." A whole new school of thought based on scientific management is gradually being built up.... Conditions change rapidly. The history of reduction or extirpation of many species and subspecies is a long one and often well documented, though the sources of information are widely scattered and difficult to come by.

John C. Phillips, *North American Big Game* (Scribner's, 1939)

The specter of disappearing large mammals was recognized by John C. Phillips and others nearly a century ago. Restoration of big game had begun well before the 1939 publication of his *North American Big Game*, but it appeared to be a relatively unenlightened process. Until the middle of the twentieth century, large mammal restoration was as simple as moving Minnesota black bears to Arkansas, moving white-tailed deer from one state to another, or moving elk throughout North America. Such translocations were just part of a common practice that not only attempted to reestablish a variety of taxa to vacant range but included the introduction of foreign wildlife species. This was an era when the homogenization of faunas (and myriad other problems caused by exotic species) was simply not a concern. Unenlightened wildlife management was more concerned with increasing the opportunities for an avid hunting public. Thus it should not be surprising to learn that Texas supports a diverse array of African antelope, New Zealand has red deer, and Scandinavia has a North American beaver population.

With respect to large mammal restoration, the twenty-first century will be distinguished from the past by the detailed planning that precedes the first capture and relocation. Not all of the stories in this section are optimistic views of repatriated carnivores roaming the landscape. Where the prey base or public perception creates unsuitable ecological and social conditions, restoration planning must be suspended and new efforts initiated. Without such planning, however, agencies would waste huge sums in personnel costs, equipment, and public trust. In this sense, a restoration effort that fails may be worse than one that is never attempted at all. Whereas wolves, wolverines, cougars, elk, and grizzly bears might have a shot at restoration in the expansive and relatively unpopulated sections of the Pacific Northwest (Carroll et al. in chapter 1) and Desert Southwest (Dugelby et al. in chapter 3), upstate New York offers an entirely different restoration potential (Paquet et al. in chapter 2). Although elk have been returned to several eastern locations, many of these efforts have failed for unknown reasons. In chapter 4, McClafferty and Parkhurst offer a blueprint that will enable the state of Virginia to avoid many of the pitfalls experienced by others who have rushed headlong into a vision of elk in the East.

All four of these chapters build on the best available information concerning the historical distribution of extirpated fauna by examining the ecological requirements of their target species and exploring the geographic possibilities for their return. Two of the chapters address the restoration of entire large mammal communities—thus pushing the envelope of traditional restoration practice. All are blueprints for those considering the return of an extirpated large mammal. All should be viewed as the first step in a long and complicated process.

Chapter 1

Is the Return of the Wolf, Wolverine, and Grizzly Bear to Oregon and California Biologically Feasible?

CARLOS CARROLL, REED F. NOSS,
NATHAN H. SCHUMAKER, AND PAUL C. PAQUET

Carnivores are indicators of ecosystem function and can serve as keystones in the top-down regulation of ecosystems (Terborgh et al. 1999). Although the strength of top-down processes varies widely among species and ecosystems (Noss et al. 1996), it is probably more prevalent than many ecologists have assumed (Terborgh et al. 1999; Crooks and Soulé 1999). Wide-ranging carnivores may serve as "bioassays" of emergent landscape characteristics such as connectivity and give us information on the optimal size and arrangement of reserves. Viability analysis of carnivore species may highlight potential reserve areas that are not targeted in other biodiversity assessments such as gap analysis (Scott et al. 1993).

The restoration of mammalian carnivore species to portions of their former range, either by restoration of habitat or through active reintroduction, presents new challenges. Besides the inevitable sociopolitical difficulties, large and medium-sized carnivores may be particularly sensitive to landscape configuration because of their low population densities and large area requirements. In these species, population processes operate on a regional scale. Thus, regional-scale habitat models can be useful management tools for prioritizing restoration efforts. Multispecies conservation strategies have many advantages over single-species strategies (Noss et al. 1997). The extent to which restoration efforts for a particular species will enhance viability of the broader carnivore guild can be assessed by considering the major factors—such as topography, forest structure, and risk of human-induced mortality—that limit their distribution and abundance.

We used predictive habitat models to develop a carnivore restoration strategy for Oregon and northern California for three species: the gray wolf (*Canis lupus*), grizzly bear (*Ursus arctos*), and wolverine (*Gulo gulo*)—species currently extirpated from most or all of the region. Natural range expansion from adjacent states or existing refugia may be possible for at least the wolf and wolverine, and reintroduction programs have been proposed for all three species. Knowledge of the current amount and configuration of habitat can help us to identify core habitat areas and dispersal routes for these species and predict the viability of restored populations.

The three carnivore species considered here differ in the degree to which they tolerate human-associated landscape change and direct persecution. Contrasts in behavior, demographic characteristics, and population or metapopulation structure result in differing levels of ecological resilience (Weaver et al. 1996). All three species avoid humans. The wolf is highly resilient demographically, but its social structure increases the area requirements for viable populations. In mountainous portions of the western United States, the wolf may be especially vulnerable because its avoidance of rugged terrain brings it into greater proximity to human settlements (Paquet et al. 1996). The wolverine has the lowest reproductive output despite its relatively small size. The grizzly bear's limited dispersal abilities make it the most vulnerable species at the metapopulation level (Weaver et al. 1996).

The three species differ as well in the ecological roles they have played in Pacific coastal ecosystems and the potential impact of their restoration on current ecosystem dynamics. Bears may be an important link between riparian and upland systems—especially in regions, such as the Pacific coast, with anadromous fish populations. Nutrient input from salmon may be a key factor in productivity of coastal forest ecosystems (Bilby et al. 1996). The grizzly bear in south-central Alaska redistributed 40 percent of the salmon entering a coastal stream and was the conduit for 20 percent of the nitrogen uptake in adjacent forests (Hilderbrand et al. 1999). Storer and Tevis (1955:17) comment: "Its numbers multiplied by its average daily metabolic requirement must have made the grizzly an outstanding factor in the total food consumption by mammals . . . [and] a dominant element in the original native biota of California." The historic ecological role of the wolf in the Pacific Northwest is unknown. Wolf reintroduction appears to have strongly affected vertebrate communities in the Greater Yellowstone Ecosystem (Smith et al. 1999), and we might expect similar effects in portions of the Pacific states. The ecological influence of wolverine populations in the lower 48 states is almost entirely unknown.

The historical factors leading to range contraction and extirpation differ among the three species, but all three were affected by predator control programs during the late 1800s and early 1900s (Schullery and Whittlesey 1999). To the extent that today's land management agencies are more tolerant of predators, the absence of these species from much of their former range is not an inevitable consequence of current human population density and land use. In eastern Europe, Italy, and China, populations of large carnivores coexist with much higher levels of human population density (Mattson 1990). In this chapter we present an approach for evaluating the biological feasibility of restoring populations of wolf, grizzly bear, and wolverine to areas within their former range in Oregon and California. We adapt habitat suitability models developed for these species in the Rocky Mountains, show potential core areas in a combined study region, and estimate potential population size.

Wolf

The historical distribution and abundance of the wolf in the Pacific coastal states is uncertain. In California, wolves were probably most abundant on the northern coast, where elk (*Cervus elaphus*) were abundant, and in the northeastern corner of the state where they were found until 1922 (Grinnell et al. 1937; Schmidt 1991). Early extirpation from northern coastal California may have been due to human settlement patterns, including the gold rush of the 1850s, that briefly made the area one of the most densely populated in the western United States. Wolves were historically common in western Oregon (Bailey 1936), as well as east of the Cascades Range (Young and Goldman 1964). Most museum specimens were collected from the western foothills of the Cascades; the last wolf bounty in Oregon was awarded there in 1946 (Verts and Carraway 1998). Wolves reportedly persisted in the Oregon Cascades even after they were extirpated from the Rocky Mountain region (Young and Goldman 1964). While only scattered wolf reports exist from the latter half of the twentieth century, wolves have recently been documented dispersing into Oregon from the rapidly growing Idaho population (see chapter 6 in this volume).

Generally wolves locate their home ranges in areas where adequate prey are available and human interference is low (Mladenoff et al. 1995). The primary limiting factor for wolves has not been habitat degradation or prey depletion but direct persecution through hunting, trapping, and predator control programs. As human tolerance of large predators increases, however, wolves are well equipped to recolonize remaining areas of their former range. Because wolves reach sexual maturity at an early age and have large litters, the wolf has a high level of ecological resilience compared with other large carnivores (Weaver et al. 1996). The species' flexible social structure allows pack size, fecundity, and dispersal to respond to shifts in population density and prey abundance (Fuller 1989; Boyd et al. 1995; Weaver et al. 1996). Nonetheless, wolves were eliminated in areas of the western United States where grizzly bears persisted, suggesting that these compensatory mechanisms have limits. Population densities of wolves are usually far lower than population densities of sympatric grizzly bears. And as social animals, wolves are more susceptible to predator control than solitary animals.

Human activities affect wolf distribution and wolf survival (Thiel 1985; Fuller et al. 1992; Mladenoff et al. 1995; Paquet 1993; Paquet et al. 1996). In Wisconsin (Mladenoff et al. 1995) and Minnesota (Fuller et al. 1992), wolves selected areas with low human population density. The absence of wolves in human-dominated areas may reflect high levels of human-caused mortality, displacement resulting from behavioral avoidance, or some combination of both (Fuller et al. 1992; Mech and Goyal 1993). Roads, by increasing human access, negatively affect wolf populations at local, landscape, and regional scales (Fuller 1989; Thurber et al. 1994; Mladenoff et al. 1995). Wolves may avoid densely roaded areas because of traffic volume (Thurber et al. 1994), or their absence may be a direct result of mortality associated with roads (Van Ballenberghe et al. 1975). Even in areas where wolf harvest is prohibited, 80 to 95 percent of mortality is often anthropogenic (Fuller 1989; Mech 1989; Paquet 1993; Pletscher et al. 1997). Wolves in mountainous regions often concentrate their activities in forest valleys where snow conditions and prey availability are optimal (Paquet 1993; Paquet et al. 1996; Singleton 1995). Topography has not been incorporated in previous models (Mladenoff et al. 1995) developed in the north-central United States, due to the flatter terrain typical of that region.

Ungulates such as elk, deer (*Odocoileus virginianus* and *O. hemionus*), moose (*Alces alces*), and bighorn sheep (*Ovis canadensis*) make up the bulk of the wolf diet (Mech 1970; Fuller 1989), although they may take smaller prey such as snowshoe hare (*Lepus americanus*) and beaver (*Castor canadensis*). Ungulate biomass (Keith 1983; Fuller 1989), density, and species diversity (Corsi et al. 1999) are important habitat factors. In a review of studies from several regions, for example, prey density explained 72 percent of the variation in wolf density (Fuller 1989). A smaller core area can support a viable wolf population if prey biomass per unit area is high (Fritts and Carbyn 1995; Wydeven et al. 1995).

Grizzly Bear

The current distribution of the grizzly bear in the Pacific coastal states is confined to a small remnant population in the North Cascades of Washington, where 17 records have been confirmed in recent decades (Almack et al. 1993). Distribution was widespread in Oregon except in the arid east (Bailey 1936). The last grizzly bear in the state was killed in northeastern Oregon in 1931 (Verts and Carraway 1998). As many as 10,000 grizzly bears occurred throughout California except in the northeast and the Mojave Desert (Storer and Tevis 1955). Although the California grizzly may have been most common in chaparral rather than dense forest, it was also abundant in oak woodlands, river valleys, and mixed-hardwood forests (Storer and Tevis 1955). In the 1850s, a settler counted 40 bears feeding along the Mattole River in coastal northwestern California (Grinnell et al. 1937). Despite its initial abundance, the species was extirpated from northern California by 1902. The last known grizzly in the state was killed in the southern Sierra Nevada in 1924 (Storer and Tevis 1955).

The grizzly bear has a combination of life history traits that contribute to its low resilience in the face of human encroachment (Bunnell and Tait 1981). Its low lifetime reproductive potential (as few as three female young per adult female) makes population viability sensitive to small declines in adult survivorship (Weaver et al. 1996). Although subadult males often disperse two home-range diameters (about 70 km), successful long-distance dispersal between subpopulations has not been recorded in the western United States.

The range of the grizzly has become increasingly fragmented (Craighead and Vyse 1996), exacerbating the demographic and genetic risks associated with small, isolated populations. Craighead and Vyse (1996) have compared the viability of bear populations on islands of varying size and conclude that isolated populations require at least 1000 bears for long-term persistence. Mattson and Reid (1991) found a similar size threshold for European brown bear populations. Roads and their traffic cause direct mortality, disrupt bear behavior, create barriers to movement (Archibald et al. 1987; McLellan and Shackleton 1988), and increase poaching and removal of habituated bears (Mattson et al. 1987; Weaver et al. 1996).

Although the grizzly is an omnivore, its resilience is limited by its seasonally high caloric needs (Weaver et al. 1996). The species was widespread in western ecosystems, and its diet reflects this distribution. Key foods range from soft mast (drupes, berries) and hard mast (acorns, whitebark pine nuts) to fish, vertebrate carrion, and insects (Storer and Tevis 1955; Mattson and Reid 1991).

Wolverine

Because wolverines exist at low densities and inhabit remote areas, it is difficult to judge whether the Pacific Northwest supports reproducing populations or just dispersing individuals. In California, the last confirmed specimens were collected from the Sierra Nevada in the early 1900s (Grinnell et al. 1937). More recent unconfirmed reports originate from the southern Sierra Nevada (Barrett et al. 1994). In Oregon, specimens were confirmed in the Blue Mountains of eastern Oregon in 1986 and 1992 and from Steens Mountain in southeastern Oregon in 1973 (Verts and Carraway 1998). Specimens were collected from the central Oregon Cascades in 1965, 1969, and 1973 (Verts and Carraway 1998). The 1969 specimen was a female, suggesting the possibility of reproduction. It seems clear, however, based on the sparse evidence, that wolverines are scarce in Oregon and California relative to the Rocky Mountains. Although little is known about the wolverine's habitat needs and distribution in the Pacific Northwest, fragmentation of the landscape by roads and human development may hinder natural recovery there as in other regions (Carroll et al. 2001).

Because the wolverine's diet includes unpredictably distributed resources such as carrion, it has larger home-range requirements than equivalent-sized carnivores. Carrion use may link wolverines to other carnivores, such as wolves, that are much reduced in the western United States. Moreover, wolf poisoning campaigns have eradicated local wolverine populations as well in some regions (Banci 1994).

Female wolverines mature at three years of age and produce less than one kit per year until death at six to eight years (Copeland 1996). Large area requirements and low reproductive rates make the wolverine especially vulnerable to human-induced mortality and habitat alteration. Populations probably cannot sustain annual rates of human-induced mortality greater than 7 or 8 percent, a rate lower than that usually caused by trapping (Gardner 1985; Banci 1994; Weaver et al. 1996). Areas closed to trapping such as Yellowstone National Park and the Canadian mountain parks in Alberta and British Columbia appear to be wolverine refugia (Hatler 1989; Buskirk 1999).

Although the wolverine's long-range dispersal abilities (greater than 200 km in Idaho; Copeland 1996) may facilitate its persistence, females tend to settle closer to their place of birth (Banci 1994). The large home-range sizes of Idaho wolverine (a mean of 384 km^2 in females) relative to those in Canada and Alaska suggest more limited food or denning resources (Copeland 1996). Female wolverines must leave their kits for lengthy foraging trips. In the lower 48 states, they often select natal dens in alpine areas where snow tunnels in talus can provide thermoregulatory benefits and safety from predators (Magoun and Copeland 1998). Natal dens in Alaska, by contrast, appear less limited by topography or human settlement (Magoun and Copeland 1998).

Habitat Models

Using habitat quality to predict carnivore distribution is especially challenging because there is much to learn about the link between habitat and demography. Habitat models such as the habitat suitability index (HSI) system were developed primarily for site-level planning and may be poor templates for regional evaluations. Such conceptual models rely on qualitative relationships derived from expert opinion and published studies. Empirical models, in contrast, base predictions on statistical analyses of species occurrence data. Regional-scale empirical models have been used to predict range expansion of wolves in the north-central and northeastern United States (Mladenoff et al. 1995, 1999), as well as grizzly bear distribution in Idaho and Montana (Merrill et al. 1999; Mace et al. 1999; Boyce and MacDonald 1999).

For most carnivores, regional species occurrence data, primarily sighting and trapping records, are not systematically collected. If the biases are accounted for and the results checked against independent data, models built from occurrence data may be more reliable than conceptual models. When telemetry information is available for part of the region, mesoscale empirical models developed from these data may be adapted to predict distribution at the regional scale—although the reliability of the original data is offset by our ignorance of emergent regional-scale factors. When demographic data are available, spatial variation in survival and fecundity can be estimated and the results used to predict the distribution of regional sources and sinks—although such demographic data are rare.

Given the diversity of ecosystem types found within our analysis area, habitat models must be broad in order to be accurate across the region. Models that might have high predictive power in the east may extrapolate poorly across the western United States. We used three approaches to predict habitat quality for wolf, grizzly bear, and wolverine: sighting and trapping records were used to develop an empirical model for the wolverine; grizzly bear habitat was predicted by adapting previously published models (Merrill et al. 1999; Mace et al. 1999); wolf habitat was predicted with a conceptual model in which individual elements were derived from statistical analysis of telemetry data (P. Paquet, unpublished data). All models used habitat data at a resolution of 1 km². Although our habitat maps are static models, they can form the basis for dynamic models that predict whether a patch of suitable habitat is large and connected enough to remain occupied by a species.

Wolf

We mapped predicted habitat value for gray wolf by combining an empirical model relating wolf distribution to topography (Paquet et al. 1996) with variables suggested by other studies. The three primary model components were prey density, prey accessibility, and security from human disturbances.

PREY DENSITY. Although ungulate population data are available for some areas (such as California and Idaho), comparable data are not available for the entire region. In order to build a seamless map of ungulate abundance, we developed a linear regression model using deer abundance data and forage availability based on "tasseled-cap greenness," a transformation of Landsat Thematic Mapper satellite imagery (Crist and Cicone 1984). We compared the regression results with deer harvest data from Oregon and ungulate abundance estimates from Idaho.

The deer regression model was developed from the average 1990–1996 deer population in northern and central California (CDFG 1998). Although a linear regression using data from all seven units was significant ($p = 0.02$; $R^2 = 0.69$; DF = 5), we chose a robust regression model that excluded the unit with highest deer density to avoid overestimating prey abundance ($p < 0.01$; $R^2 = 0.85$; DF = 4). The regression equation was deer/km² = 2.2789 + 0.0533 × greenness.

Deer harvest data for Oregon were correlated with greenness ($p < 0.01$; $R^2 = 0.4088$; DF = 65). Although elk population estimates were not available on a statewide level for California, elk abundance is low compared with deer. Oregon elk harvest was not correlated with greenness ($p > 0.10$). In Idaho, however, the only state where abundance estimates for all ungulate species were available, wolf population estimates derived from ungulate abundance data were within 6 percent of those derived using the deer/greenness model.

Although the greenness model may not accurately predict abundance of particular species as prey community composition changes in interior ecosystems, it may be more robust as an estimator of total prey biomass.

PREY ACCESSIBILITY. Rugged terrain may make prey less available to wolves (Paquet et al. 1996). The relationship between prey accessibility and slope was modeled as the power equation $Y = 28.18405 \times 0.931377^x$, where $x =$ slope in degrees. This equation was developed from wolf radiotelemetry data from four study areas with different topography (Riding Mountain National Park, Manitoba; Pukaskwa National Park, Ontario; Jasper National Park, Alberta; and the central Canadian Rockies) (Paquet et al. unpublished data).

SECURITY. We incorporated road and human population density into a composite habitat-effectiveness metric representing level of security or lack of human presence (Merrill et al. 1999). GIS data for roads, trails, and railroads at the 1:100,000 scale (USGS, unpublished data) were grouped into expected use classes, weighted, and examined at the 1-km² resolution. Paved highways were weighted two to three times more heavily than unpaved roads, and trails were weighted at 0.35 that of unpaved roads (Merrill et al. 1999).

Local human population density was derived from census blocks, which range in size from 1 to more than 100 km². The effects of population centers over distance were derived from population data interpolated by using an inverse distance weighting algorithm (Merrill et al. 1999).

COMBINING HABITAT FACTORS. To determine the likelihood that wolves or other wide-ranging carnivores will inhabit a region, we developed a GIS function (FOCALSEARCH) that sums total mortality risk within a variable-size moving window with a diameter based on habitat productivity. Wolf pack territories may range in size from 100 km² to more than 2000 km² depending on prey productivity (Wydeven et al. 1995; Paquet et al. 1996).

PREDICTING WOLF POPULATION SIZE. We used a two-step process for estimating potential wolf population size (Mladenoff and Sickley 1999). This method assumes that although regional wolf distribution is most limited by human-caused mortality, prey density determines abundance in otherwise suitable areas. Once mortality risk was approximated using a moving window, we delineated areas where survival rates would be high enough to permit wolf persistence. Although the habitat effectiveness values represent relative rather than absolute mortality risk, thresholds can be based on areas of known wolf distribution. In our region, we used current wolf distribution in the Rocky Mountains (Houts 2000) to select the 25 percent of the region with lowest mortality risk as core habitat and the second most favorable quartile as peripheral habitat. These thresholds provide a means of ranking areas based on proportion of core habitat, which is known to influence population persistence (Haight et al. 1998).

Once areas of core and peripheral habitat were delineated, we could estimate the number of wolves expected to inhabit the area. Here we used an equation relating wolf density to prey density (Fuller 1989, 1995; Mladenoff et al. 1999):

$$\text{wolf density}/1000 \text{ km}^2 = 4.19 \times \text{DEPU}/\text{km}^2$$

where DEPU (deer-equivalent prey units) was estimated from our deer/greenness regression model. Besides evaluating the relative levels of mortality risk, we evaluated the proportion of an area in public ownership as a factor influencing the feasibility of wolf restoration.

Grizzly Bear

We mapped predicted habitat for grizzly bear by adapting a model previously developed from Idaho sighting data (Merrill et al. 1999) and incorporated tasseled-cap greenness in order to measure habitat productivity (Mace et al. 1999; Gibeau 2000). This model was similar to the wolf model but lacked the negative effect of rugged terrain. Mortality risk was summed within a variable-size moving window whose total productivity was similar to that of grizzly bear home ranges in the Rocky Mountains (Merrill et al. 1999). Because grizzly bears do not show a high ability to compensate demographically for increased mortality (Weaver et al. 1996), we used this scaled total mortality risk as the principal predictor of whether the species could persist. We did not attempt to incorporate greenness values further in predicting potential population size.

Wolverine

We combined wolverine occurrence records (Natural Heritage Database, unpublished data) with GIS data on vegetation, topography, climate, and human-use variables (Carroll et al. 2001). We used multiple logistic regression to compare habitat variables at sighting locations with those at random points (Hosmer and Lemershow 1989). A large set of alternative multivariate models was constructed and evaluated with the Bayesian information criterion (BIC) (Schwarz 1978). We used the coefficients from the final multivariate model to calculate a resource selection function (RSF) (Manly et al. 1993; Boyce and McDonald 1999) representing the relative probability of wolverine occurrence at a location. The final model incorporated annual snowfall (Daly et al. 1994), interpolated human population density (Merrill et al. 1999), road density, and cirque habitat for wolverine dens (Hart et al. 1997).

Results

Wolf

If wolves successfully disperse to the southern Cascades and Modoc Plateau, this region may well hold the largest subpopulation in the Pacific coastal states (190–470 wolves) (Figure 1.1). Northeastern Oregon could support a population of about 100. Although southeastern Oregon has low human use, it is arid and may only support a low-density population. Coastal areas may support small populations, but a lower proportion of core habitat may make persistence even less probable than in arid areas. The central and southern Sierra Nevada region, in contrast, has a high proportion of core habitat, but low prey densities, rugged terrain, and distance from existing wolf populations reduce the likelihood of occupation. Among the larger habitat areas, the southern Oregon Cascades and Modoc Plateau have a relatively high proportion of public land (Table 1.1), making these areas most feasible for restoration.

Table 1.1. GAP Land Management Classes in Potential Wolf Population Areas

Region	Area (km²)	Park/wilderness	Other public	Private
SE Oregon	18,100	14.4%	50.8%	34.8%
NE Oregon	7,800	18.1	31.6	50.3
N Oregon Cascades	6,700	28.8	48.4	22.8
S Cascades	14,600	13.6	58.2	28.2
Modoc Plateau	33,700	6.8	57.7	35.5
Sierra Nevada	15,400	54.2	42.2	3.6
North coast	7,300	0.6	9.7	89.7
Central coast	7,800	0.9	32.0	67.1
South coast	6,100	16.4	40.2	43.4
California coast	8,100	9.0	20.5	70.5

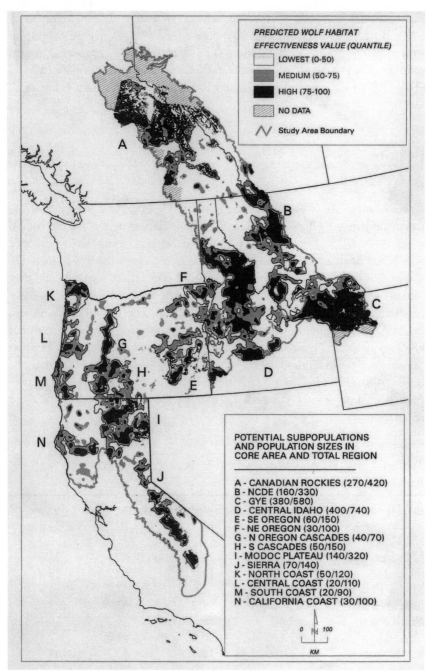

Figure 1.1. Habitat for the gray wolf in the northwestern United States and Rocky Mountains as predicted by a regional-scale habitat model.

Grizzly Bear

Besides the large area of habitat in the proposed central Idaho reintroduction area, a smaller cluster of habitat patches is evident in the Klamath region (northwestern California and southwestern Oregon) and the southern Sierra Nevada (Figure 1.2). Based on population estimates from other areas in the

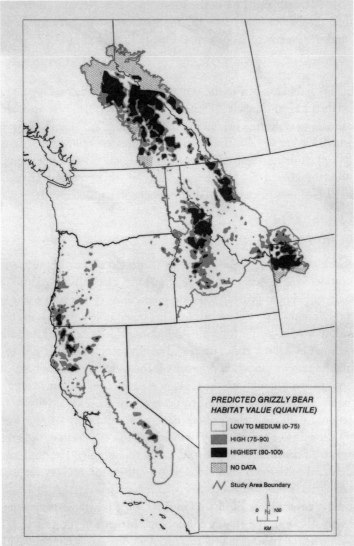

Figure 1.2. Habitat for the grizzly bear in the northwestern United States and Rocky Mountains as predicted by a regional-scale habitat model.

lower 48 states (USFWS 1993; Pease and Mattson 1999), the Klamath region could support 163 to 367 grizzly bears over 20,900 km^2 and the southern Sierra Nevada could support 48 to 108 bears over 6100 km^2. (Preliminary results of dynamic viability models [Carroll et al., unpublished data] suggest that this figure underestimates the potential size of the Sierra Nevada population.) The percentage of highest-security core habitat (90–100 decile) is 20.7 in the Klamath and 13.1 in the southern Sierra Nevada. Both values are lower than those in central Idaho (33.6), the Greater Yellowstone Ecosystem (43.3), and the Northern Continental Divide Ecosystem (42.7).

Wolverine

High-quality habitat occurs in the central and southern Sierra Nevada (22,000 km^2) and the Oregon Cascades (11,200 km^2) (Figure 1.3). These areas are comparable in size to wolverine habitat in central Idaho (32,200 km^2), the northern Continental Divide (14,600 km^2), and the Greater Yellowstone Ecosystem (20,200 km^2). Although wolverine demography is still poorly understood in the lower 48 states, based on home range sizes in other regions (Banci 1994; Copeland 1996) these areas might each support 50 to 100 animals.

Discussion

The accuracy of regional habitat models is limited by our imprecise knowledge of underlying species/habitat relationships at this scale. Although the models are unlikely to provide accurate predictions of the future size of carnivore populations, they are useful tools for qualitative comparisons between regions. The models could form the foundation for testable hypotheses that can be refined as habitat restoration and natural recolonization proceed.

Obstacles to reintroduction or recolonization vary by species because each species has its own habitat needs, ecological resilience, and area and connectivity requirements. They are similar, however, in their sensitivity to human development. Although the greater ecological resiliency of the wolf will probably allow it to occupy a larger portion of the region than the grizzly or wolverine, wolf distribution is constrained by topography. Thus mountainous areas such as the Klamath Mountains and southern Sierra Nevada are better habitat for grizzly bear and wolverine than for wolves.

Habitat areas for large carnivores in the Pacific coastal states are generally smaller than those in the Rocky Mountains. Moreover, the scattered distribution of potential habitat makes metapopulations less attainable. Although its extensive area requirements may help explain why the wolf was the first of the three species to be extirpated, its exceptionally high dispersal ability make its natural restoration more probable. Comparison of occupied and predicted

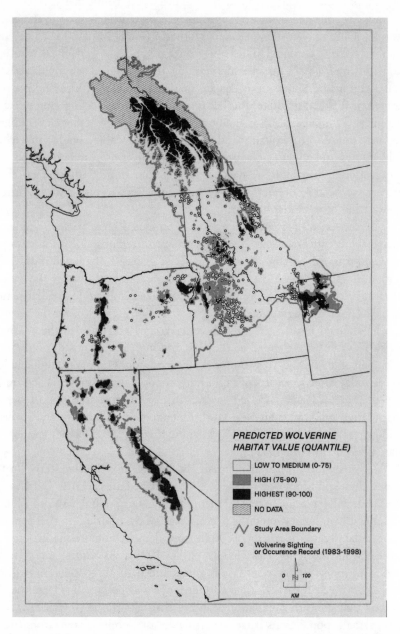

Figure 1.3. Habitat for the wolverine in the northwestern United States and Rocky Mountains as predicted by a regional-scale habitat model.

habitat for the wolverine suggests the importance of connectivity. Wolverines are rare or extirpated from what appears to be extensive habitat in the Sierra Nevada while smaller habitat patches close to Rocky Mountain source populations remain occupied. This suggests that the rescue effect (Brown and Kodric-Brown 1977) may be maintaining these smaller subpopulations or that sightings of dispersing males are masking the loss of breeding populations. It is difficult to judge whether range contraction is a lingering effect of trapping (Grinnell et al. 1937) and predator control, or whether current habitat is no longer of the quality and size to support viable populations. More field studies on habitat associations of extant wolverine populations and surveys in potential habitat are needed. Despite our limited knowledge of the species, we need to develop management guidelines in order to halt further degradation of potential habitat.

The conservation importance of core protected areas versus semideveloped habitat differs among the three species. In regions such as northwestern California where core areas are small, for example, predicted mean population sizes for grizzly bear may be comparable to those in the Rocky Mountains, but high variance in population size results in higher extinction risk. Although demographic data for the wolverine are scarce, the species may also be affected by these factors. A century ago, wolverine in the southern Sierra Nevada were trapped as they left the high mountains during winter (Grinnell et al. 1937). Areas buffering Sierran mountain parks are likely less secure today after a century of rapid human population growth. The limited size of protected areas has led several writers to stress the importance of cooperative planning among land management agencies and private landowners (Salwasser et al. 1987; Mladenoff et al. 1995; Paquet and Hackman 1995; Boyd et al. 1995).

As wolves recolonize an increasingly human-inhabited western United States, they will often occur outside core protected areas (Fritts and Carbyn 1995) but will continue to rely on them for long-term population persistence. Peripheral wolf habitat presents a "conservation conundrum" in that semideveloped landscapes can support high prey densities but result in human-caused wolf mortality (Mladenoff et al. 1997). As Fritts and Carbyn (1995:26) note: "Reintroduction programs and natural repopulation will reestablish wolves in parts of their former range and potentially create the situations that concern population-viability theorists the most—that is, relatively small populations that are isolated or semi-isolated from other populations."

The smaller size and greater isolation of high-quality core habitat in the Pacific states highlight the importance of metapopulation connectivity for the survival of these semidisjunct populations. Simulation models suggest that connectivity can counteract the negative effects of small area, as wolves dispersing from protected areas strengthen peripheral populations (Haight et al. 1998). Regional planning that incorporates core, buffer, and dispersal habitat can increase the effective size of reserves and allow wolves to expand into the semideveloped landscape matrix (Fritts and Carbyn 1995). Translocation of animals may be necessary for some of the more isolated areas; others will benefit from the restoration of regional-scale linkages. Even if wolves successfully recolonize coastal Oregon and other areas with limited core habitat, they are likely to remain vulnerable to extirpation. A threat analysis that integrates information on the location of critical habitat with data on human population growth and landscape change can help in the prioritization of conservation efforts (Carroll et al., unpublished data). Although our results suggest a high potential for restoring large carnivores in the Pacific Northwest, current development trends may foreclose options for carnivore restoration unless steps are taken soon to protect critical habitat.

Acknowledgments

This study was funded primarily by a grant from the Foundation for Deep Ecology; we received additional support from Defenders of Wildlife. The information in this chapter has been funded in part by the U.S. Environmental Protection Agency, has been subjected to the agency's peer and administrative review, and has been approved for publication as an EPA document.

Literature Cited

Almack, J. A., W. L. Gaines, P. H. Morrison, J. R. Eby, R. H. Naney, G. F. Wooten, S. H. Fitkin, and E. R. Garcia. 1993. North Cascades grizzly bear ecosystem evaluation. Final report. Denver: Interagency Grizzly Bear Committee.

Archibald, W. R., R. Ellis, and A. N. Hamilton. 1987. Responses of grizzly bears to logging truck traffic in the Kimsquit river valley, British Columbia. *International Conference on Bear Research and Management* 7:251–257.

Bailey, V. 1936. The mammals and life zones of Oregon. *North American Fauna* 55:1–416.

Banci, V. 1994. Wolverine. In L. F. Ruggiero, K. B. Aubry, S. W. Buskirk, L. J. Lyon, and W. J. Zielinski, eds., *The Scientific Basis for Conserving Forest Carnivores: American Marten, Fisher, Lynx, and Wolverine.* General Technical Report RM-254. Fort Collins: USDA Forest Service Rocky Mountain Forest and Range Experiment Station.

Barrett, R., R. Golightly, and T. E. Kucera. 1994. California wolverine. In C. G. The-lander, ed., *Life on the Edge: A Guide to California's Endangered Natural Resources.* Santa Cruz: Biosystems Books.

Bilby, R. E., B. R. Fransen, and P. A. Bisson. 1996. Incorporation of nitrogen and car-bon from spawning coho salmon into the trophic systems of small streams: Evi-dence from stable isotopes. *Canadian Journal of Fisheries and Aquatic Sciences* 53:164–173.

Boyce, M. S., and L. L. MacDonald. 1999. Relating populations to habitats using resource selection functions. *Trends in Ecology and Evolution* 14:268–272.

Boyd, D. K., P. C. Paquet, S. Donelon, R. R. Ream, D. H. Pletscher, and C. C. White. 1995. Transboundary movements of a recolonizing wolf population in the Rocky Mountains. In L. N. Carbyn, S. H. Fritts, and D. R. Seip, eds., *Ecology and Conservation of Wolves in a Changing World.* Edmonton: Canadian Circumpolar Institute, University of Alberta.

Brown, J. H., and A. Kodric-Brown. 1977. Turnover rates in insular biogeography: Effect of immigration on extinction. *Ecology* 58:445–449.

Bunnell, F. L., and D. E. N. Tait. 1981. Population dynamics of bears—implications. In C. W. Fowler and F. D. Smith, eds., *Dynamics of Large Mammal Populations.* New York: Wiley.

Buskirk, S. W. 1999. Mesocarnivores of Yellowstone. In T. W. Clark, A. P. Curlee, S. C. Minta, and P. M. Kareiva, eds., *Carnivores in Ecosystems: The Yellowstone Expe-rience.* New Haven: Yale University Press.

California Department of Fish and Game (CDFG). 1998. An assessment of mule and black-tailed deer habitats and populations in California. Unpublished report. Sacramento.

Carroll, C., R. F. Noss, and P. C. Paquet. 2001. Carnivores as focal species for conser-vation planning in the Rocky Mountain region. *Ecological Applications* 11: 961–980.

Copeland, J. P. 1996. Biology of the wolverine in central Idaho. M.S. thesis, Univer-sity of Idaho, Moscow.

Corsi, F., E. Dupre, and L. Boitani. 1999. A large-scale model of wolf distribution in Italy for conservation planning. *Conservation Biology* 13:150–159.

Craighead, L., and E. R. Vyse. 1996. Brown/grizzly bear metapopulations. In D. R. McCullough, ed., *Metapopulations and Wildlife Conservation.* Washington, D.C.: Island Press.

Crist, E. P., and R. C. Cicone. 1984. Application of the tasseled cap concept to sim-ulated thematic mapper data. *Photogrammetric Engineering and Remote Sensing* 50:343–352.

Crooks, K. R., and M. E. Soulé. 1999. Mesopredator release and avifaunal extinctions in a fragmented system. *Nature* 400:563–566.

Daly, C., R. P. Neilson, and D. L. Phillips. 1994. A statistical-topographic model for mapping climatological precipitation over mountainous terrain. *Journal of Applied Meteorology* 33:140–158.

Fritts, S. H., and L. N. Carbyn. 1995. Population viability, nature reserves, and the outlook for gray wolf conservation in North America. *Restoration Ecology* 3:26–38.

Fuller, T. K. 1989. Population dynamics of wolves in north-central Minnesota. *Wildlife Monographs* 105:1–41.

———. 1995. Comparative population dynamics of North American wolves and African wild dogs. In L. N. Carbyn, S. H. Fritts, and D. R. Seip, eds., *Ecology and Conservation of Wolves in a Changing World.* Edmonton: Canadian Circumpolar Institute, University of Alberta.

Fuller, T. K., W. E. Berg, G. L. Radde, M. S. Lenarz, and G. B. Joselyn. 1992. A history and current estimate of wolf distribution and numbers in Minnesota. *Wildlife Society Bulletin* 20:42–55.

Gardner, C. L. 1985. The ecology of wolverines in southcentral Alaska. M.S. thesis, University of Alaska, Fairbanks.

Gibeau, M. L. 2000. A conservation biology approach to management of grizzly bears in Banff National Park, Alberta. Ph.D. dissertation, University of Calgary, Alberta.

Grinnell, J., J. S. Dixon, and J. M. Linsdale. 1937. *Fur-Bearing Mammals of California: Their Natural History, Systematic Status, and Relations to Man.* Vol. 2. Berkeley: University of California Press.

Haight, R. G., D. J. Mladenoff, and A. P. Wydeven. 1998. Modeling disjunct gray wolf populations in semi-wild landscapes. *Conservation Biology* 12:879–888.

Hart, M. M., J. P. Copeland, and R. L. Redmond. 1997. Mapping wolverine habitat in the Northern Rockies using a GIS. Poster presented at the fourth annual conference of the Wildlife Society, Snowmass, Colo.

Hatler, D. F. 1989. A wolverine management strategy for British Columbia. Wildlife Bulletin B-60. Victoria, B.C.: Wildlife Branch, Ministry of Environment.

Hilderbrand, G. V., T. A. Hanley, C. T. Robbins, and C. C. Schwartz. 1999. Role of brown bears (*Ursus arctos*) in the flow of marine nitrogen into a terrestrial ecosystem. *Oecologia* 121:546–550.

Hosmer, D. W., and S. Lemeshow. 1989. *Applied Logistic Regression.* New York: Wiley.

Houts, M. E. 2000. Modeling gray wolf habitat in the Northern Rocky Mountains using GIS and logistic regression. Master's thesis, University of Kansas, Lawrence.

Keith, L. B. 1983. Population dynamics of wolves. In L. N. Carbyn, ed., *Wolves in Canada and Alaska.* Canadian Wildlife Report 45. Ottawa.

Mace, R. D., J. S. Waller, T. L. Manley, K. Ake, and W. T. Wittinger. 1999. Landscape evaluation of grizzly bear habitat in western Montana. *Conservation Biology* 13:367–377.

Magoun, A. J., and J. P. Copeland. 1998. Characteristics of wolverine reproductive den sites. *Journal of Wildlife Management* 62:1313–1320.

Manly, B. F. J., L. L. McDonald, and D. L. Thomas. 1993. *Resource Selection by Animals.* New York: Chapman & Hall.

Mattson, D. J. 1990. Human impacts on bear habitat use. *International Conference on Bear Research and Management* 8:35–56.

Mattson, D. J., R. R. Knight, and B. M. Blanchard. 1987. The effects of development and primary roads on grizzly bear habitat use in Yellowstone National Park, Wyoming. *International Conference on Bear Research and Management* 7:259–273.

Mattson, D. J., and M. M. Reid. 1991. Conservation of the Yellowstone grizzly bear. *Conservation Biology* 5:364–372.

McLellan, B. N., and D. M. Shackleton. 1988. Grizzly bears and resource extraction industries: Effects of roads on behaviour, habitat use, and demography. *Journal of Applied Ecology* 25:451–460.

Mech, L. D. 1970. *The Wolf: The Ecology and Behavior of an Endangered Species.* Garden City: Natural History Press.

———. 1989. Wolf population survival in an area of high road density. *American Midland Naturalist* 121:387–389.

Mech, L. D., and S. M. Goyal. 1993. Canine parvovirus effect on wolf population change and pup survival. *Journal of Wildlife Diseases* 22:104–106.

Merrill, T., D. J. Mattson, R. G. Wright, and H. B. Quigley. 1999. Defining landscapes suitable for restoration of grizzly bears (*Ursus arctos*) in Idaho. *Biological Conservation* 87:231–248.

Mladenoff, D. J., and T. A. Sickley. 1999. Assessing potential gray wolf restoration in the northeastern United States: A spatial prediction of favorable habitat and potential population levels. *Journal of Wildlife Management* 62:1–10.

Mladenoff, D. J., T. A. Sickley, R. G. Haight, and A. P. Wydeven. 1995. A regional landscape analysis and prediction of favorable gray wolf habitat in the northern Great Lakes region. *Conservation Biology* 9:279–294.

———. 1997. Causes and implications of species restoration in altered ecosystems: A spatial landscape projection of wolf population recovery. *Bioscience* 47:21–31.

Mladenoff, D. J., T. A. Sickley, and A. P. Wydeven. 1999. Predicting gray wolf landscape recolonization: Logistic regression models vs. new field data. *Ecological Applications* 9:37–44.

Noss, R. F., H. B. Quigley, M. G. Hornocker, T. Merrill, and P. C. Paquet. 1996. Conservation biology and carnivore conservation in the Rocky Mountains. *Conservation Biology* 10:949–963.

Noss, R. F., M. A. O'Connell, and D. D. Murphy. 1997. *The Science of Conservation Planning.* Washington, D.C.: Island Press.

Paquet, P. C. 1993. Summary reference document—ecological studies of recolonizing wolves in the Central Canadian Rocky Mountains. Unpublished report by John/Paul and Associates. for Canadian Parks Service, Banff, Alberta.

Paquet, P. C., and A. Hackman. 1995. Large carnivore conservation in the Rocky Mountains. Toronto and Washington, D.C.: World Wildlife Fund Canada and World Wildlife Fund U.S.

Paquet, P. C., J. Wierzchowski, and C. Callaghan. 1996. Effects of human activity on gray wolves in the Bow River Valley, Banff National Park, Alberta. In J. Green, C. Pacas, S. Bayley and L. Cornwell, eds., *A Cumulative Effects Assessment and Futures Outlook for the Banff Bow Valley.* Prepared for the Banff Bow Valley Study. Ottawa: Department of Canadian Heritage.

Pease, C. M., and D. J. Mattson. 1999. Demography of the Yellowstone grizzly bears. *Ecology* 80:957–975.

Pletscher, D. H., R. R. Ream, D. K. Boyd, M. W. Fairchild, and K. E. Kunkel. 1997. Population dynamics of a recolonizing wolf population. *Journal of Wildlife Management* 61:459–465.

Salwasser, H., C. Schonewald-Cox, and R. Baker. 1987. The role of interagency cooperation in managing for viable populations. In M. Soulé, ed., *Viable Populations for Conservation.* New York: Cambridge University Press.

Schmidt, R. H. 1991. Gray wolves in California: Their presence and absence. *California Fish and Game* 77:79–85.

Schullery, P., and L. H. Whittlesey. 1999. Greater Yellowstone carnivores: A history of changing attitudes. In T. W. Clark, A. P. Curlee, S. C. Minta, and P. M. Kareiva, eds., *Carnivores in Ecosystems: The Yellowstone Experience.* New Haven: Yale University Press.

Schwarz, G. 1978. Estimating the dimension of a model. *Annals of Statistics* 6:461–464.

Scott, J. M., F. Davis, B. Csuti, R. Noss, B. Butterfield, C. Groves, H. Anderson, S. Caicco, F. D'Erchia, T. C. Edwards Jr., J. Ulliman, and R. G. Wright. 1993. Gap analysis: A geographic approach to the protection of biological diversity. *Wildlife Monographs* 123:1–41.

Singleton, P. H. 1995. Winter habitat selection by wolves in the North Fork of the Flathead River Basin, Montana and British Columbia. M.S. thesis, University of Montana, Missoula.

Smith, D. W., W. G. Brewster, and E. E. Bangs. 1999. Wolves in the greater Yellowstone ecosystem: Restoration of a top carnivore in a complex management environment. In T. W. Clark, A. P. Curlee, S. C. Minta, and P. M. Kareiva, eds., *Carnivores in Ecosystems: The Yellowstone Experience.* New Haven: Yale University Press.

Storer, T. I., and L. P. Tevis Jr. 1955. *California Grizzly.* Berkeley: University of California Press.

Terborgh, J., J. A. Estes, P. Paquet, K. Ralls, D. Boyd-Heger, B. J. Miller, and R. F. Noss. 1999. The role of top carnivores in regulating terrestrial ecosystems. In M. E. Soulé and J. Terborgh, eds., *Continental Conservation: Scientific Foundations of Regional Reserve Networks*. Washington, D.C.: Island Press.

Thiel, R. P. 1985. Relationship between road densities and wolf habitat suitability in Wisconsin. *American Midland Naturalist* 113:404.

Thurber, J. M., R. O. Peterson, T. R. Drummer, and S. A. Thomasma. 1994. Gray wolf response to refuge boundaries and roads in Alaska. *Wildlife Society Bulletin* 22:61–68.

U.S. Fish and Wildlife Service (USFWS). 1993. *Grizzly Bear Recovery Plan (and Summary)*. Bethesda: U.S. Department of Agriculture.

Van Ballenberghe, V., W. Erickson, and D. Byman. 1975. Ecology of the timber wolf in northeastern Minnesota. *Wildlife Monographs* 43:1–43.

Verts, B. J. 1975. New records for three uncommon mammals in Oregon. *Murrelet* 56:22–23.

Verts, B. J., and L. N. Carraway. 1998. *Land Mammals of Oregon*. Berkeley: University of California Press.

Weaver, J. L., P. C. Paquet, and L. F. Ruggiero. 1996. Resilience and conservation of large carnivores in the Rocky Mountains. *Conservation Biology* 10:964–976.

Wydeven, A. P., R. N. Schultz, and R. P. Thiel. 1995. Monitoring of a recovering gray wolf population in Wisconsin, 1979–1991. In L. N. Carbyn, S. H. Fritts, and D. R. Seip, eds., *Ecology and Conservation of Wolves in a Changing World*. Edmonton: Canadian Circumpolar Institute, University of Alberta.

Young, S. P., and E. A. Goldman. 1964. *Wolves of North America*. New York: Dover.

Chapter 2

Feasibility of Timber Wolf Reintroduction in Adirondack Park

PAUL C. PAQUET, JAMES R. STRITTHOLT, NANCY L. STAUS,
P. J. WILSON, S. GREWAL, AND B. N. WHITE

During the last two centuries, the primary limiting factor for timber wolves (*Canis lupus*) has been human-caused mortality through hunting, trapping, and predator control (Young 1946; Smith et al. 1999). As antipredator sentiment and the economic importance of the livestock industry wanes, wolves may continue to recolonize portions of their former range. In recent years, natural and human-assisted wolf colonization has been successful in several North American locations. The most effective, to date, has been the restoration of eastern timber wolves in the Upper Great Lakes region of the United States (Fuller 1995). Establishment of a second population in the eastern United States is a wolf recovery goal (USFWS 1992). Adirondack Park, New York, has been identified as a potential recovery site for the eastern timber wolf (Mladenoff and Sickley 1998) along with Maine, New Hampshire, and Vermont.

Although some form of wolf existed in New York State before and during early European settlement, the regional abundance and distribution of timber wolves are poorly understood. De Kay (1842) described two varieties of the common American gray wolf (*Lupus occidentalis*) inhabiting New York State. The variety corresponding most closely to the modern description of timber wolves was considered rare (De Kay 1842). In the mid-1800s, the range of wolflike canids in New York State was reduced to mountainous and forested areas and the counties along the St. Lawrence River. The last reported wolf in New York was killed and mounted in 1893 (Adirondack Museum, Catalog No. 79.10.1).

The suitability of Adirondack Park to sustain wolves, however, has not been evaluated adequately (Henshaw 1982; USFWS 1992). A thorough evaluation requires an understanding of wolf population dynamics and the ecological relations between wolves and their prey (primarily ungulates) and other predators. An assessment of human influences is required, as well, because historic and ongoing human activities modify and often constrain the wolf's inherent behavioral ecology (see also chapter 13 in this volume). The ultimate factor determining population viability for wolves is human attitude (Boitani 1982; Fritts and Carbyn 1996). Regional planning can simplify coexistence by identifying spatial refugia or core areas with a level of protection sufficient to buffer populations against conflicts with humans (Mladenoff et al. 1995, 1997; Mladenoff and Sickley 1998; Boitani et al. 1997; Carroll et al. 1999, 2001). Planning can also identify optimal locations of buffer zones and corridors that will expand the effective size of core areas by allowing use of semideveloped lands while reducing the probability of human-caused mortality. Human/wolf conflict is most likely to occur in areas of highly productive habitat with above-average human use, in spatial buffers between large core habitat areas and zones of high human use, or in zones likely to experience human occupation in the future.

Predicting precisely how wolves will respond to a new environment is impossible. We can, however, assess the feasibility of wolf reintroduction by constructing spatially explicit models based on field research. The recent development of such models using geographic information systems (GIS) suggests that the wolf's habitat preference and movements are predictable at some spatial scales. (See Mladenoff et al. 1995, 1999; Paquet et al. 1996; Alexander et al. 1996; Boitani 1997; Mladenoff and Sickley 1998; Massolo and Meriggi 1998; Haight et al. 1998; Carroll et al. 2001.) Extensive field-work has confirmed the efficacy of several of these theoretical models (Alexander et al. 1996; Mladenoff et al. 1999; chapter 13 in this volume). By generalizing site-specific empirical models that reflect local conditions, we can identify potential wolf habitat in areas where humans have extirpated wolves and where suitable habitat persists (Mladenoff and Sickley 1998).

Although GIS wolf models have not been developed for Adirondack Park, predictive habitat models have been developed for the northeastern United States. Harrison and Chapin (1997) and Mladenoff and Sickley (1998) have identified 14,000 to 16,000 km² of land in the park as potential wolf habitat—less than a previous estimate of 24,280 km² (USFWS 1992).

We constructed a spatially explicit timber wolf habitat suitability model that considered landscape linkages and the region's ability to support a wolf population. In this chapter we use the suitability model to identify potential wolf habitat within Adirondack Park and nearby areas. We then estimate the potential number of wolves the park could support based on prey available within areas suitable for wolf occupation. Finally, we use the estimates of habitat and prey populations to assess the potential of Adirondack Park to sustain permanently a reintroduced population of gray wolves.

Methods

The relatedness of the original Adirondack wolf, the current eastern timber wolf, and other North American wolves is uncertain. Thus, we carried out a preliminary genetic assessment of eastern timber wolves using current and historical samples of DNA (see Wilson et al. 1999, 2000 for complete methodological details and results). Informed by this assessment, we constructed five submodels based on biophysical and cultural factors (Figure 2.1). Because snow influences habitat use by wolves, we modeled winter and summer seasons independently. Summer was defined as 15 April to 15 September and

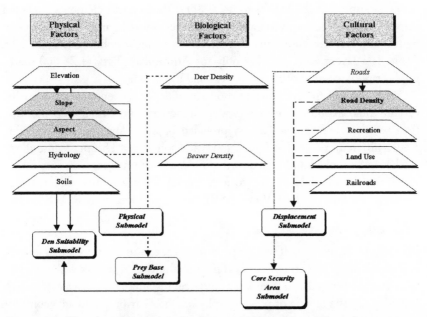

Figure 2.1. Flowchart illustrating development of five submodels used to assess potential suitability of wolf habitat in Adirondack Park, New York, 1999. Shaded rectangles are surfaces that we derived from primary data.

winter as 16 September to 14 April. These periods reflect denning activity in summer and pup travel with adults during winter. We used the submodels to assess habitat potential, habitat fragmentation, and connectivity and to simulate movements of wolves in the region. Integrating the submodels allowed us to examine how wolves might use Adirondack Park now and in pristine conditions. We defined "pristine conditions" as the absence of human activity on the landscape. Though speculative, this retrospective view helps us understand how humans influence the distribution and viability of regional wolf populations. Future conditions, based on current rates and types of development, can also be considered with and without mitigation of human activity.

We derived biological attributes from radiotelemetry and snow-tracking data collected in the Rocky Mountains of the United States and Canada; Riding Mountain National Park, Manitoba; Pukaskwa National Park, Ontario; and the Apennine Mountains, Italy. These areas differ considerably in landscape complexity, prey abundance, prey diversity, and extent of human development. Adirondack Park contains elements representative of all four study areas. We supplemented these empirically derived attributes with habitat descriptions in the relevant ecological literature. When information was lacking or equivocal, we used expert opinion to modify model attributes. We derived spatially explicit biophysical and cultural attributes from data sets at scales ranging from 1:24,000 to 1:250,000. Biophysical elements included slope, aspect, elevation, vegetation, and water. Human influences included roads, railways, structures, and activities such as snowmobiling (Figure 2.1). Some data layers were available only for Adirondack Park (1:24,000 roads, for example), whereas others were regional in coverage.

We evaluated the certainty of model results with a sensitivity analysis by varying the contribution of each model component to the final outcome. Criteria based on information theory provided objective measures for selecting among different models (Burnham and Anderson 1998). This procedure allows models to be ranked and scaled so that estimates of precision can account for uncertainty as to which is the "correct" model. We used three GIS software packages: Arc/Info and ArcView (ESRI 1996) and Imagine (ERDAS 1998).

Landscape Submodel

We created the ideal wolf landscape surface that assumed no human activity in the past or present (null model). The probability that wolves will use a certain area is expressed as a function of physical resources (slope, aspect, water). Included are the effects of physiography on the distribution, size, geometry, and juxtaposition of habitat patches and behavioral responses of wolves to the natural physical environment. The model output expresses the probability that a given pixel is of high survival value to wolves.

Snow accumulation was included in the physical model for winter. Because we could not accurately predict precipitation over the park, we assumed that most November through March precipitation would fall as snow. To estimate where snowfall would have its greatest influence on wolves, we used mean monthly precipitation data (1961–1990) from the PRISM data set (Daly et al. 1997) for the months of November through March. This information was then factored into the slope/aspect model by subtracting the snowfall score from the combined slope/aspect score.

Prey-Base Submodel

We assumed that the distribution and density of prey, as well as land cover, were modified by human activity. We also assumed that without humans all prey were accessible to wolves. Our prey-base submodel was based on 1996 white-tailed deer (*Odocoileus virginianus*) estimates and the estimated number of beaver (*Castor canadensis*) colonies per kilometer of shoreline from 1993 to 1994 (N.Y. Department of Environmental Conservation, unpublished data). Deer density ranged from 0.5 to 8.7 deer/km^2 and beaver density from 0.09 to 0.38/km^2. We classified densities of 0.5 to 0.9 deer/km^2 as unsuitable, 1.0 to 1.3 as low, 1.4 to 1.8 as moderate, 1.9 to 2.9 as high, and 3.0 to 8.7 as very high. Densities of 0.09 beaver/km^2 were classified as low, 0.16 as moderate, and 0.34 to 0.38 as high.

Habitat Displacement Submodel

We superimposed human developments and zones of human activity over the landscape and prey-base submodels. The resulting probability surface predicts wolf landscape avoidance due to biophysical features, human disturbance, and habitat loss. Environmental security implies that high-quality habitat satisfies day-to-day needs in the presence of low-level, nonlethal disturbance. When high levels of stress are imposed on wolves in otherwise good habitat, the wolves abandon or avoid that area. Therefore, landscape factors such as isolation, steep topography, and inaccessibility were important model components.

We used roads and land use to predict exclusion of wolves. Adirondack Park contains about 8200 km of roads passable by two-wheel-drive vehicles. Road density was calculated for each 1-km^2 cell and then generalized using a round, 5-km^2 moving GIS window. Wolf suitability was rated as high (road densities of 0–0.23 km/km^2), medium (0.23–0.45 km/km^2), low (0.45–0.6 km/km^2), and unsuitable (more than 0.6 km/km^2).

Den Suitability Submodel

Denning wolves prefer deep, well-drained soils near water. The presence of beaver also influences the location of den sites (Carbyn 1983; D. Smith, pers.

comm.). We used soil depth, drainage characteristics, distance to water, and availability of beaver to create a combined suitability score. Potential den habitat was rated as poor (1–9), fair (10–12), and good (13–15).

Core Security Area Submodel

In human-dominated landscapes, wolf survival depends largely on reduced contact with people. As a proxy for human activity, we used a 1-km buffer (Chapman 1977; Singleton 1995; Paquet et al. 1996) on either side of roads passable by two-wheel-drive vehicles. Within this 2-km band, wolves were excluded. Unaffected areas were considered core security sites.

Least-Cost Pathways

Differences in landscape features, human activity, prey distribution, and habitat quality affect wolf movement and dispersal. Roads, railways, and large water bodies are potential barriers to wolf movement. An obstacle's permeability depends on its structure, physical location, and level of human activity. Specific values for all landscape elements were expressed within the model as coefficients. The values combine to create a probability surface with a variable resistance to the movement of wolves.

We used ArcView least-cost path analysis to assess seasonal landscape connectivity in Adirondack Park. The analysis simulated wolf movements by calculating the relative resistance of the landscape measured in individual pixels. Higher costs reflect increased environmental resistance to movement. Simulated animals selected travel routes with an optimal combination of security (avoid towns), habitat quality (select areas of abundant prey), and energy efficiency (avoid deep snow). We simulated seasonal dispersal from four destinations: the three largest core security areas and a peripheral area in the northern section of the park. Cardinal points (north, south, east, and west) on the park's boundary were used as target destinations.

Criteria for Evaluating Wolf Restoration

From the literature, our own studies, and discussions with other researchers, we identified nine conditions that characterize viable populations of timber wolves. These criteria reflect our belief that the goal of wolf reintroduction is to establish packs, which are the basic social and biological units of the species.

MINIMUM POPULATION SIZE. The USFWS (1992) concluded that an isolated population of eastern timber wolves would be viable if it occupied a minimum contiguous area of 25,906 km^2 and averaged at least one wolf per 129 km^2 (that is, 201 wolves). Using North American density estimates of

0.25 to 4.0 wolves/100 km (Ballard et al. 1987:25; Paquet et al. 1996), we calculated the range of wolf numbers that could occur in potential core habitat within Adirondack Park.

PREY BASE. Wolves require a prey biomass of at least 100 kg/km^2 (Keith 1983; Fuller 1989; Messier 1994, 1995). Adult prey species equivalents are approximately 0.25 moose/km^2, 1.0 deer/km^2, and 4.0 beaver/km^2. Several prey species combined can provide the needed biomass (Mladenoff et al. 1995; Paquet et al. 1996).

CORE HABITAT SECURITY. Wolves require high-quality habitat with fewer than 1000 people or 1000 events per month (Paquet et al. 1996). In human-dominated landscapes, habitat quality and human activity interact to create a dynamic tension between wolf occupation and abandonment. Sometimes wolves sacrifice security for the benefits associated with productive habitat. Conversely, wolves are easily displaced from poor-quality habitats when exposed to low levels of human activity. As a result, the areal distribution of wolf populations may be patchy.

HABITAT CONNECTIVITY. Wolves need travel networks that link populations through dispersal (Paquet et al. 1996). Travel corridors improve genetic exchange, reduce local extinction probabilities, and enhance survival of dispersers.

DEN SITES. Sustained reproduction depends on suitable den and rendezvous sites protected from human intrusion.

MORTALITY. Annual sustained mean mortality (natural causes, hunting, trapping, highway collision, railway collision) should be less than 30 percent of the adult population (Fuller 1989). Wolves can probably sustain higher mortality for short periods (one or two years).

HUMAN OCCUPATION. Permanent human densities should be lower than 0.4 people/km^2 in core wolf habitat (Mladenoff et al. 1995). Wolves can tolerate higher human densities but prefer low-density areas (L. Boitani, pers. comm.; T. Fuller, pers. comm.).

ROAD DENSITY. Many studies have found that wolf populations have a low probability of persistence in areas with road densities greater than 0.58 to 0.72 km/km^2 (Thiel 1985; Jensen et al. 1986; Mech et al. 1988; Mech 1989; Fuller 1989; Mladenoff et al. 1995). Mladenoff et al. (1995, 1999) suggest that suitable wolf habitat is restricted to areas with road density not exceeding 0.45 km/km^2 in the overall pack area and 0.23 km/km^2 in the pack's core areas (areas containing sensitive den and rendezvous sites). In areas where road access increases exposure of wolves to people with vehicles and guns, the density of traversable roads should be less than 0.6 km/km^2 within the entire home range of a wolf pack (Mladenoff et al. 1998; Merrill 2000). Where wolves are protected, the density of low-speed paved roads and railways should be less than 1.42 km/km^2 within the entire home range (Paquet et al. 1996; Merrill 2000).

SPEED LIMITS AND TRAFFIC VOLUME. Speed limits on roads and railroads in wolf habitat should be less than 70 km/hour (Merrill 2000), and traffic volume on highways should be less than 2000 vehicles a day (J. Bertwistle, pers. comm.; D. Smith, pers. comm.).

Results

In this section we review the findings of our genetic assessment of eastern timber wolves, report on the results of wolf habitat suitability and landscape connectivity models, and summarize estimates of wolf populations that Adirondack Park could potentially support.

Genetics

According to recent genetic research (Wilson et al. 2000), the characteristics described by De Kay (1842) are consistent with two overlapping types of wolves: the proposed eastern Canadian wolf (*C. lycaon*) and the timber wolf (*C. lupus*). Mitochondrial DNA (mtDNA) and microsatellite DNA analyses of the eastern Canadian wolf and the red wolf (*C. rufus*) show a close relationship. The previously unidentified wolflike canid now living in the Adirondacks apparently represents a hybrid between the eastern Canadian wolf (*C. lycaon*) and the coyote (*C. latrans*).

Physical Landscape Suitability

Topographic constraints (such as steep slopes) prevent wolves from using only a small portion of the park. Given the somewhat gentle terrain of the Adirondacks, much of the park area is physically suitable for wolf habitation. Such terrain, however, is also conducive to human access and activity.

Den Sites

Comparing den site suitability under "pristine" conditions with present conditions showed a 52 percent loss of suitable den sites due to human displacement. A disproportionate amount of this loss (71 percent) was from the best denning sites.

Prey Base

Though we have a reasonable understanding of the distribution of prey in the park, we are more confident of the location of prey than their abundance. Overall, 44 percent of the park has low prey suitability, 30 percent moderate, and 26 percent high. Populations of white-tailed deer and beaver appear more than adequate to sustain wolves. Only high-prey-base areas in northwestern Adirondack Park, however, are sufficiently secure for long-term wolf survival. Most of the high-density prey areas in the eastern portion of the park lack adequate security for wolves. Wolves attracted to these areas would likely be killed or displaced by human disturbances. Further, New York State lacks a significant moose (*Alces alces*) population. Thus potential wolf density is likely lower in New York than in other regions of North America where several abundant prey species occur. In the future, recovering moose populations could augment the prey base.

Core Security Areas

Using only security from humans as a criterion, we identified 420 landscape patches totaling 23,172 km^2. We then eliminated areas less than 150 km^2 as too small to support a pack of timber wolves. This left 18 core security areas greater than 150 km^2 (Table 2.1). Our estimate is liberal because distances between polygons, condition of the matrix separating polygons, existence of barriers, and polygon shape were not accounted for. Long and narrow areas have a high perimeter/interior ratio, for example, which provides little protection from disturbance.

Based on our summer model (Figure 2.2), 6 of the 18 core areas were ranked as good habitat, 7 as fair, and 5 as poor (Table 2.1). Half of this combined area is not available to wolves, however, because less topographically complex areas have been developed for human use. Without humans and settlements, wolves would prefer most of the combined area. Den habitat was found in all 18 core security areas. Core areas in the northwestern quarter of the park contained the largest contiguous sites. Considering core security areas collectively, 60 percent of the land area is classified as having low prey suitability, 14 percent moderate, and 26 percent high. All the sites classified as high are in the northwestern core areas (most or all of areas 5, 4, 6, 9, and 13). Most high-quality prey areas are outside core security. About 33 percent of the core security areas are in private ownership (Table 2.1).

During winter, the most suitable wolf habitat was concentrated in the northeastern quarter of the park with moderately good conditions in the

Table 2.1. Size and ownership of 18 Core Security Areas Mapped for Adirondack Park

Core security area	Area (ha)	Percent private	Percent easement	Percent public
1	19,792	93.90	0	6.10
2	15,435	82.27	0	17.73
3	20,477	38.21	0	61.79
4	94,085	39.50	39.44	21.06
5	15,675	24.84	69.93	5.23
6	41,035	65.54	29.11	5.35
7	15,051	4.60	0	95.40
8	167,265	33.49	2.52	63.99
9	197,263	35.66	4.96	59.38
10	49,808	42.91	0	57.09
11	20,594	3.34	0	96.66
12	17,588	10.52	0	89.48
13	55,210	39.05	3.48	57.47
14	202,998	16.65	2.15	81.20
15	52,846	16.05	0	83.95
16	41,073	19.86	0	80.14
17	46,342	3.33	0	96.67
18	28,238	19.71	0	80.29
TOTALS	1,100,775	32.75	8.42	58.83

northwestern quarter. If we exclude areas with road densities greater than 0.6 km/km^2, a large portion of the best winter habitat is eliminated.

Snowmobile Trails

Access provided by winter trails may lead to harassment or killing of wolves. Conversely, wolves travel more efficiently on compacted routes such as deer and snowmobile trails (Mech 1970; Paquet et al. 1996). The total length of registered snowmobile trails in Adirondack Park is 2381 km with a concentration in core area 6. Area 6 also had the best habitat suitability of all 18 core areas. Accordingly, the likelihood of winter encounters with wolves would be high in this region.

Trains

With one possible exception, railroads pose little threat to wolves in Adirondack Park. The park has four tracks with 520 km of rail line leading to Ausable Forks, Newton Falls, Tahawus, and Lake Placid. The only track that could pose a threat to wolves passes through the Ha-De-Ron-Dah Wilder-

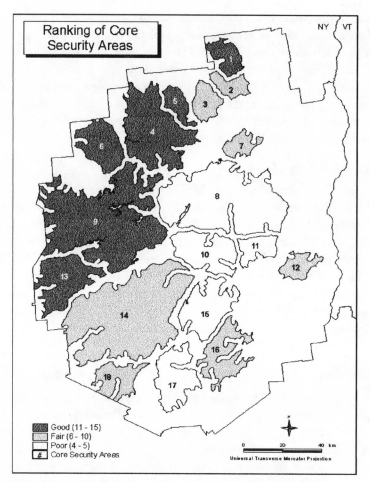

Figure 2.2. Location and ranking of potential core wolf habitat in Adirondack Park, New York.

ness, the Lake Canada Wilderness, and Five Ponds Wilderness. The threat posed by railroads is associated with location and traffic frequency. If tracks in high-quality wolf habitat are seldom used, they are of little concern. Although we could not get traffic information from railroad companies operating in the park, biologists in the region reported a low frequency of use.

Connectivity

Least-cost pathway analysis showed that much of Adirondack Park will not provide secure wolf movement. Patchy core areas would force wolves across an inhospitable matrix where human activity is widespread, especially during summer. Connectivity between the park and the surrounding region is even less likely because the Adirondack Park is isolated from the nearest potential timber wolf habitat by about 250 km (Harrison and Chapin 1997; Mladenoff and Sickley 1998). Geographic barriers such as the St. Lawrence River and Lake Champlain may isolate other suitable areas in New York from wolves in Canada and potential habitat in Maine and New Hampshire. Even using the most optimistic conditions, the prospect of wolves penetrating the matrix surrounding Adirondack Park is very low.

Estimated Wolf Population

Adirondack Park comprises 11,008 km^2 of potential core habitat (38 percent good habitat, 31 percent fair habitat, and 30 percent poor habitat) that can support a winter wolf density of 85 to 288 wolves. Because the biomass of prey in the park is low compared with other North American wolf habitats, we believe the higher estimate is unrealistic. A refined prey biomass estimate (Keith 1983; Fuller 1989) of 1.8 to 2.5 white-tailed deer/km^2 (N.Y. Department of Environmental Conservation, unpublished data) in potential core habitat produces an estimate of 132 to 154 wolves. This number would likely decline as predation reduces deer numbers to a lower level (Ballard et al. 2001).

Discussion

The biophysical and human elements that determine the habitability of potential wolf habitat vary throughout Adirondack Park. The sum of partially connected habitat patches combined with larger contiguous habitats could support about 150 wolves. The potential core habitat, however, is less than 45 percent of the area necessary to maintain long-term viability of an isolated wolf population. Moreover, predicted population size does not meet the minimum criterion of 201 (USFWS 1992).

Although corridors link high-quality habitats within the park, human activity and developments could impede access to areas that might otherwise be occupied by wolves. Connections outside the park are tenuous at best. At worst, the distance to the nearest wolf population is too great and the matrix habitat too harsh for one to expect regular exchange. Further, regional road density is even higher outside of the park. A reintroduced wolf population would create a long-term management problem if it expanded into agricultural lands. But fewer packs on less land would invite random population problems.

If the annual mortality of wolves in Adirondack Park exceeds 30 percent, population persistence may be in doubt—especially during establishment phases of reintroduction when high mortality is expected. As wolves seek home ranges, travel patterns are certain to be exaggerated (Boyd et al. 1996; chapter 6 in this volume). Highway mortality was the major cause of low survival for reintroduced lynx in New York (Brocke et al. 1991). In the Bow Valley of Banff National Park, recolonizing wolves has had poor success due to highway and railway deaths (Paquet et al. 1996). In Yellowstone National Park, road-related deaths are an important cause of mortality for reintroduced wolves (D. Smith, pers. comm.). If timber wolf restoration proceeds in Adirondack Park, we believe the initial population will need to be augmented until packs have established.

Natural landforms and the condensed arrangement of potential habitats in some core areas of the park may make wolves highly susceptible to human disturbance. In less physiographically complex environments such as the Great Lakes region, multiple travel routes link blocks of wolf habitat (D. Mech, pers. comm.; D. Mladenoff, pers. comm.). Because safe alternative routes are available, destruction or degradation of one or two routes is not usually critical. But wolves living in Adirondack Park probably could not avoid valley bottoms or use other travel routes without affecting their security. Therefore, tolerance of disturbance may be lower than in other human-dominated environments (Minnesota, Wisconsin) where wolves can avoid disturbed sites without jeopardizing their own survival.

The presence of coyote hybrids in New York has important implications for the potential reintroduction of timber wolves into the Adirondacks (Wilson et al. 1999). It is unclear whether coyotes interbred with wolves in New York or whether hybridization between eastern Canadian wolves and coyotes in Ontario was followed by a southern migration into New York. Even if conditions are right for establishment of timber wolves in Adirondack Park, the issue of which canid species originally occupied the area is unresolved. Historical accounts and genetic data from historic samples suggest the eastern Canadian wolf (*C. lycaon*)/red wolf (*C. rufus*) was common in New York State before extirpation. Although rare, timber wolves (*C. lupus*) may have also been present (De Kay 1842). Recent genetic evidence strongly suggests red wolves were endemic and the current dominant canid is a coyote hybrid (Wilson et al. 1999).

Whether the relocation of wolves into the Adirondacks could result in cohesive packs that avoid interbreeding with sympatric hybrid coyotes is unclear. Ongoing research on eastern Canadian wolves and red wolves may provide insights. An analysis of Algonquin wolves and neighboring Frontenac Axis canids has shown limited gene flow outside the park (Wilson et al. 2000). Conversely the red wolf in Alligator River, North Carolina, has hybridized with coyotes (B. Kelley, pers. comm.).

Conclusions

The current ecological and social conditions in Adirondack Park are probably unsuitable for timber wolves. Although our analysis suggests that the park has sufficient habitat to support a small population of wolves, regional conditions are not conducive to sustaining wolves over the long term. And given current trends in regional development, we anticipate a deterioration in potential wolf habitat. Because most development occurs in areas preferred by wolves, human activity will increase the risk of death and injury for wolves, reduce opportunities for wolves to move freely, displace or alienate wolves from preferred range, and interrupt normal periods of activity.

The taxonomic identity of the original Adirondack canid remains a mystery. We believe that if timber wolves were never present (or existed only in low numbers), then introduction of timber wolves is inappropriate. Today coyote hybrids are the predominant predator of white-tailed deer. From an ecological perspective, the functional role of a keystone predator may be more important than its genetic makeup.

If wolf reintroduction proceeds, a comparison of the ecological differences among timber wolves, eastern Canadian wolves, eastern coyotes, and hybrids should be undertaken. If the reintroduction of eastern Canadian wolves is intrinsically important because the species existed in New York and was extirpated by humans, then the feasibility of maintaining a population of *C. lycaon* must be addressed.

Without a commitment by governments to protect wolves inside and outside Adirondack Park, we doubt a reintroduction of timber wolves could succeed. The call for large reserves—especially when they are isolated from other areas with similar habitat—is a reflection of the spatial needs of inherently rare wide-ranging carnivores. All else being equal, large populations are less vulnerable to extinction than small populations (Jones and Diamond 1976; Pimm et al. 1988, 1993; Berger 1990; Schoener and Spiller 1992). In Adirondack Park, regional isolation would expose reintroduced timber wolves to the perils that threaten survival of all small populations. Without a source of timber wolves to augment the local population, wolves in the park would be subject to demographic and genetic problems that would depress reproduction and accelerate their mortality.

Literature Cited

Alexander, S., C. Callaghan, P. Paquet, and N. Waters. 1996. GIS predictive model for habitat use by wolves (*Canis lupus*). *CD Conference Proceedings. GIS '96—10 Years of Excellence*, March 19–23. Fort Collins: GIS World Inc.

Ballard, Warren B., Jackson S. Whitman, and Craig L. Gardner. 1987. Ecology of an exploited wolf population in south-central Alaska. *Wildlife Monographs* 98:1–54.

Ballard, Warren B., D. Lutz, T. W. Keegan, L. H. Carpenter, and J. C. deVos, Jr. 2001. Deer-predator relationships: A review of recent North American studies with an emphasis on mule and black-tailed deer. *Wildlife Society Bulletin* 29:99–115.

Berger, J. 1990. Persistence of different-sized populations: An empirical assessment of rapid extinctions in bighorn sheep. *Conservation Biology* 4:91–98.

Boitani, L. 1982. Wolf management in intensively used areas of Italy. In F. H. Harrington and P. C. Paquet, eds., *Wolves of the World.* Park Ridge, N.J.: Noyes.

———. 1995. Ecological and cultural diversities in the evolution of wolf-human relationships. In L. N. Carbyn, S. H. Fritts, and D. R. Seip, eds., *Ecology and Conservation of Wolves in a Changing World.* Edmonton: Canadian Circumpolar Institute, University of Alberta.

Boitani, L., F. Corsi, and E. Dupre. 1997. Large scale approach to distribution mapping: The wolf in the Italian peninsula. Oral presentation to the annual meeting of the Society for Conservation Biology, Victoria, B.C.

Boyd, D. K., P. C. Paquet, S. Donelon, R. R. Ream, D. H. Pletscher, and C. C. White. 1996. Transboundary movements of a recolonizing wolf population in the Rocky Mountains. In L. N. Carbyn, S. H. Fritts, and D. R. Seip, eds., *Ecology and Conservation of Wolves in a Changing World.* Edmonton: Canadian Circumpolar Institute, University of Alberta.

Brocke, R. H., K. A. Gustafson, and L. B. Fox. 1991. Restoration of large predators: Potentials and problems. In D. J. Decker, M. E. Krasny, G. R. Goff, C. R. Smith, and D. W. Gross, eds., *Challenges in the Conservation of Biological Resources: A Practioner's Guide.* Boulder: Westview.

Burnham, K., and D. Anderson. 1998. *Model Selection and Inference: A Practical Information-Theoretic Approach.* New York: Springer-Verlag.

Carbyn, L. N. 1983. Wolf predation on elk in Riding Mountain National Park, Manitoba. *Journal of Wildlife Management* 47:963–976.

Carroll, C., P. C. Paquet, and R. F. Noss. 1999a. *Modeling Carnivore Habitat in the Rocky Mountain Region: A Literature Review and Suggested Strategy.* Toronto: World Wildlife Fund.

Carroll, C., R. F. Noss, and P. C. Paquet. 1999b. *Carnivores as Focal Species for Conservation Planning in the Rocky Mountain Region.* Toronto: World Wildlife Fund.

———. 2001. Carnivores as focal species for conservation planning in the Rocky Mountain region. *Ecological Applications.*

Chapman, R. C. 1977. The effects of human disturbance on wolves (*Canis lupus*). M.S. thesis, University of Alaska, Fairbanks.

Daly, C., G. H. Taylor, and W. P. Gibson. 1997. The PRISM approach to mapping precipitation and temperature. *Tenth Conference on Applied Climatology*. Reno: American Meteorological Society.

De Kay, J. D. 1842. *Natural History of New York*. New York: D. Appleton & Co. and Wiley Putnam.

ERDAS IMAGINE. 1998. *Version 8.2*. Atlanta: ERDAS, Inc.

ESRI. 1996. *Arc-Info Version 7.0*. Redlands, Calif.: Environmental Systems Research Institute.

Fritts, S. H., and L. N. Carbyn. 1996. Population viability, nature reserves, and the outlook for gray wolf conservation in North America. *Restoration Ecology* 3:26–38.

Fuller, T. K. 1989. Population dynamics of wolves in north-central Minnesota. *Wildlife Monographs* 105:1–41.

———. 1995. Comparative population dynamics of North American wolves and African wild dogs. In L. N. Carbyn, S. H. Fritts, and D. R. Seip, eds., *Ecology and Conservation of Wolves in a Changing World*. Edmonton: Canadian Circumpolar Institute, University of Alberta.

Haight, R. G., D. J. Mladenoff, and A. P. Wydeven. 1998. Modeling disjunct gray wolf populations in semi-wild landscapes. *Conservation Biology* 12:879–888.

Harrison, D. J., and T. G. Chapin. 1997. An assessment of potential habitat for eastern timber wolves in the northeastern United States and connectivity with occupied habitat in southeastern Canada. Working Paper 7. New York: Wildlife Conservation Society.

Henshaw, R. E. 1982. Can the wolf be returned to New York? In F. H. Harrington and P. C. Paquet, eds., *Wolves of the World*. Park Ridge, N.J.: Noyes.

Jensen, W. F., T. K. Fuller, and W. L. Robinson. 1986. Wolf (*Canis lupus*) distribution on the Ontario-Michigan border near Sault Ste. Marie. *Canadian Field Naturalist* 100:363–366.

Jones, H. L., and J. Diamond. 1976. Short-term base studies of turnover in breeding bird populations on the California Channel Islands. *Condor* 78:526–549.

Keith, L. B. 1983. Population dynamics of wolves. In L. N. Carbyn, ed., *Wolves in Canada and Alaska*. Canadian Wildlife Report 45. Ottawa: Government of Canada.

Massolo, A., and A. Meriggi. 1998. Factors affecting habitat occupancy by wolves in northern Apennines (northern Italy): A model of habitat suitability. *Ecography* 21:97–107.

Mech, L. D. 1970. *The Wolf: The Ecology and Behavior of an Endangered Species*. Garden City: Natural History Press.

———. 1989. Wolf population survival in an area of high road density. *American Midland Naturalist* 121:383–389.

Mech, L. D., S. H. Fritts, G. Radde, and W. J. Paul. 1988. Wolf distribution in Minnesota relative to road density. *Wildlife Society Bulletin* 16:85–88.

Merrill, S. B. 2000. Road densities and gray wolf, *Canis lupus,* habitat suitability: An exception. *Canadian Field Naturalist* 114:312–313.

Messier, F. 1994. Ungulate population models with predation: A case study with the North American moose. *Ecology* 75:478–488.

———. 1995. On the functional and numerical responses of wolves to changing prey density. In L. N. Carbyn, S. H. Fritts, and D. R. Seip, eds., *Ecology and Conservation of Wolves in a Changing World.* Edmonton: Canadian Circumpolar Institute, University of Alberta.

Mladenoff, D. J., and T. A. Sickley. 1998. Assessing potential gray wolf restoration in the Northeastern United States: A spatial prediction of favorable habitat and population level. *Journal of Wildlife Management* 62:1–10.

Mladenoff, D. J., T. A. Sickley, R. G. Haight, and A. P. Wydeven. 1995. A regional landscape analysis and prediction of favorable gray wolf habitat in the northern Great Lakes region. *Conservation Biology* 9:279–294.

———. 1997. Causes and implications of species restoration in altered ecosystems: A spatial landscape projection of wolf population recovery. *BioScience* 47:21–31.

Mladenoff, D. J. , T. A. Sickley, and A. P. Wydeven. 1999. Predicting gray wolf landscape recolonization: Logistic regression models vs. new field data. *Ecological Applications* 9:37–44.

Paquet, P. C., J. Wierzchowski, and C. Callaghan. 1996. Effects of human activity on gray wolves in the Bow River Valley, Banff National Park, Alberta. In J. Green, C. Pacas, S. Bayley, and L. Cornwell, eds., *A Cumulative Effects Assessment and Futures Outlook for the Banff Bow Valley.* Prepared for the Banff Bow Valley Study. Ottawa: Department of Canadian Heritage.

Pimm, S. L., H. L. Jones, and J. Diamond. 1988. On the risk of extinction. *American Naturalist* 132:757–785.

Pimm, S. L., J. Diamond, T. M. Reed, G. J. Russell, and J. Verner. 1993. Times to extinction for small populations of large birds. *Proceedings of the National Academy of Sciences of the United States of America* 90:10871–10875.

Schoener, T. W., and D. A. Spiller. 1992. Is extinction rate related to temporal variability in population size? An empirical answer for orb spiders. *American Naturalist* 139:1176–1207.

Singleton, P. H. 1995. Winter habitat selection by wolves in the North Fork of the Flathead River Basin, Montana and British Columbia. M.S. thesis, University of Montana, Missoula.

Smith, D. W., W. G. Brewster, and E. E. Bangs. 1999. Wolves in the Greater Yellowstone ecosystem: Restoration of a top carnivore in a complex management environment. In T. W. Clark, A. P. Curlee, S. C. Minta, and P. M. Kareiva, eds., *Carnivores in Ecosystems.* New Haven: Yale University Press.

Thiel, R. P. 1985. The relationship between road density and wolf habitat suitability in Wisconsin. *American Midland Naturalist* 113:404–407.

U.S. Fish and Wildlife Service (USFWS). 1992. *Recovery Plan for the Eastern Timber Wolf*. St. Paul: USDI Fish and Wildlife Service.

Wilson, P. J., S. Grewal, R. C. Chambers, and B. N. White. 1999. Genetic characterization and taxonomic description of New York canids. In P.C. Paquet, J. R. Stritholt, and N. Staus, eds., *Wolf Reintroduction Feasibility in the Adirondack Park: A Report to the Adirondack Park Citizens Action Committee and Defenders of Wildlife*. Adirondack Park: Conservation Biology Institute.

Wilson, P. J., S. Grewal, I. D. Lawford, J. Heal, A. G. Granacki, D. Pennock, J. B. Theberge, M. T. Theberge, D. Voigt, W. Waddell, R. E. Chambers, P. C. Paquet, G. Goulet, D. Cluff, and B. N. White. 2000. DNA profiles of the eastern Canadian wolf and the red wolf provide evidence for a common evolutionary history independent of the gray wolf. *Canadian Journal of Zoology* 78:2156–2166.

Young, S. D. 1946. *The Wolf in North American History*. Caldwell, Idaho: Caxton Printers.

Chapter 3

Rewilding the Sky Islands Region of the Southwest

BARBARA L. DUGELBY, DAVE FOREMAN, RURIK LIST,
BRIAN MILLER, JACK HUMPHREY, MIKE SEIDMAN,
AND ROBERT HOWARD

The creation of islandlike protected areas such as national parks fails to conserve all elements of biological diversity, especially large mammals (Newmark 1985). In fact, large mammals are particularly susceptible to extirpation from isolated reserves (Newmark 1985). Yet keystone species, especially top carnivores, are essential to the maintenance of biological diversity and promotion of long-term ecosystem integrity (Soulé and Noss 1998). There is increasing evidence that many ecosystems are regulated from the top by large carnivores, that ecosystems may collapse or be radically altered without them, and that diversity and resilience will be lost as a result (Terborgh et al. 1999).

Rewilding—a large-scale approach to nature protection (Soulé and Noss 1998)—calls for large, connected core reserves with their full complement of native species. The central goal of rewilding is to maintain or restore ecologically viable populations of large carnivores and other keystone species (Soulé and Terborgh 1999; Foreman et al. 2000a). The Wildlands Project was organized in 1992 with the mission of restoring the ecological integrity of North America. One of our primary goals is to design and implement wildlands networks across the North American continent. Wildlands network designs place a strong emphasis on large mammals, especially large carnivores, due to their focal roles: "Focal species are organisms used in planning and managing reserves because their requirements for survival represent

65

factors important to maintaining ecologically healthy conditions" (Miller et al. 1999:82). Further, they support the overall goals of creating the Sky Islands Wildlands Network (SIWN): restoration of all large carnivores, ungulates, and other species native to the region; restoration of watersheds, streams, and riparian forests; restoration of a natural fire regime; restoration and protection of historical connectivity for native wide-ranging species; elimination or control of exotic species; protection of all remaining native forests and woodlands; and restoration of natural forest conditions. In this chapter we discuss the use of large mammals in the design and management of the proposed Sky Islands Wildlands Network (Foreman et al. 2000b).

The Greater Sky Islands Region

"To my mind," wrote Aldo Leopold, "these live oak-dotted hills fat with side oats grama, these pine-clad mesas spangled with flowers, these lazy trout streams burbling along under great sycamores and cottonwoods, come near to being the cream of creation" (1937:118). The landscape that so enthralled Leopold was where the Rocky Mountains and the Sierra Madre meet, where the plants and animals of the Neotropics mingle with those of the Nearctic, where jaguar (*Panthera onca*) and grizzly (*Ursus arctos*) hunted the same ridges, where elk (*Cervus elaphus*) and javelina (*Tayassu tajacu*) grazed and rooted cheek to jowl, and where northern goshawks (*Accipiter gentilis*) took thick-billed parrots (*Rhynchopsitta pachyrhyncha*) on the wing. Southwestern New Mexico, southeastern Arizona, northwestern Chihuahua, and north-eastern Sonora are a landscape of wonder, beauty, and wildness—and of great biological diversity. The greater Sky Islands region is globally important for the lessons it taught Leopold, for its role in launching the wilderness preservation movement, and for its biodiversity.

The Sky Islands Wildlands Network is intended to protect and restore the region's ecological integrity. It is embedded in an ecoregion of about 7 million hectares that extends from the Mogollon Rim in east-central Arizona and west-central New Mexico south to the northern Sierra Madre Occidental in Chihuahua and Sonora, Mexico (Figure 3.1). The Sky Islands cover nearly 4 million hectares at the center of the region (McLaughlin 1994). Weldon Heald coined the term "sky islands" in 1967 to denote mountain ranges that are isolated from each other by lower topography (Warshall 1994; McLaughlin 1994). Since basins may inhibit the movement of certain woodland and forest species just as saltwater seas isolate plants and animals on oceanic islands, the 40 ranges in the Sky Islands system may be thought of as an archipelago (Warshall 1994).

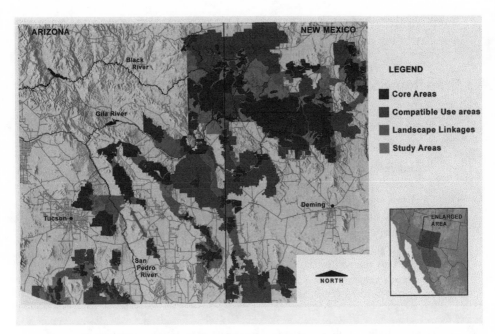

Figure 3.1. The proposed Sky Islands Wildlands Network.

Although numerous local influences play a role, the great diversity of the Mogollon Highlands/Sky Islands/northern Sierra Madre Occidental network stems from its location, topography, geological history (Warshall 1994), and incorporation of the Neotropic and Holarctic floristic provinces and the Neotropic and Nearctic faunal realms. The lowest gap in the continental cordillera between northern Canada and the Isthmus of Tehuantepec in southern Mexico is in the Sky Islands—a low pass that encouraged the mixing of eastern and western species, thus adding to the diversity. Some temperate species reach the southern limit of their ranges here while it is the northern limit for many tropical species (Felger and Wilson 1994). The Sky Islands region supports at least 104 mammal species and more than half of North America's bird species. The Sky Islands are the northern limit of 14 plant families and 4 bird families and the southern limit of 7 bird families. The geographic limits of 30 bird, 35 reptile, and 15 mammal species also occur here (Warshall 1994).

The Sky Islands, as well as the Mogollon Highlands and Sierra Madre, are covered primarily by oak and pine-oak woodlands and, above 3000 m, by conifer forest (McLaughlin 1994). The White Mountains, Mogollon Mountains, and Black Range (all with elevations over 3000 m) in the northern part of the region support spruce-fir forests and subalpine grasslands typical of the southern Rocky Mountains. Mixed conifer forests also flourish in the high country of the northern Sierra Madre Occidental.

Rewilding and Top-Down Regulation

The SIWN design is based on rewilding and focal species planning. Rewilding emphasizes keystone species, particularly large carnivores (Soulé and Noss 1998:22):

> Studies are demonstrating that the disappearance of large carnivores often causes ecosystems to undergo dramatic changes, many of which lead to biotic simplification and species loss (Mills et al. 1993). On land, these changes are often triggered by exploding ungulate populations. For example, deer, in the absence of wolves and cougars, have become extraordinarily abundant and emboldened in many rural and suburban areas throughout the United States, causing both ecological and economic havoc (McShea et al. 1977; Nelson 1997; McLaren and Peterson 1994).

Top carnivores such as puma (*Puma concolor*) encourage herding in elk and deer (*Odocoileus* spp.). In the absence of top carnivores, herd cohesiveness relaxes and individuals forage alone and tend to become stationary (Brown 1997). It has been suggested that the extermination of top predators increases ungulate densities and promotes selective foraging, which alters vegetation structure and composition (Stacy and Berger 2000). This can result in overgrazing, disruptions of nutrient flow and soil formation, and desertification in the southwestern United States (Savory 1990; Fleischner 1994). In the Tropics, Terborgh and colleagues (Terborgh 1992) have found that the absence of carnivore control of herbivores—tapir (*Tapirus bairdii*), monkeys, rodents, insects—can precipitate a rapid loss of plant species diversity. Construction of the Lago Guri reservoir in Venezuela created islands devoid of large predators such as jaguar, puma, and harpy eagle (*Harpia harpyja*). On such islands the recruitment of many canopy tree species has come to a halt because herbivores and seed predators are superabundant.

Top carnivores normally prey upon smaller predators and inhibit their foraging. Several studies have suggested that "demographic release" of mesopredators such as the house cat (*Felis cattus*), fox (*Vulpes* sp.), raccoon (*Procyon lotor*), and opossum (*Didelphis virginiana*) causes declines in songbirds and other small prey (Soulé et al. 1988; Palomares et al. 1995; Coté and Sutherland 1997; Terborgh et al. 1999). In isolated remnants of scrub habitat in southern California, Crooks (1997; Crooks and Soulé 1999) found that coyotes (*Canis latrans*) can limit feral house cat populations. In some situations the absence of top predators can lead to intense competition among former prey species for space or food, sometimes allowing one species of competitor to eliminate others (Paine 1966; Terborgh et al. 1999).

Three arguments support rewilding with an emphasis on large predators:

> First, the structure, resilience, and diversity of ecosystems are often maintained by "top-down" ecological (trophic) interactions initiated by top predators (Terborgh 1988; Terborgh et al. 1999). Second, wide-ranging predators usually require large protected landscapes for foraging, seasonal movement, and other needs; they justify bigness. Third, connectivity is also required because core protected areas typically are not large enough in most regions; they must be linked to insure long-term viability of wide-ranging species. (Soulé and Noss 1998)

In short, rewilding uses large predators and their prey to restore ecosystem integrity throughout a system of large interconnected reserves. Focal species analysis identifies additional high-value habitats and addresses three questions: How much area is needed? What is the quality of habitat? And in what configuration should we design components of a reserve network?

Miller et al. (1999) warn against selecting endangered species as focal species in reserve design; but if they are employed, they should be used judiciously and not exclusively. As focal species we recommend those that have viability and recovery targets that are linked to restoration of the Sky Islands Wildlands Network (Table 3.1). Accordingly, the large mammals that make

Table 3.1. Focal species of Sky Islands Wildlands Network

Focal species	*Umbrella*	*Keystone*	*Flagship*	*HQI[a]*	*WQI[b]*	*Prey*
CARNIVORES						
Mexican wolf						
(*Canis lupus baileyi*)	X	X	X		X	
Grizzly bear						
(*Ursus arctos horribilis*)	X		X	X	X	
Black bear						
(*Ursus americanus*)	X		X	X		
Jaguar						
(*Felis onca*)	X		X	X	X	
Mountain lion						
(*Felis concolor*)	X	X	X			

(continued)

Table 3.1. *Continued*

Focal species	Umbrella	Keystone	Flagship	HQI[a]	WQI[b]	Prey
UNGULATES						
Bighorn sheep (*Ovis canadensis* and *Ovis canadensis mexicana*)	X		X	X	X	
Bison (Mex.) (*Bison bison*)	X	X	X			X
Elk (*Cervus elaphus*)	X		X			X

[a]Habitat quality indicator
[b]Wilderness quality indicator

up this group embody these attributes: keystone species, umbrella species, flagship species, habitat quality indicator, wilderness quality indicator, and prey. A final criterion for selected species: they should be important both in the United States and Mexico. For example, the golden eagle (*Aquila chrysaetos*) is the national bird of Mexico and an important conservation flagship in Mexico even more so than in the United States.

Mexican Wolf

The Mexican wolf (*Canis lupus baileyi*) is an umbrella species, keystone species, flagship species, and wilderness quality indicator that was probably extirpated from the Sky Islands region in the late 1980s or early 1990s. Its preferred habitat is believed to be Madrean evergreen woodland, open pine forest, piñon-juniper woodland, and grasslands above 1800 m (Leopold 1959; McBride 1980; Brown et al. 1984; Johnson et al. 1992). As reintroduced wolves sort out their relationship with mountain lions, wolves will likely concentrate in more open and less rugged terrain (Leopold 1959; Brown et al. 1984; Parsons, pers. comm.). The reintroduction and protection of the wolf aids in achieving the SIWN goals of recovering native species and restoring connectivity.

As a wide-ranging species that maintains large territories and disperses long distances, the Mexican wolf is an umbrella species. Its recovery will require a metapopulation within the SIWN. At the very least, breeding packs should be present in the Gila/Apache NF complex, the Chiricahua/Peloncillo/Animas/Sierra San Luis complex, the Galiuro/Aravaipa complex, the Tumacacori/Pajarito/Buenos Aires complex, the Santa Rita/Canelo

Hills/Huachuca complex (Johnson et al. 1992), and several locations in the northern Sierra Madre. The wolf is also a keystone species because it is an important top-down regulator (Stacy and Berger 2000) that affects regional ecosystem processes. The presence of wolves will promote grizzly bear reintroduction by providing carcasses for scavenging (Mattson, pers. comm.).

The wolf's high profile in both the United States and Mexico and its strong public support make it a classic conservation flagship. Public opinion in Arizona, New Mexico, and Texas is in favor of wolf reintroduction (Taugher 1995). Although Mexican wolves do not require wilderness for habitat, they are vulnerable to vehicle-assisted poaching and do not usually persist where road density is more than 0.58 km/km^2 (Thiel 1985; Jensen et al. 1986). Therefore, wolves need secure roadless cores and extensive areas of low road density (Johnson et al. 1992)—which makes them a wilderness quality indicator for cultural, not ecological, reasons.

A priority in the conservation efforts for the Mexican wolf must be the evaluation of recent reports of wild wolves in Mexico. The goal for wolf recovery should be a viable linked population throughout the entire region. The SIWN will facilitate safe passage of wolves from the Gila/Apache to the Chiricahuas and Peloncillos and other linked regions.

MANAGEMENT RECOMMENDATIONS. To reduce further poaching we recommend:

1. Strategic road closures
2. Additional releases in remote parts of the Gila and Aldo Leopold Wilderness Areas, Blue Range Primitive Area, and Ladder Ranch
3. Effective investigations of wolf poaching
4. Reduction or elimination of livestock grazing in wolf release areas
5. Review of hunting regulations with respect to the danger of related activities disturbing wolves

International coordination should expand, perhaps through participation in PROFAUNA (a Mexican nonprofit conservation organization). Such collaborations should include searches for wolves in central Chihuahua and building public support for wolf restoration in Mexico.

Grizzly Bear

The grizzly bear is an umbrella species, flagship species, habitat quality indicator, and wilderness quality indicator. Although it is not present in the SIWN region—the last Mexican grizzly was killed in northwestern Chihuahua in 1960 (Brown 1985) and in New Mexico and Arizona in the mid-1930s—it was once present throughout the region and abundant in some areas. In 1973, the New Mexico–Arizona Section of The Wildlife Society resolved to support grizzly reintroduction (Brown 1985). A study for the U.S. Forest Service in 1974 showed that the Gila Wilderness Area is a viable reintroduction site (Erickson 1974).

Grizzly populations require large, intact, high-quality areas to maintain viability (Craighead et al. 1995). As a result, the umbrella of grizzly habitat protection will benefit many other species. As the most powerful and imposing carnivore south of the Arctic in North America, the grizzly is a classic flagship species. The grizzly also serves as a habitat quality indicator due to its demanding habitat requirements. Only an ecosystem of high quality and integrity can support grizzlies in the long run (Mattson et al. 1996). Finally, the grizzly is a wilderness quality indicator for the SIWN. Fragmentation and edge effects increase human-related mortality, which tends to be higher on the boundary of protected areas (Woodroffe and Ginsberg 1998). The grizzly requires large wilderness areas to avoid interaction with people. In the Northern Continental Divide Ecosystem, habitat use by grizzly bears declined where total road density reached 0.8 km/km^2 and open road density exceeded 0.5 km/km^2 (Mace and Manley 1993).

MANAGEMENT RECOMMENDATIONS. If grizzly bears are to be reintroduced, some roads may need to be closed in order to provide large blocks of habitat that reduce human contact and support large-scale ecosystem processes such as fire (Willcox 1998). Black bear hunting should be closed in occupied grizzly range to avoid mortality resulting from mistaken identity.

Black Bear

The black bear (*Ursus americanus*) is an umbrella species, flagship species, and habitat quality indicator. It aids in achieving the goals of protecting and restoring native forests and forest connectivity. The black bear has expansive area requirements and uses a variety of forest and woodland habitats (LeCount 1977). It is a popular species in North America and thus fills the role of flagship species as well.

Because the black bear has seasonally dynamic food habits that are habitat specific, it is an indicator of habitat quality and diversity. Poorly planned human activities can turn black bears into garbage feeders and panhandlers—behavior that may lead to lethal control (Morgan 2000). Core wilderness areas would be a buffer against such human influences. Despite a ban on black bear hunting in the Mexican portion of the reserve network, the species is found only in areas of difficult access and low human population density.

MANAGEMENT RECOMMENDATIONS. Bear hunting regulations should be closely monitored to ensure that female harvest does not cause population declines (Powell et al. 1997). Studies in the Sky Islands region should focus on the black bear's spatial requirements, distribution, metapopulation linkages, and vulnerability to harvest.

Jaguar

The jaguar is an umbrella species, flagship species, and wilderness quality indicator. It promotes the goals of carnivore recovery, protecting and restoring connectivity through Madrean woodland and riparian areas, and protecting and restoring riparian areas. Modern jaguar occurrences in the United States are likely the result of dispersal from Sonora, Mexico (R. Valdéz, pers. comm.). A reserve system that protects jaguars is an umbrella for many other species. The jaguar has large area requirements (10–90 km^2), and in the Sky Islands region it prefers Madrean woodland and riparian areas (Mondolfi and Hoogesteijn 1986; Lorenzana and López 2000)

Although research has found the jaguar to be a keystone in subtropical and tropical America, it is probably not a classic keystone at the northern edge of its range in the SIWN because of its historic low population density in the region. Considered by many to be one of most beautiful and mysterious animals in the world, the jaguar is an ideal conservation flagship. After recent sightings in New Mexico and Arizona, public interest in the jaguar is high. Even a group of ranchers, the Malpai Borderlands Group, has developed a jaguar protection campaign. The jaguar will prey on livestock, however, and suffers mortality from local ranchers where cattle-ranching is a dominant land use (Lopez et al. 2000). For security, it requires large wilderness cores without cattle and is a wilderness quality indicator.

MANAGEMENT RECOMMENDATIONS. The status of the jaguar in northern Chihuahua and Sonora should be examined to determine actual and potential dispersal corridors to the United States. Efforts should be made to improve legal protection for the Mexican source population and to acquire or otherwise protect the land that supports this northernmost population. Moreover, habitat suitability analyses should focus on areas in Arizona and New Mexico that might support breeding females.

Research in Mexico by Miller (pers. comm.) shows that jaguars are easily disturbed by human presence. Jaguars in western Mexico have a narrower food niche than pumas and are more restricted to large prey (such as white-tailed deer, *Odocoileus virginianus,* and peccary, *Tayassu tajacu),* than are pumas (Núñez et al. 2000). Even with full protection of the jaguar in the United States, the species remains highly threatened by poaching. Success-ful recovery of the jaguar may require limiting access to canyons and Madrean woodland in mid- and lower-elevation portions of Sky Islands. The protected Cienega Creek corridor under I-10 east of Tucson appears to be suitable for jaguar movement (Terborgh, pers. comm.). Protection of this corridor and an increase in javelina will encourage jaguar recovery in the United States.

Mountain Lion

The mountain lion (*Puma concolor*) is an umbrella species, keystone species, and flagship species. It offers top-down regulation of ungulate prey that can be enhanced by restoring connectivity among various ecosystems. The inter-connected landscape that a viable population requires would benefit many other species. The mountain lion is of value to hunters and nonhunters alike—characteristics of a potential conservation flagship. Logan et al. (1996) have suggested that harvest-free areas of at least 2600 km^2 would support long-term population stability.

MANAGEMENT RECOMMENDATIONS. The Gila and Aldo Leopold Wilder-ness areas including parts of the Gila National Forest and the Ladder Ranch are large enough to serve as a suitable harvest-free refuge. Regional harvest regulations should be examined and altered if necessary to assure that self-sustaining populations continue to serve their keystone roles, and can colo-nize vacant range.

Desert Bighorn and Rocky Mountain Sheep

The bighorn sheep (*Ovis canadensis*) is a flagship species, habitat quality indi-cator, and wilderness quality indicator. Successful recovery and management will promote connectivity among mountain ranges. Both subspecies are flag-ships because they are popular with the general public, hunters, and wildlife managers. In fact, New Mexico and Arizona wildlife agencies are reintro-ducing bighorns to suitable mountain ranges in the SIWN. One of these reintroductions took place on the Pedro Armenderas ranch owned by Ted Turner. Bighorn are vulnerable to diseases carried by domestic sheep and require access to pathogen-free habitat (see chapter 11 in this volume). They also have demanding habitat requirements for parturition and feeding

(Valdez and Krausman 1999). The bighorn is present or has been reintroduced to several mountain ranges in the region. They are vulnerable to
poaching and roadkill when moving from one range to another. For these
reasons bighorn are habitat quality indicators. And because they require core
wilderness habitat they are wilderness quality indicators as well.

MANAGEMENT RECOMMENDATIONS. We recommend that regional reintroductions continue and that landscape connections within an emerging
metapopulation should be enhanced and protected. Because bighorn are vulnerable to human disturbances including livestock, roads, and fences
(SEMARNAP 2000), habitat-fragmenting influences in the SIWN should
be minimized.

American Bison

The bison (*Bison bison*) is a flagship species and prey species with expansive
spatial needs in grassland landscapes. The recovery of bison in northern Mexico and the southwestern United States would protect prairie dogs and ecological grassland processes that have traditionally been disrupted by cattle.
Also, the border-inhabiting herd may be genetically unique (Trefethen 1975)
and could provide an important food source for wolves and grizzly bears
(Carbyn and Trottier 1987; Craighead et al. 1995). (Although it has been
questioned whether the American bison historically ranged in Mexico, the
evidence of its presence in Chihuahua is indisputable. The free-ranging herd
recently fenced in in New Mexico presents the possibility, although remote,
that the bison found today in the region are genetically descended from the
bison of the southern Great Plains.)

MANAGEMENT RECOMMENDATIONS. The Janos Antelope Wells herd, which
has resided mostly on a ranch in the United States since the summer of 1998,
could form the nucleus of a free-ranging population. Studies should determine the exact status and genetic history of the herd on the New Mexico–Chihuahua border and develop a compensation program for potential
damage to fences.

Elk

The elk (*Cervus elaphus*) is a flagship species and an important prey species.
Viable elk will facilitate carnivore recovery and reflect an interconnected
matrix of grasslands and forests across a broad range of elevations. The success of elk restoration throughout North America is evidence of strong popular support and ecological resilience. The expansion of elk restoration efforts

throughout the SIWN, including Mexico, is not likely to be controversial. Elk are important prey for Mexican wolf, mountain lion, grizzly bear, and, potentially, jaguar.

MANAGEMENT RECOMMENDATIONS. Managers should work with hunters, outfitters, and ranchers to replace cattle ranching with trophy elk hunting permits on the Gila and Apache National Forests (Hess 1998). Elk should be reestablished wherever feasible in Sky Islands ranges in northern Mexico.

Conclusions

Compared to most single-species initiatives, this proposal is a radical but essential step toward regional landscape restoration (Figure 3.1) and takes a larger-scale view than is traditional (see chapter 16 in this volume). The SIWN is driven by ecological goals. We anticipate its implementation to occur over a period of decades.

Wildlands conservation can be an effective, ethical response to many of the threats to our quality of life—such as lifestyle changes due to sprawling cities and boom-or-bust economic cycles. Protecting native ecosystems is a traditional tool for conserving sensitive species, but ecosystem protection is now recognized as a significant factor in ensuring the quality of our own life and economic security for generations to come. When large natural ecosystems are protected, neighboring communities receive direct economic benefits. These benefits are possible because communities near protected areas offer a clean, healthy, highly desirable setting that invites long-term economic investment from those who value nature. In the long term, economic health is a true by-product of ecological health.

The social challenges presented by the implementation of the SIWN are considerable—but not insurmountable. Even though the social and economic opportunities associated with wildlands preservation have demonstrated their merit, rural residents and government agencies tend to resist such ambitious changes. Achieving full implementation of the SIWN and similar networks will require changes in social attitudes about property rights, government regulations, and accepting responsibility for actions that degrade nature. Modifying these attitudes will require a broad public outreach program involving local farmers and ranchers, businesspeople, residents, schools, and government agencies. We must move beyond the pervasive negative mindset that prevents us from restoring natural landscapes while developing healthy local economies.

Literature Cited

Brown, D. E. 1985. *The Grizzly in the Southwest.* Norman: University of Oklahoma Press.

Brown, D. E., D. M. Gish, R. T. McBride, G. L. Nunley, and J. F. Scudday. 1984. *The Wolf in the Southwest: The Making of an Endangered Species.* Tucson: University of Arizona Press.

Brown, D. E., and N. B. Carmony, eds. 1995. *Aldo Leopold's Southwest.* Albuquerque: University of New Mexico Press.

Brown, J. S. 1997. *The Ecology of Fear: Optimal Foraging, Game Theory, and Trophic Interactions.* Abstracts of oral and poster papers. Acapulco: Seventh International Theriological Congress.

Carbyn, L. N., and T. Trottier. 1987. Responses of bison on their calving grounds to predation by wolves in Wood Buffalo National Park. *Canadian Journal of Zoology* 65:2072–2078.

Côté, I. M., and W. J. Sutherland. 1997. The effectiveness of removing predators to protect bird populations. *Conservation Biology* 11:395–405.

Craighead, J. J., J. S. Sumner, and J. A. Mitchell. 1995. *The Grizzly Bears of Yellowstone: Their Ecology in the Yellowstone Ecosystem, 1959–1992.* Washington, D.C.: Island Press.

Crooks, K. 1997. Tabby go home: House cat and coyote interactions in southern California habitat remnants. *Wild Earth* 7(4):60–63.

Crooks, K., and M. E. Soulé. 1999. Mesopredator release and avifaunal extinctions in a fragmented system. *Nature* 400:563–566.

Erickson, L. R. 1974. Livestock utilization of a clear-cut burn in northeastern Oregon. M.S. thesis, Oregon State University.

Estes, J. A., N. S. Smith, and J. F. Palmisano. 1978. Sea otter predation and community organization in the western Aleutian Islands, Alaska. *Ecology* 59:822–833.

Felger, R. S., and M. F. Wilson. 1994. Northern Sierra Madre Occidental and its apachian outliers: A neglected center of biodiversity. In L. F. DeBano, P. F. Ffolliott, A. Ortega-Rubio, G. J. Gottfried, R. H. Hamre, and C. B. Edminster, tech. coords., *Biodiversity and Management of the Madrean Archipelago: The Sky Islands of Southwestern United States and Northwestern Mexico.* General Technical Report RM-GTR-264. Tucson: USDA Forest Service, Rocky Mountain Forest and Range Experiment Station.

Fleischner, T. L. 1994. Ecological costs of livestock grazing in western North America. *Conservation Biology* 8:629–644.

Foreman, D., B. Dugelby, J. Humphrey, B. Howard, and A. Holdsworth. 2000. The elements of a wildlands network conservation plan: An example from the Sky Islands. *Wild Earth,* special issue, 10(1):17–30.

Foreman, D., K. Daly, B. Dugelby, R. Hanson, R. E. Howard, J. Humphrey, L. Klyza Linck, R. List, and K. Vacariu. 2000. *Sky Islands Wildlands Network Conservation Plan.* Tucson: Wildlands Project.

Frey, J. K. 1998. *Natural History Characteristics of Focal Species for the Sky Island/Greater Gila Reserve Design.* Albuquerque: Sky Island Alliance.

Hess, K. Jr. 1998. *Incentive-Based Conservation for the Sky Island Complex: A Draft Report to the Wildlands Project on Livestock, Elk, and Wolves.* Tucson: Wildlands Project.

Jensen, W. F., T. K. Fuller, and W. L. Robinson. 1986. Wolf, *Canis lupus,* distribution on the Ontario-Michigan border near Sault Ste. Marie. *Canadian Field Naturalist* 100(3):363–366.

Johnson, T., D. C. Noel, and L. Z. Ward. 1992. *Summary of Information on Four Potential Mexican Wolf Reintroduction Areas in Arizona.* Tucson: Arizona Game and Fish Department.

Lambeck, R. J. 1997. Focal species: A multi-species umbrella for nature conservation. *Conservation Biology* 11:849–856.

LeCount, A. L. 1977. Some aspects of black bear ecology in the Arizona chaparral. *International Conference on Bear Research and Management* 4:175–179.

Leopold, A. 1937. Conservationist in Mexico. *American Forests* 43:118–120.

———. 1949. *A Sand County Almanac.* Oxford: Oxford University Press.

———. 1959. *Wildlife of Mexico: The Game Birds and Mammals.* Berkeley: University of California Press.

Logan, K. A., L. L. Sweanor, T. K. Ruth, and M. G. Hornocker. 1996. *Cougars of the San Andres Mountains, New Mexico.* Project W-128-R. Sante Fe: New Mexico Department of Game and Fish.

López, C. A., D. E. Brown, and G. Lorenzana. 2000. El jaguar en Sonora: ¿Desapareciendo o solamente desconocido? *Especies* 3:19–23.

Lorenzana Piña, G., and C. A. López González. 2000. Carnivore distribution, activity and habitat association in central Sonora. In *Carnivore 2000: Proceedings and Agenda.* Denver: Defenders of Wildlife.

Lowe, C. H. 1985. *Arizona's Natural Environment.* Tucson: University of Arizona Press.

Mace, R. D., and T. L. Manley. 1993. *South Fork Flathead River Grizzly Bear Project: Progress Report for 1992.* Kalispell: Montana Department of Fish, Wildlife and Parks.

Mattson, D. J., S. Herrero, R. G. Wright, and C. M. Pease. 1996. Designing and managing protected areas for bears: How much is enough? In R. G. Wright, ed., *National Parks and Protected Areas: Their Role in Environmental Protection.* Cambridge, Mass.: Blackwell Science.

McBride, R. T. 1980. *The Mexican Wolf (*Canis lupus baileyi*): A Historical Review and*

Observations in Its Status and Distribution. Washington, D.C.: U.S. Fish and Wildlife Service.

McLaren, B. E., and R. O. Peterson. 1994. Wolves, moose and tree rings on Isle Royale. *Science* 266:1555–1558.

McLaughlin, S. P. 1994. An overview of the flora of the Sky Islands, southeastern Arizona: Diversity, affinities, and insularity. In L. F. DeBano, P. F. Ffolliott, A. Ortega-Rubio, G. J. Gottfried, R. H. Hamre, and C. B. Edminster, tech. coords., *Biodiversity and Management of the Madrean Archipelago: The Sky Islands of Southwestern United States and Northwestern Mexico.* General Technical Report RM-GTR-264. Tucson: USDA Forest Service, Rocky Mountain Forest and Range Experiment Station.

McShea, W. J., H. B. Underwood, and J. H. Rappole. 1997. The Science of Overabundance: Deer Ecology and Population Management. Washington, D.C.: Smithsonian Institution Press.

Miller, B., R. Reading, J. Strittholt, C. Carroll, R. Noss, M. Soulé, O. Sanchez, J. Terborgh, D. Brightsmith, T. Cheeseman, and D. Foreman. 1999. Focal species in the design of reserve networks. *Wild Earth* 8(4):81–92.

Mills, L. S., M. E. Soulé, and D. F. Doak. 1993. The history and current status of the keystone species concept. *BioScience* 43:219–224.

Mondolfi, E., and R. Hoogesteijn. 1986. Notes on the biology and status of the jaguar in Venezuela. In S. D. Miller and D. D. Everett, eds., *Cats of the World: Biology, Conservation, and Management.* Washington, D.C.: National Wildlife Federation.

Morgan, C. 2000. Bear safe Washington: An education program to alleviate conflict between humans and bears. In *Carnivore 2000: Proceedings and Agenda.* Denver: Defenders of Wildlife.

Nelson R. 1997. *Heart and Blood: Living with Deer in America.* New York: Knopf.

Newmark, W. D. 1985. Legal and biotic boundaries of western North American national parks: A problem of congruence. *Biological Conservation* 33:197–208.

Núñez, R., B. Miller, and F. Lindzey. 2001. Food habits of jaguars and pumas in Jalisco, Mexico. *Journal of Zoology* 252:373–379.

Palomares, F., P. Gaona, P. Ferreras, and M. Debiles. 1995. Positive effects on game species of top predators by controlling smaller predator populations: An example with lynx, mongooses, and rabbits. *Conservation Biology* 9:295–305.

Paine, R. T. 1966. Food web complexity and species diversity. *American Naturalist* 100:65–75.

Powell, R. A., J. W. Zimmerman, and D. E. Seaman. 1997. *Ecology and Behavior of North American Black Bears: Home Ranges, Habitat, and Social Organization.* New York: Chapman and Hall.

Quigley, H. B. 1987. Ecology and conservation of the jaguar in the Pantanal Region, Mato Grosso do Sul, Brazil. Ph.D. dissertation, University of Idaho, Moscow.

Rabinowitz, A. R. 1986. Ecology and behaviour of the jaguar (*Panthera onca*) in Belize, Central America. *Journal of Zoology* 210:149–159.

Savory, A. 1990. Wolf, man and cow—The essential alliance. Paper presented to the Arizona Wolf Symposium, 23–24 March 1990. Tempe: Preserve Arizona's Wolves and Arizona Chapter of the Wildlife Society.

SEMARNAP. 2000. *Proyecto para la conservación, manejo y aprovechamiento del borrego cimarrón (*Ovis canadensis*) en México.* Mexico City: Secretaría de Marina, Recursos Naturales y Pesca.

Soulé, M. E., and R. F. Noss. 1998. Rewilding and biodiversity as complementary goals for continental conservation. *Wild Earth* 8(3):18–28.

Soulé, M. E., D. T. Bolger, A. C. Alberts, J. Wright, M. Sorice, and S. Hill. 1988. Reconstructed dynamics of rapid extinctions of chaparral-requiring birds in urban habitat islands. *Conservation Biology* 2:75–92.

Soulé, M. E., and J. Terborgh. 1999. *Continental Conservation: Scientific Foundations of Regional Reserve Networks.* Washington, D.C.: Island Press.

Stacey, P. B., and J. Berger. 2000. How extinctions of grizzly bears and wolves can affect the diversity of riparian birds. In *Carnivores 2000: Proceedings and Agenda.* Denver: Defenders of Wildlife.

Taugher, M. 1995. Rural residents back N.M. wolf release. *Albuquerque Journal,* 6 December.

Terborgh, J. 1988. The big things that run the world—A sequel to E. O. Wilson. *Conservation Biology* 2:402–403.

Terborgh, J. 1992. Maintenance of diversity in tropical forests. *Biotropica* 24:283–292.

Terborgh, J., J. A. Estes, P. Paquet, K. Ralls, D. Boyd, B. Miller, and R. Noss. 1999. The role of top carnivores in regulating terrestrial ecosystems. In M. Soulé and J. Terborgh, eds., *Continental Conservation.* Washington, D.C.: Island Press.

Thiel, R. P. 1985. Relationship between road densities and wolf habitat suitability in Wisconsin. *American Midland Naturalist* 113:404–407.

Trefethen, J. V. 1975. *An American Crusade for Wildlife.* Alexandria, Va.: Boone and Crocket Club.

Valdez, R., and P. R. Krausman. 1999. *Mountain Sheep of North America.* Tucson: University of Arizona Press.

Warshall, P. 1994. The Madrean Sky Island Archipelago: A planetary overview. In L. F. DeBano, P. F. Ffolliott, A. Ortega-Rubio, G. J. Gottfried, R. H. Hamre, and C. B. Edminster, tech. coords., *Biodiversity and Management of the Madrean Archipelago: The Sky Islands of Southwestern United States and Northwestern Mexico.* General Technical Report RM-GTR-264. Tucson: USDA Forest Service, Rocky Mountain Forest and Range Experiment Station.

———. 1995. Southwestern Sky Island Ecosystems. In E. T. LaRoe, G. S. Farris, C. E. Pucket, P. D. Doran, and M. J. Mac, eds., *Our Living Resources: A Report to the*

Nation on the Distribution, Abundance, and Health of U.S. Plants, Animals, and Ecosystems. Washington, D.C.: U.S. Department of Interior, National Biological Service.

Willcox, L. 1998. *A Sense of Place: An Atlas of Issues, Attitudes and Resources in the Yellowstone to Yukon Ecoregion.* Canmore, Alberta: Yellowstone to Yukon Conservation Initiative.

Woodroffe, R., and J. R. Ginsberg. 1998. Edge effects and the extinction of populations inside protected areas. *Science* 280:2126–2128.

Chapter 4

Using Public Surveys and GIS to Determine the Feasibility of Restoring Elk to Virginia

JULIE A. MCCLAFFERTY AND JAMES A. PARKHURST

Prior to European settlement, the eastern elk (*Cervus elaphus canadensis*) in Virginia was most abundant in the central and western mountainous regions (McKenna 1962). It was extirpated from Virginia in 1855 (Wood 1943), and for the next 62 years wild elk were not known to exist within the commonwealth. In 1917, the Virginia Game Commission transported about 150 Rocky Mountain elk (*C. elaphus nelsoni*) from Yellowstone National Park and released them in small groups across the state (Wood 1943). Because little was known about the habitat requirements of elk at that time, the locations of many releases were ill advised and most releases failed. Two small herds (smaller than 100 elk each) survived in west-central Virginia, but soon they conflicted with agricultural interests. Virginia's first regulated elk season opened in 1922, and many additional elk were poached by frustrated farmers (Wood 1943; McKenna 1962). By 1960, when legal elk hunting was discontinued, both herds had declined. They vanished in 1970 (Gwynne 1977).

In response to a renewed public interest in Virginia elk, we began a restoration feasibility study in 1997. To be successful, elk restoration must be based on specific information about the elk's historic range, habitat requirements, interspecific relations, socioeconomics, public relations, and management. Our first objective was to define suitable elk habitat and determine if it existed in Virginia. Our second objective was to determine the socioeconomic costs, benefits, and other ramifications of elk restoration in Virginia.

In this chapter we review the methods and results of this feasibility study and discuss its local and regional implications on elk and other large mammal restoration projects.

Methods

We performed this feasibility study in two stages, first examining only the biological aspects of elk restoration and then qualifying that assessment with a socioeconomic analysis. In the first step, we assessed potential elk habitat to determine if Virginia has the necessary biological/ecological foundation to support a restored elk herd. To keep this step as objective and quantitative as possible, only the physical condition of existing habitat was considered. In the second step, we built upon this concept of biological feasibility to include socioeconomic factors such as public attitudes and human population trends.

Biological Assessment

We assessed the biological feasibility of elk restoration in two stages. First we identified potential study areas using major roads (USGS 1985) as an indicator of development and habitat fragmentation. Using USGS digital line graphs (DLGs) and Map and Image Processing Software (MIPS ver 5.6, TNT-MicroImages, Inc., Lincoln, Nebraska), we created a 150-m buffer on each side of every Class 1 or 2 road. Potential habitat areas were identified as any polygon falling outside these buffers. Only polygons greater than 25,000 ha (approximately the maximum size of Virginia's previous restoration area) and adjacent to at least one other polygon greater than 25,000 ha (a combined area of more than 50,000 ha, or 500 km^2; Witmer 1990) were extracted for further consideration.

Second we assessed habitat conditions within each extracted study area by using a habitat suitability index (HSI) model (McClafferty 2000) developed according to guidelines prepared by the U.S. Fish and Wildlife Service (USFWS 1981). Similar models developed by Wisdom et al. (1986), Beyer (1987), Thomas et al. (1988), Long (1996), and Van Deelen et al. (1997) were useful references during model development, but because these models were designed for other geographic regions and operational scales they were not directly applicable. We executed our HSI model using MIPS for geographic information systems (GIS) processing. The model assesses the four basic life requisites of elk (forage, cover, water, and space) using four variables: landscape composition (HE_C), landscape interspersion (HE_I), water availability (HE_W), and road density (HE_R). Each variable was averaged across the landscape to give a single score for each variable in each polygon ranging from zero (nonviable) to 1 (optimal). (See McClafferty 2000 for a thorough description of these variables.)

The first two variables (landscape composition and interspersion) employed a digital land-cover map that consisted of 1991–1993 Landsat Thematic Mapper satellite imagery collected at a resolution of 30 m by 30 m and classified into seven land cover categories: deciduous forest, coniferous forest, deciduous shrub, herbaceous, open water, disturbed, and coastal wetland (Morton 1998). These we collapsed into three categories: forest, open (shrub and herbaceous), and other (considered nonhabitat). The composition variable measured the amount of forest versus open land in a given portion of the landscape; interspersion measured the accessibility (proximity) of these critical habitat types from each location within the polygons.

We obtained road density and water availability data from 1:100,000 DLGs (USGS 1985). To measure road density, we converted Class 3 and 4 roads (improved local, residential, and rural roads) to MIPS raster maps and calculated density at each location within a polygon. We measured water availability by using the permanent (nonintermittent) water features from the hydrographic databases and calculating the linear distance from each point in a polygon to the nearest water source.

We combined the four habitat variables using a simple geometric mean:

$$HSI = (HE_C \times HE_I \times HE_W \times HE_R)^{(1/4)}$$

Given this relationship, a high score on one variable will tend to compensate low scores. But a variable with a low score influences the final index more than variables with high scores. We calculated HSI scores at each pixel in a polygon and then averaged across each polygon to obtain the final polygon score.

Interpreting Model Output

To compare study areas, we converted each final HSI score into habitat units by multiplying the HSI by the size (in hectares) of the study area (USFWS 1981). This process allowed us to compare directly the habitat availability in study areas with different HSI scores and different areas—for example, a 25,000-ha polygon with an average HSI score of 0.6 would be equivalent functionally to a hypothetical polygon of 15,000 ha of optimal habitat. This measure is similar to Lidicker's (1988) ratio of optimal to marginal patch area (ROMPA). We then extrapolated habitat units to an estimate of expected elk population size (based on an "optimal" elk density of 1 elk per 125 habitat units). This figure is the average density of established elk herds in the East (Michigan: 1 elk/230 ha, SEAFWA 1997; Pennsylvania: 1 elk/147 ha, Forbes and Ferrence 1999; Arkansas: 1 elk/202 ha, SEAFWA 1997) multiplied by 0.65 (the average HSI score we obtained for Virginia). Given that both habitat units and population estimates are "per area" estimates, either can be used to directly compare study areas.

Each study area received two ranks: one according to estimated population size (higher population size corresponding to higher rank), the other on the effect of study area shape (higher area/perimeter ratio corresponding to higher rank) (McClafferty 2000). A study area's final rating for overall biological feasibility was the lower of these two values.

Socioeconomic Assessment

We assessed socioeconomic feasibility at statewide and regional scales (McClafferty 2000). An informational brochure and mail survey were sent to 2400 randomly selected Virginia residents stratified by county. The brochure explained the purpose and approach of the feasibility study, a history of elk in Virginia, the life history of elk, and a summary of basic elk habitat requirements. The survey contained 60 questions that assayed participant demographics, knowledge, attitudes, and interest regarding elk and elk restoration. We administered the survey in spring 1999 using a modified version (McClafferty 2000) of Dillman's (1978) total design method; we assessed nonresponse bias through 50 telephone interviews with nonrespondents during summer 1999. Our primary objective in conducting this survey was to build a foundation for greater public involvement by obtaining preliminary information on citizens' knowledge and opinions and identifying the issues we would have to address later in decision-making processes.

We conducted four stakeholder workshops to assess the socioeconomic parameters surrounding elk restoration in the regions where potential elk habitat had been identified (Figure 4.1). We invited representatives of three major stakeholder groups: agricultural producers (foresters, livestock producers, orchardists, vegetable growers, grain farmers), government agencies (U.S. Forest Service, National Park Service, Virginia Department of Transportation, police), and users (recreation, research, environmental, economic development). We strove to get 20 to 25 participants, equally divided among the three groups, at each meeting.

At the start of each workshop, participants were informed that no decisions about elk restoration would be made during the meeting. Instead they were to discuss what they considered necessary in the feasibility study for their area. Following a slide show about the history of elk, elk restoration, habitat needs, life history, and an overview of this feasibility study, participants received instruction in nominal group technique (Moore 1987). Over

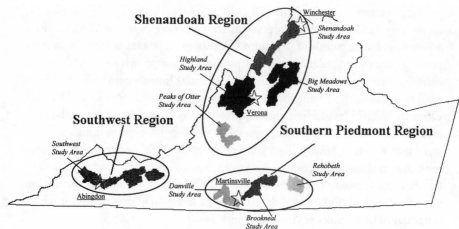

Figure 4.1. Location map of study areas (italics, color-coded according to biological feasibility) and regional stakeholder workshops (underlined text, stars) held during summer and fall 1999 as part of a feasibility assessment of elk restoration in Virginia. Dark colored study sites indicate high biological feasibility, medium gray sites indicate medium biological feasibility, and light gray sites indicate low biological feasibility.

the course of the day, they were asked to list all possible benefits of elk restoration to their area, then list all possible costs or detriments of elk restoration to their area, and finally to suggest how these issues might be reconciled. We asked all participants to complete pre- and postmeeting surveys to help us assess the workshop's effectiveness in meeting their expectations (McClafferty 2000).

Results and Discussion

The two-step process that we used in our analysis worked very well, and the socioeconomic assessment complemented the biological assessment in many ways. In this section, we first review the results of these two phases individually and then discuss how we combined them to produce our final recommendations.

Identification of Potential Elk Habitat

The HSI model identified eight areas in three regions of Virginia as potential habitat for elk: one in southwestern Virginia, four in the Shenandoah Mountains, and three in the Southern Piedmont (Figure 4.1). The area in the Southwest Region encompasses 209,974 ha and portions of nine counties in public (26 percent) and private (73 percent) ownership. This elongated area, containing 126,975 habitat units (estimated carrying capacity: 976), received a potential habitat rank of "medium." Although this area is capable of supporting elk, its irregular shape and predominance of private lands present complex management issues, including potential human conflicts.

The areas in the Shenandoah Region ranged from 67,994 ha at Peaks of Otter to 222,875 ha at Highland. Approximately 33 to 50 percent of the land area in these sites is in public ownership (national forest, national park, or state management lands). Two sites received a "high" biological feasibility rank due to their large, contiguous layout, low road density, and high potential carrying capacity (more than 700 elk). The other two sites in this region were smaller, had higher road densities, were irregularly shaped, and received potential habitat ranks of "medium" and "low."

The three sites in the Southern Piedmont Region were small (under 90,000 ha) and mostly in private ownership (more than 95 percent). Despite having high landscape composition and interspersion values, these sites received potential habitat rankings of "low" or "medium" due to high road densities and small size. Although only one of these three sites appeared capable of supporting a viable elk population (estimated carrying capacity: more than 400), their close proximity to each other enhanced the potential for the region as a whole to fulfill the habitat needs of elk and perhaps function as a metapopulation (Gilpin 1996). Such potential is dampened, however, by a high probability of human/elk conflicts on private lands (owing to crop depredation).

Because the HSI model used only data from Virginia, our habitat evaluation terminated at the state boundary. With databases for West Virginia and Kentucky, the number and size of study areas identified in Virginia would have been much larger. The extensive Monongahela National Forest across the West Virginia border and the Cumberland Plateau along the Kentucky border (where an elk restoration is currently under way as explained in chapter 5) are examples of such areas. Thus our estimate of potential elk habitat in Virginia is conservative.

Based strictly on biological parameters, the Shenandoah Region appears to have the greatest restoration potential—given that two of the four areas ranked "high" in this first phase of study. The Southwest and Southern Piedmont regions ranked lower because of the predominance of private land and habitat that is more patchy, discontinuous, and more heavily settled.

Other Biological Considerations

Identifying potential elk habitat is only one aspect of assessing the biological feasibility of restoration. An HSI model alone is insufficient for evaluating biological implications such as (1) disease and parasite transmission (from elk to indigenous wildlife, domestic stock, and humans or from indigenous wildlife and domestic stock to elk); (2) interactions between elk and native fauna (competition with deer and effects on songbirds, small mammals, herps, and other fauna through induced habitat modification); and (3) impacts of elk on native flora (intensified grazing pressure, forest browsing, trampling). Clearly each of these issues might jeopardize present health, economic, and ecological conditions. Precautionary protocols (Nettles and Corn 1998; chapter 7 in this volume) have been developed to avoid disease problems stemming from restoration. And with the implementation of elk population management strategies, most troublesome issues could be prevented. But a thorough management plan addressing these issues must be developed and set in place before any restoration is attempted—an approach that has seldom been applied in the eastern United States.

Statewide Socioeconomic Assessment

Most respondents (61 percent) agreed that "reintroducing elk into Virginia is a good idea"; 14 percent disagreed and 25 percent were undecided ("neutral" or "don't know"). When asked a similar question later in the survey without the "neutral" or "don't know" options, support increased marginally to 68 percent whereas opposition more than doubled to 32 percent. Clearly those who initially responded strongly in either direction did not change their minds during the survey. But most of those (65–69 percent) who initially were undecided later indicated they would oppose elk restoration if forced to take a position.

Approximately half of the respondents (51 percent) indicated they were confident of their knowledge of elk; the other half were uncertain. Interestingly, respondents who indicated confidence in their knowledge were more likely to agree that elk restoration was a good idea than those who were uncertain. Further, confident respondents expressed polar attitudes ("strongly agree" or "strongly disagree"), whereas uncertain respondents expressed moderate attitudes ("agree," "neutral," "disagree," "don't know"). Only 6 percent of confident respondents were neutral about elk in Virginia (versus 44 percent of uncertain respondents). These findings mirror those of Lauber and Knuth (1997, 1998) concerning moose restoration in New York and those of Lohr et al. (1996) and Bright and Manfredo (1996) concerning wolf restoration in New Brunswick and Colorado.

The propensity for undecided and uncertain respondents to oppose elk restoration creates an interesting dilemma. The less people know or care about elk restoration, the more likely they are to oppose it. At first glance, this observation may lead one to conclude that education programs would increase public support for restoration. This, however, is not entirely the case. Although public education programs are important tools to maintain support and let interested parties know what to expect from a restoration, they have only limited power to change opinions or create new attitudes in ambivalent minds. Attitudes toward elk restoration are based, not only on a person's knowledge of elk, but also that person's values, interests, and past experiences—factors that education alone cannot alter (Kellert 1991; Bright and Manfredo 1996). People accept information that reinforces what they already believe, and information considered valid and useful by one stakeholder may be interpreted differently or even ignored by others (Kellert 1991; Lauber and Knuth 1998). Respondents who supported elk restoration in our study found the natural history and ecological information we provided in brochures more helpful than did opponents. Opponents wanted more information about the downside of elk restoration—a topic mentioned only briefly in the brochure. Changing people's attitudes may be possible only when communication and understanding are encouraged among conflicting groups (Kellert 1991; Lauber and Knuth 1998) and when education programs are long-term and broad-based.

Upon examining respondents' motivations, we found that supporters and opponents of restoration act upon very different motives (Table 4.1). Those who support restoration based their decision primarily on ethical and ecological values; four of the top five reasons to support elk restoration were unrelated to the desire for direct personal gain. These included the desire to restore Virginia's natural history, return elk to historical native range, and increase Virginia's biodiversity. Factors involving direct personal impact (recreational opportunities, economic benefits) were less important. Interestingly, the opportunity to hunt elk was the least important reason to support restoration. Opponents of elk restoration, by contrast, clearly made a need-based decision. The top five reasons given for opposing elk restoration focused on safety (automobile accidents, disease transmission) or economic issues (such as crop damage). Value-based factors such as inhumane treatment of elk or competition with deer ranked low among opponents' motivations. Kellert (1980, 1991) has classified such value-based motives as ecologistic, humanistic, or moralistic; he classed need-based motives as negativistic or utilitarian.

This pattern in motivations may be common. In New Brunswick and Michigan, the primary reasons offered to support wolf restoration were that wolves were present historically and that they are endangered (value-based). Opposition to wolf restoration centered on potential loss of livestock due to wolf predation and concern for personal safety (need-based) (Kellert 1991;

Table 4.1. Comparison of Perceived Positive and Negative Aspects of Elk Restoration

Survey respondents	Workshop participants
TOP FIVE REASONS TO SUPPORT ELK RESTORATION:	TOP FIVE POTENTIAL BENEFITS:
1. To restore natural heritage	1. Economic/cultural benefits of tourism
2. Elk were present historically	2. Provision of new/enhanced recreational opportunities
3. Opportunity to see elk	3. Positive impact on wildlife/habitat management
4. To increase biodiversity	4. Educational opportunity
5. Elk have a right to live in native range	5. Fostering local biodiversity
TOP FIVE REASONS TO OPPOSE ELK RESTORATION:	TOP FIVE POTENTIAL CONCERNS:
1. Will cause automobile/elk collisions	1. Private property damages due to elk depredation
2. Will increase agricultural damage	2. Ecosystem impacts
3. Not a good use of funds	3. Economic costs of restoration/ management program
4. Disease transmission to other wildlife	4. Public safety on roads
5. Disease transmission to livestock	5. Disease transmission/introduction

Note: These findings are based on the responses of Virginia's general public (left column) during a mail survey conducted in spring 1999 and local stakeholder groups (right column) in regions where potentially suitable elk habitat was found.

Lohr et al. 1996). Restoration of moose in New York was supported because it would help restore the local ecosystem and because moose belong in their native habitat (value-based); opposition focused on financial cost and other competing concerns (need-based) (Lauber and Knuth 1997). Finally, Bright and Manfredo (1996) found that the "symbolic existence beliefs" and emotional responses elicited by wolves largely determined attitudes toward wolf restoration in Colorado. People with strong "existence beliefs" and positive emotional responses supported wolf restoration; those with weaker "existence beliefs" and negative emotional responses opposed it.

Burton (1990) cites three categories of motivations: those universal to the human species (needs), those that are cultural (values), and those that are transitory (interests). Interests refer to a person's social, political, and economic objectives and can be negotiated or traded for gains in other interests. Values and needs, however, are much more persistent. Needs are universal and are usually pursued by all means possible when their fulfillment is threatened. Values are deeply held, culturally ingrained ideas, habits, or beliefs that are resistant to compromise or change. When conflict arises about values, people often respond to opposing arguments in nonrational, noncompromising ways (Burton 1990). Peripheral perceptions about a management agency or another interest group reinforce a person's values and can further divide people's attitudes. A direct link between public attitudes toward restoration and attitude toward a management agency (Virginia Department of Game and Inland Fisheries) arose in our survey and in the proposal to restore moose to New York (Lauber and Knuth 1999).

The motivations of supporters and opponents of elk restoration in Virginia follow Maslow's need-hierarchy theory (Maslow 1970 as cited in Muchinsky 1990). Lower-order needs (physiological needs, safety) must be met before higher-order needs (social needs, self-actualization) become primary motivators. This may explain why opponents (need-based concerns) were unable to consider value-based motives worthy or justifiable. If elk restoration threatens a person's safety or economic livelihood, ecologistic and moralistic factors will be less important.

Regional Socioeconomic Assessment

Regional workshop participants cited five major benefits of elk restoration (Table 4.1): tourism, recreation, management, education, and biodiversity. The potential for economic return from tourism topped the list of benefits at three of the four meetings (Abingdon, Verona, Winchester). Participants in Martinsville ranked creation of new/enhanced recreational opportunities as most important. Other benefits listed were the enhancement of biodiversity, positive effects of elk on current management programs, and educational opportunities presented by a restored elk herd.

This list of potential benefits differs markedly from those expressed by mail survey respondents. Workshop participants highlighted the potential economic returns from elk (such as tourism) whereas survey respondents promoted the less tangible benefits (such as fostering biodiversity). Only in Martinsville was a noneconomic benefit (enhanced recreation) ranked as most important. Although biodiversity and ecosystem enhancement were discussed as important at all workshops, they consistently ranked lower than economic or recreational benefits. This reversal of the survey results reflects the close proximity of workshop participants to proposed release sites versus the geographic distance between much of the general public and those sites (Creighton 1981).

Workshop participants cited five major concerns (Table 4.1): damage to private property, impacts on the local ecosystem, economic costs of restoration, the risk of elk/vehicle collisions, and the possibility of disease transmission among elk, other wildlife, livestock, and humans. Among both workshop participants and the general public, direct negative consequences outweighed value-based consequences.

Many workshop participants expressed concern about inequitable distribution of costs and benefits in the community. Agricultural producers, in particular, believed the burden of supplying habitat and forage to elk would fall unfairly onto them. Benefits from restoration, by contrast, would probably spread over the community at large, bypassing agricultural producers. Many participants suggested that this constituted a property rights infringement—in that private landowners would be expected to provide for a public good—and a demoralization of the agricultural community in that other community interests would take precedence over the needs of producers. Participants were steadfast in one opinion: should elk restoration proceed, an inequitable distribution of costs and benefits must be avoided.

Participants in Abingdon and Winchester resolved to oppose elk restoration in their areas. Participants in Verona believed that economic concerns could be addressed by securing funding before restoration proceeds. Participants suggested that expedient damage compensation and economic incentives (easements, tax adjustments) were necessary. Martinsville participants recommended that a small experimental herd be studied before taking on a large-scale restoration project. There were other resolutions as well: a need for further research on elk biology and landscape ecology; identification of favorable elk habitat in the context of current landownership patterns; a cost/benefit analysis; a review of elk management in other eastern states; and education programs on the history and biology of elk, the objectives of the program, the conflicts that could arise, and the proposed management of elk and elk/human conflicts.

Participants in Abingdon and Winchester had more negative attitudes toward elk restoration than participants in Verona or Martinsville. Those from Abingdon and Winchester feared crop and property damage caused by elk and thought this risk outweighed all potential benefits. These areas, important dairy and apple-producing regions, have already incurred heavy damage from white-tailed deer (*Odocoileus virginianus*) (VDGIF 1994; West 1997). Verona and Martinsville participants believed it would be possible to minimize risks through careful and attentive management so long as the major concerns were addressed beforehand.

The public involvement strategy that guides the decision-making process on restoration must be able to resolve the inequity issue and the "values versus needs" dispute. In the words of Creighton (1981:26): "If people see themselves as standing to gain or lose something they value as result of an agency action (particularly if they are losing while others are gaining), the agency must demonstrate the equity of its decision-making process." To foster productive discussions and bring opposing groups together, the process must be open, informative, and noncompetitive. By encouraging people to recognize that all interested parties have equal standing and that the purpose of discussion is to learn about alternative interests and create mutually agreeable solutions—not to demoralize or defeat the opposing side—animosity can be reduced and consent, if not consensus, may be reached (Curtis and Hauber 1997). Having initiated this process, we believe the regional workshops were a good first step. But public involvement efforts must be continued throughout the decision-making and restoration processes.

The Final Steps

In the final steps of our feasibility study we performed a cost/benefit analysis, conducted a risk analysis, and then analyzed all the information for each study area in a matrix. Using data from Pennsylvania (Shafer et al. 1993; SEAFWA 1997; Lord et al. 1999) and Michigan (Parker 1990; SEAFWA 1997; Matthews 1999) on management costs, property damage, elk/vehicle collisions, tourism revenues, and hunting revenues, we calculated a cost/benefit ratio of 1:1.73 (McClafferty 2000). Using a three-step risk analysis, we then evaluated human population trends, landownership patterns (private versus public land), and land use trends (number, acreage, and type of farmlands) in and around potential elk ranges. Although projected risk varied by region, it was lowest in the southern Shenandoah Valley (McClafferty 2000). When we examined the biological results, public response, and economic/risk analysis results for each study area (McClafferty 2000), only the Highland site received a "high" feasibility rank.

Conclusion

Our HSI model represents an objective, systematic procedure to assess potential elk habitat over large geographic areas and helped us to identify biologically feasible restoration areas. (It has yet to be validated and field-tested on restored elk populations.) Although the model was designed to assess suitability for elk in Virginia at a landscape scale, it could be adapted for use at different geographic scales for elk, other ungulates, and large mammals elsewhere. Our estimates of potential elk population size and density were used only as theoretical guides to compare areas and should not necessarily be viewed as management goals. Desired population size must be based on management goals, socioeconomic factors, and habitat quality. Our estimates might be useful as starting points, but managers must set their own targets according to site-specific conditions and objectives.

We found that most Virginians lack sufficient knowledge to form an opinion about elk restoration. Many simply are not interested. But among those who were interested enough to express an opinion, most supported restoration. Nonetheless, one must be careful not to base a final decision on the preferences of ill-informed stakeholders. Public input and education should be part of a process that also considers biological feasibility, economic ramifications, and risk of elk/landowner conflicts.

Moralistic, humanistic, and ecologistic values were statewide motivating factors among supporters of elk restoration. The desire to return the elk to its historic range, reverse human wrongs, and restore native biodiversity—all were important to supporters of restoration. Local supporters shared many of these attitudes, but they were more interested in the direct economic and recreational benefits of restoration. Opponents of elk restoration did not display this divergence in motivation. Potentially negative effects on the economic livelihood and safety of the residents of Virginia were of concern, as was the equitable distribution of economic costs and benefits of restoration throughout the community.

Our assessment revealed many concerns and benefits associated with a restored elk herd, and we suspect the same issues would be raised for other ungulates as well. Because many costs and benefits of restoration fall directly on local communities, their concerns must be satisfied before a final decision is made. The ultimate success of a restoration program depends on the level of support within the community—and that support depends in turn on how well community interests are reconciled and the fairness of the public involvement process that produced the reconciliation. Our study has laid a firm foundation of ecological information and public involvement for elk restoration in Virginia. With continued attention to these concerns, we come one step closer to restoration success.

Acknowledgments

We thank D. F. Stauffer, S. L. McMullin, and J. S. Fraser (Department of Fisheries and Wildlife Sciences, Virginia Polytechnic Institute and State University) and W. M. Tzilkowski (School of Forest Resources, Pennsylvania State University) for their critical reviews. We are indebted to M. Cartwright (Arkansas Game and Fish Commission), R. Grimes (Kentucky Department of Fish and Wildlife Resources), and M. Knox (Virginia Department of Game and Inland Fisheries) for providing us with important information. This study was funded by grants from the Rocky Mountain Elk Foundation and the National Science Foundation.

Literature Cited

Beyer, D. E. 1987. Population and habitat management of elk in Michigan. Ph.D. dissertation, Michigan State University, Lansing.

Bright, A. D., and M. J. Manfredo. 1996. A conceptual model of attitudes: A case study of wolf reintroduction. *Human Dimensions of Wildlife* 1:1–21.

Burton, J. 1990. *Conflict: Human Needs Theory.* New York: St. Martin's Press.

Creighton, J. L. 1981. *The Public Involvement Manual.* Cambridge, Mass.: Abt Books.

Curtis, P. D., and J. R. Hauber. 1997. Public involvement in deer management decisions: Consensus versus consent. *Wildlife Society Bulletin* 25:399–403.

Dillman, D. 1978. *Mail and Telephone Surveys: The Total Design Method.* New York: Wiley.

Forbes, B., and G. Ferrence. 1999. The elk in Pennsylvania. Web page: http://www.iup.edu/ferenc/elk.htm.

Gilpin, M. 1996. Metapopulations and wildlife conservation: Approaches to modeling spatial structure. In D. R. McCullough, ed., *Metapopulations and Wildlife Conservation.* Washington, D.C.: Island Press.

Gwynne, J. V. 1977. Elk stocking in Virginia. Unpublished report. Richmond: Virginia Department of Game and Inland Fisheries.

Kellert, S. R. 1980. American's attitudes and knowledge of animals. *Transactions of the North American Wildlife Conference* 45:111–124.

———. 1991. Public views of wolf restoration in Michigan. *Transactions of the North American Wildlife and Natural Resources Conference* 56:152–161.

Lauber, T. B., and B. A. Knuth. 1997. Fairness in moose management decision-making: The citizens' perspective. *Wildlife Society Bulletin* 25:776–787.

———. 1998. Refining our vision of citizen participation: Lesson from a moose reintroduction proposal. *Society and Natural Resources* 11:411–424.

———. 1999. Measuring fairness in citizen participation: A case study of moose management. *Society and Natural Resources* 11:19–37.

Lidicker, W. Z. Jr. 1988. Solving the enigma of microtine "cycles." *Journal of Mammalogy* 69:225–235.

Lohr, C., W. B. Ballard, and A. Bath. 1996. Attitudes toward gray wolf reintroduction to New Brunswick. *Wildlife Society Bulletin* 24:414–420.

Long, J. R. 1996. Feasibility assessment for the reintroduction of North American elk into Great Smoky Mountains National Park. M.S. thesis, University of Tennessee, Knoxville.

Lord, B. E., C. H. Strauss, and W. M. Tzilkowski. 1999. Economic impact of elk viewing in rural Pennsylvania. Unpublished report presented at the fourth annual Eastern States Elk Management Workshop, Clam Lake, Wisc.

Maslow, A. H. 1970. *Motivation and Personality.* 2nd ed. New York: Harper & Row.

Matthews, G. 1999. Status report of the Michigan elk herd. Unpublished report presented at the fourth annual Eastern States Elk Management Workshop, Clam Lake, Wisc.

McClafferty, J. A. 2000. An assessment of the biological and socioeconomic feasibility of elk restoration in Virginia. M.S. thesis, Virginia Polytechnic Institute and State University, Blacksburg.

McKenna, G. 1962. Elk in Virginia. *Virginia Wildlife* 23:6–7.

Moore, C. M. 1987. *Group Techniques for Idea Building.* London: Sage.

Morton, D. M. 1998. Land cover of Virginia from Landsat thematic mapper imagery. M.S. thesis, Virginia Polytechnic Institute and State University, Blacksburg.

Muchinsky, P. M. 1990. *Psychology Applied to Work: An Introduction to Industrial and Organizational Psychology.* 3rd ed. Pacific Grove, Calif.: Brooks/Cole.

Nettles, V. F., and J. L. Corn. 1998. *Model Health Protocol for Importation of Wild Elk* (Cervus elaphus) *for Restoration.* Athens, Ga.: Southeastern Cooperative Wildlife Disease Study.

Parker, L. R. 1990. Feasibility assessment for the reintroduction of North American elk, moose, and caribou into Wisconsin. Unpublished report. Madison: Wisconsin Department of Natural Resources.

SEAFWA. 1997. Impacts of reintroducing North American elk to the southeastern United States: A report of the SEAFWA's Ad Hoc Committee investigating elk reintroduction. Unpublished report. N.p.: Southeastern Association of Fish and Wildlife Agencies.

Shafer, E. L., R. Carlane, R. W. Gulin, and H. K. Cordell .1993. Economic amenity values of wildlife: Six case studies in Pennsylvania. *Environmental Management* 17:669–682.

Thomas, J. W., D. A. Leckenby, M. Henjum, R. J. Pedersen, and L. D. Bryant. 1988. *Habitat-Effectiveness Index for Elk on Blue Mountain Winter Ranges.* General Technical Report PNW-GTR-218. Portland: U.S. Forest Service, Pacific Northwest Research Station.

U.S. Fish and Wildlife Service (USFWS). 1981. *Standards for the Development of Habitat Suitability Index Models.* Ecological Services Manual 103. Washington, D.C.: U.S. Fish and Wildlife Service, Division of Ecological Services.

U.S. Geological Survey (USGS). 1985. *Digital Line Graphs from 1:100,000-Scale*

Maps: Data Users Guide 2. Reston, Va.: U.S. Department of the Interior, National Mapping Program.

Van Deelen, T. R. 1997. *Use of Simulations to Estimate the Growth of an Elk Population Introduced into Southern Illinois.* Urbana: Illinois Natural History Survey.

VDGIF. 1994. *Deer Damage in Virginia.* House Document 19. Richmond: Virginia Department of Game and Inland Fisheries.

West, B. C. 1997. Deer damage in Virginia: Implications for management. M.S. thesis, Virginia Polytechnic Institute and State University, Blacksburg.

Wisdom, M. J., L. R. Bright, C. G. Carey, W. W. Hines, R. J. Pedersen, D. A. Smithey, J. W. Thomas, and G. W. Witmer. 1986. *A Model to Evaluate Elk Habitat in Western Oregon.* Publication R6-F & WL-216-1986. Portland: U.S. Department of Agriculture, U.S. Forest Service, Pacific Northwest Region.

Witmer, G. W. 1990. Re-introduction of elk in the United States. *Journal of the Pennsylvania Academy of Science* 64:131–135.

Wood, R. K. 1943. The elk in Virginia. M.S. thesis, Virginia Polytechnic Institute and State University, Blacksburg.

PART II

Practice

Successful large-mammal restoration efforts have occurred worldwide, but the process by which species were returned is often sketchy. That entire species have been restored at statewide scales (see Gassett in case 1) is as much a testament to the resilience of most large mammals as it is to our often haphazard attempts to bring them back. Many of the critical hands-on aspects of reintroduction are not only tedious but potentially dangerous to handled and handler alike. Others are very technical in nature, expensive, or otherwise an inconvenience to the manager. It is little wonder that many restorations were poorly documented or did not use the best available technology. Nonetheless, the likelihood of success is enhanced when problems are solved *before* they occur and procedures for dealing with the inevitable mishaps along the way are at least considered beforehand. Detailed records of ongoing and future restoration efforts are critical if we are to learn from our mistakes.

The momentous return of elk to Kentucky has, in many instances, been an exercise in reinventing the wheel (Larkin et al. in chapter 5). With limited literature on the subject, many aspects of elk transport and release were trial and error. Despite significant losses early in the process, the result has been the creation of what appears to be the nucleus of a self-sustaining herd. Now that both of Kentucky's native ungulates are well on their way to recovery—if not overabundance (case 1)—what's next? Will restoration of a large carnivore soon be considered part of every wildlife manager's strategic plan?

That the wolf is once again a classic keystone species in Yellowstone was made possible by intensive planning and detailed documentation of the process. As a result, future stewards of wolf reintroduction now have a tool chest that includes the best release method for different ecological and geo-

graphic settings (Fritts et al. in chapter 6) as well as detailed husbandry procedures that will help to ensure that each wolf arrives at the moment of
release in the best physical condition possible (Johnson in case 2). With current veterinary and pathology state-of-the-art, there is no reason why reintroduction stock should not be returned to the wild in top physical condition
(Gaydos and Corn in chapter 7). On the other hand, wolf restoration in the
Desert Southwest (Brown and Parsons in chapter 8) faces additional challenges due to a long history of antipredator prejudice and the need to utilize
animals that were born in captivity. This is where public involvement and
support are critical to a successful outcome.

Chapter 5

Returning Elk to Appalachia: Foiling Murphy's Law

JEFFERY L. LARKIN, ROY A. GRIMES,
LOUIS CORNICELLI, JOHN J. COX, AND DAVID S. MAEHR

Although the restoration of free-ranging elk (*Cervus elaphus*) east of the Mississippi River has a history filled with success and failure, scanty documentation and inadequate postrelease monitoring limits its usefulness as a lesson in ungulate restoration. In this chapter we examine organizational and ecological aspects of Kentucky's effort to restore free-ranging elk. A restoration effort of this magnitude requires extensive cooperation among a network of interstate partners. The complexity of such an undertaking creates logistical complications that can jeopardize success. We use the Kentucky Elk Restoration Program to illustrate the value of detailed planning, the role of unpredictable events, and the evolution of restoration protocols. Lessons learned from such an effort could enhance the success of future elk restoration efforts and, ultimately, the return of this species to many parts of its former range.

Elk in Eastern North America

As many as 10 million elk (Seton 1927) of six subspecies inhabited North America prior to European settlement (Murie 1951; Lyon and Thomas 1987). By the early 1900s, the Merriam's (*C. e. merriami*) and eastern (*C. e. canadensis*) subspecies were extinct. Whereas the Merriam's elk represented a distributional extreme and may have been more susceptible to extinction (Diamond 1984), the eastern elk inhabited as much as a third of the contiguous United States from Canada to the southeastern coastal plain, and from the eastern seaboard to the Great Plains.

There have been six attempts to restore elk east of the Mississippi River since the early 1900s (Bryant and Maser 1982; Wathen et al. 1996), but success occurred only in Pennsylvania and Michigan prior to 1990 (Witmer 1990). Both populations (Pennsylvania: 350; Michigan: 1400) are self-sustaining, and the larger herd requires hunting as a means of population control (Wathen et al. 1996). Why did two elk restoration attempts succeed while four failed? The answer to this question is problematic due to limited postrelease monitoring (hence unknown fates of released animals) and insufficient understanding of elk restoration ecology (hence managers' inability to modify restoration protocols) (Witmer 1990). The Pennsylvania herd, for example, was not studied until 60 years after release (Eveland et al. 1979).

The elk is not the only repatriated species in which fragmented documentation and limited monitoring typify postrelease management. In fact, only 27 percent of translocations specified the types of information recorded during the translocation process (Griffith et al. 1989), and most cannot be adequately evaluated (Kleiman 1996). Witmer (1990:134) suggests that "state and federal agencies . . . develop, standardize, and require permitting, data recording, and documentation of methodologies . . . thereby providing a useful data base of successful and unsuccessful [elk] re-introductions." Each reintroduction should be carefully planned (May 1991) and viewed as an exercise in adaptive management. Modification of protocols should mirror expanding knowledge and experience. Information pertaining to mortality, reproduction, age structure, habitat use, and movements will be needed to determine the success of elk reintroductions and improve the results of future attempts (Murie 1951; McCullough 1969; Witmer 1990; Wathen et al. 1996). Further, detailed and accurate demographic statistics increase the relevance of predictive population models (Maehr et al. 1999).

Although Barbour and Davis (1974) exclude elk from their accounts of Kentucky mammals, there is ample evidence that elk were widespread in the state at the time of European settlement (Funkhouser 1925). Remains in cave middens, eighteenth-century place-names, and pioneer diaries support the assertion that elk were not uncommon in Kentucky's presettlement fauna (Shoemaker 1939; Cahalane 1967; Wharton and Barbour 1991). Early explorers of the commonwealth such as Simon Kenton observed "a number of elk . . . upon the bare ridges [at May's Lick]" and at "Upper Blue Lick . . . again beheld elk and buffalo in immense numbers" (cited in McClung 1986). Accounts of large elk herds appear in journals and letters of other pioneers such as John Strader, James Yager, Thomas Walker, and Daniel Boone (Shoemaker 1939). In 1775, Dr. Thomas Walker recounted a Cumberland hunting trip where they "killed in the journey 13 buffaloes, 8 elks, 53 bears, 20 deers, 4 wild geese, and about 150 wild turkeys" (Arnow 1960). That elk were harvested less often than other large mammals during the Walker excursion suggests that they were relatively less abundant in eastern Kentucky. Nonetheless, elk herbivory may have facilitated the maintenance of barrens, glades, and other early successional habitats in an otherwise forested landscape. A growing human population spurred overharvest, habitat loss, and elk extirpation. In 1847, John J. Audubon lamented: "When we first settled in the state of Kentucky [1810], some of these animals [elk] were still to be met with; but at present we believe none are to be found within hundreds of miles of our . . . residence"(Cahalane 1967:274).

Most animal restorations occur at small spatial scales (Simberloff et al. 1999) and usually result in the release of 75 or fewer animals (Griffith et al. 1989). The Kentucky elk restoration project has targeted the release of 1800 elk over a nine-year period (Maehr et al. 1999) and has utilized stock from Arizona, Kansas, North Dakota, Oregon, and Utah. The first 484 translocated elk were radio-instrumented and monitored to document the restoration process and drive adaptive management promoting population growth.

The Restoration Process

Although the notion of elk restoration in Kentucky is not new, it has been widely accepted that elk restoration would fail due to deaths caused by the meningeal worm (*Parelaphostrongylus tenuis*) (Severinghaus and Darrow 1976; Eveland et al. 1979; Raskevitz et al. 1991). This nematode causes lethal neurological disease (Anderson et al. 1966; Samuel et al. 1992) and was implicated in the failure of four elk reintroductions in the eastern United States (Carpenter et al. 1973; Severinghaus and Darrow 1976; Raskevitz et al. 1991). Nonetheless, the survival of a captive herd in western Kentucky compelled commissioners of the Kentucky Department of Fish and Wildlife Resources (KDFWR) to reconsider the establishment of a free-ranging elk herd in the state. The protocol for elk restoration addressed these issues: minimum viable population; quality and quantity of available habitat; biological concerns including disease-testing protocol; monitoring of survival, movements, and habitat use; public concerns and benefits; project cost; and wild sources of the Rocky Mountain or Manitoban subspecies (Phillips 1997).

Funding

KDFWR estimated that the project would cost at least $1.3 million over the first three years and $20,000 to $400,000 each of the next six years. Such expenses were beyond the fiscal resources of KDFWR, so we turned to the fund-raising capabilities of the Rocky Mountain Elk Foundation (RMEF). The foundation pledged $996,000 through the early years of the project. At least 70 percent of the funding would be raised in Kentucky. In addition, KDFWR and the University of Kentucky supplemented this amount with infrastructure, equipment, and personnel, including all overhead costs.

Restoration Zone and Release Sites

We analyzed elk habitat in Kentucky by using aerial photos, topographic maps, GIS maps, and aerial surveys to identify potential restoration zones. Regions containing mixed forest and open habitat, few major highways, sparse human populations, and little row-crop activity were considered suitable elk habitat. In the early 1900s, elk reintroductions in other states led to crop depredation and eradication of elk (Witmer 1990). The only region of Kentucky extensive enough to contain a large elk population, but offering limited opportunity for elk/human conflicts, was a 1.06-million-hectare block of 14 counties in southeastern Kentucky where forest covered 93 percent of the landscape (Figure 5.1) (Phillips 1997).

Because less than 20 percent of the restoration zone is publicly owned, KDFWR sought the cooperation of private landowners and access to their land as release sites. Leases with landowners assured public access to view and hunt elk and granted research access as well. An agreement with the U.S. Forest Service (USFS) resulted in an environmental assessment of the potential for elk to colonize the Daniel Boone National Forest (DBNF) and targeted the Redbird district as the sole publicly owned release site. The other sites were on reclaimed grasslands associated with coal surface mines.

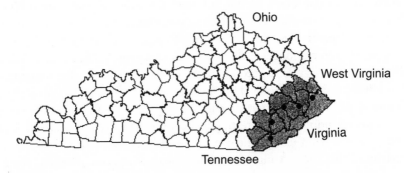

Figure 5.1. The 1.04-million-hectare Kentucky Elk Restoration Zone comprises Kentucky's 14 most southeastern counties. Release sites used during the first three years of the project are indicated with a dot.

Public Opinion

Kentucky contains more livestock than any state in the eastern United States (1.2 million), and row crops are an important economic commodity in many regions. Thus we recognized that elk restoration would be controversial in the agricultural community. Further, we recognized the need for disease testing and avoidance of source populations where brucellosis and tuberculosis occur. Our testing protocol targeted tuberculosis, brucellosis, anaplasmosis, bluetongue, Johne's disease, and vesicular stomatitis. To appease the Kentucky Cattleman's Association, KDFWR agreed to radio-collar every elk released during the first two years in order to promote a better understanding of disease in elk mortality. While this agreement added unexpected costs to the project, it avoided a public controversy and allowed an unprecedented opportunity to intensely monitor movements, survivorship, and causes of mortality.

Other agricultural concerns were addressed in two meetings with the Kentucky Farm Bureau (KFB), an organization representing 400,000 members statewide and 25,000 in the 14-county restoration zone. These meetings, attended by the state veterinarian, were open discussions that stressed restoration protocol and safety measures. During this process we pledged to prevent the establishment of elk in row-crop regions. KFB was appreciative that KDFWR discussed project plans with them before announcing restoration plans to the pubic and that we had proposed to target elk releases in a nonagricultural part of the state.

In March 1997, our presentation of the elk restoration proposal to the KDFWR Commission preceded instructions for the Wildlife Division to announce the restoration plan to the public. This was followed by a 90-day public comment period and a series of news releases describing the project. Public meetings in eight towns in the restoration zone consisted of a presentation of the proposal that addressed crop damage, disease, auto collisions, competition with other wildlife, future hunting, poaching, and dog harassment. Participants were asked to complete a questionnaire at the end of each meeting. Attendance ranged from 30 to 125 people. Support for the project among workshop participants was based on the promise for future elk hunting and viewing. Most of the 3200 statewide comments (90 percent) favored the return of elk to Kentucky. Within the 14-county restoration zone, 99 percent of the more than 1300 comments favored elk restoration. Because of this strong public support, the KDFWR Commission voted to reintroduce elk to the commonwealth.

Stock Acquisition

Once suitable habitat was located, public opinion assayed, support generated, and funding secured, KDFWR contacted elk donors. While five states have provided elk for the Kentucky restoration program, Utah has supplied more than half (418) of the animals. The Utah Division of Wildlife Resources (UDW) also developed the blueprint and supplied experienced staff for carrying out the translocation methodology. Utah's elk population is the result of a multistate restoration effort that occurred nearly a century ago. Elk were widely distributed throughout Utah until the mid-1800s before overhunting and habitat loss left only small populations in the most remote parts of the state. Utah received shipments of elk from Yellowstone National Park from 1910 to 1925. Following intrastate transplants and natural migration, elk repopulated much of their former range and colonized areas that were not previously considered elk habitat (Bryant and Maser 1982). Today Utah's population of about 60,000 elk is managed primarily through licenced hunting of both antlered and antlerless elk. Nonetheless, current harvest rates are insufficient to relieve elk/human conflicts involving damage to crops and private property. Thus Utah's role as an elk source for restoration projects serves as a supplemental harvest that is intended to help reduce conflicts with humans. With only a few exceptions, the Utah public has been generally supportive of out-of-state restocking efforts.

Capture, Disease Testing, and Transport

Elk that cause problems for private landowners are trapped by UDW to maintain forage quality and reduce crop depredation. Such partnerships are essential to future big game populations in Utah because a majority of winter range is on private land and the abundance and distribution of large carnivores are insufficient to prevent elk overabundance. Killing or relocating problem animals not only reinforces good agency/landowner relations but maintains the ecological integrity of native range.

Elk are baited with hay; a portable corral trap is constructed; then the trap is activated after elk become accustomed to entering it (Schemnitz 1994). Depending on prevailing weather conditions, up to 35 elk are captured in a single trap. Severe weather and a scarcity of natural forage produce the highest capture rates. In addition to corral traps, helicopters drive elk into holding facilities (Beasom et al. 1980) and are used as platforms for net-gunning (Schemnitz 1994). Captured elk are then loaded into modified horse trailers and transported to Hardware Ranch, an elk feeding ground established by UDW in 1949 to relieve agricultural damage and competition with mule deer

(*Odocoileus hemionus*). Elk are held in two 1-acre pens and fed approximately 5 kg of alfalfa hay/elk/day for up to 21 days. Water is provided in a trough that runs through each pen. Elk condition usually improves during their temporary confinement.

Elk are processed twice through a series of individual holding chutes and a squeeze chute. The first time they are sexed, aged, weighed, ear tagged, bled, injected with a vitamin B complex, wormed with ivermectin, and given a tuberculosis test. At this time, adults and calves are separated to prevent calves from being trampled by larger adults. Blood samples are sent to disease testing laboratories where tests are completed within a few days. After 72 hours, the tuberculosis test is checked by a federally certified veterinarian and each animal is reweighed. If determined to be disease-free and healthy, each elk is fitted with a radio-collar and loaded onto a livestock trailer. Elk are then transported nonstop approximately 3000 km to eastern Kentucky where they are released immediately upon arrival.

Postrelease Monitoring

Monitoring of radio-collars begins within 24 hours after release. Collars are monitored for mortality signals during routine fieldwork and once a week from fixed-wing aircraft. Dead animals are brought to the University of Kentucky Livestock Disease Diagnostic Center (LDDC) where a complete necropsy is conducted and cause of death determined. If an animal cannot be retrieved from the field, head and tissue samples (heart, lung, liver) and blood samples are collected and submitted to LDDC.

Large-scale movements (representing colonization progress) of each elk are determined with monthly flights. This coarse-resolution approach is useful in describing initial movements from release sites and in estimating the rate at which a reintroduced herd fills vacant range. During the first two years, 140 animals were intensively tracked at least once a week from fixed-wing aircraft (Mech 1983) to determine annual home range sizes and habitat preferences. A subset of 35 elk were fitted with collars equipped with global positioning systems (GPS) programmed to record positions every six hours during the first year.

Restoration Zone

Narrow, winding ridges, steep side-slopes, deep dendritic drainages, and narrow valleys characterize the unaltered topography of the restoration zone (McFarlan 1943). Elevations range from 244 to 488 m above mean sea level (Overstreet 1984). This region was selected because of its low human population and its distance from row crops and urban centers. Active and reclaimed surface mines, the Daniel Boone National Forest, and private forests dominate the landscape. Mountaintop removal for coal and subsequent reclamation has converted rugged topography into a landscape of flat-topped mesas and gently sloping grasslands. The restoration zone is 93 percent forested, 6 percent reclaimed surface mines, and 1 percent agriculture—primarily small residential produce gardens (Phillips 1997).

The principal vegetative cover type in eastern Kentucky is mixed-mesophytic forest characterized by approximately 30 dominant overstory tree species including American beech (*Fagus grandifolia*), yellow poplar (*Liriodendron tulipifera*), basswood (*Tilia* spp.), sugar maple (*Acer saccharum*), northern red oak (*Quercus rubra*), white oak (*Q. alba*), eastern hemlock (*Tsuga canadensis*), and yellow buckeye (*Aesculus octandra*) (Braun 1950). Common understory species include eastern redbud (*Cercis canadensis*), flowering dogwood (*Cornus florida*), spicebush (*Lindera benzoin*), and pawpaw (*Asimina triloba*). Mixed-mesophytic forests occur on moist, well-drained sites (Braun 1950). Dry ridge-tops, upper southwestern slopes, and areas with rocky shallow soils are characterized by oak (*Quercus*)–hickory (*Carya*) and oak–pine (*Pinus*) climax communities (Overstreet 1984).

Few openings of substantial size existed in eastern Kentucky prior to European settlement. Strip mining and subsequent reclamation, however, have created large (up to 1200 ha) herbaceous openings. In some regions of eastern Kentucky, the original forested landscape matrix has been reduced to woodland patches in a landscape dominated by grasslands. From a structural perspective, these early successional areas bear a striking resemblance to elk habitat in the western United States. Common plant species occurring in these openings are Kentucky-31 tall fescue (*Festuca arundinacea*), bush clover (*Lespedeza* spp.), perennial ryegrass (*Lolium perenne*), and orchardgrass (*Dactylis glomerata*).

The climate of the restoration zone is temperate humid continental (Overstreet 1984) with an average annual temperature of 13°C (Hill 1976), a mean winter temperature of 4°C, and a mean summer temperature of 24°C (McDonald and Blevins 1965). Annual precipitation averages 117 cm and is evenly distributed throughout the year (Hill 1976). Average snowfall is approximately 50.8 cm, but accumulation rarely remains for more than a few days (McDonald and Blevins 1965).

Results and Discussion

Despite initial opposition from the Kentucky Farm Bureau over potential row crop depredation and automobile collisions, no other opposition has

Table 5.1. Number of Elk Translocated from Each Source State During the First Three Years of Kentucky Elk Restoration

Source	1997–1998	1999	2000	Total
Arizona	—	—	29	29
Kansas	9	—	—	9
North Dakota	—	—	145	145
Oregon	—	—	118	118
Utah	159	143	115	417
TOTAL	168	143	407	718

materialized from other segments of the public. During the first three years of Kentucky elk restoration, 718 elk were translocated from five western states (Table 5.1). Of these, 484 were fitted with radio-collars. Monitoring efforts during the first two and a half years resulted in more than 13,000 VHF radiolocations. Although we underestimated our financial contribution by $210,000 during the first three years, we were under budget by $72,000 for the RMEF grant. KDFWR's overexpenditure was due largely to personnel costs associated with assisting western states with elk captures.

Annual mortality rates for 1998 and 1999 were 40 percent and 6 percent, respectively. Most 1998 deaths occurred as a result of capture myopathy, a condition that is largely caused by capture stress and compromised physical condition (Pond and O'Gara 1994). Although most animals suffering from capture myopathy die within two weeks of release, some animals may linger for several months before succumbing. During the 1998 translocations, adults and calves did not travel well together and many calves were injured en route to Kentucky. In subsequent years, calves and adults were separated prior to shipping. The drastic reduction in mortalities resulting from capture myopathy during 1999 compared to 1998—a reduction from 47 cases to 3— was attributed largely to the separation of age classes and a milder winter in the second year.

Other sources of mortality in translocated elk were automobile collisions (six deaths), poaching (three deaths), and removal of elk that had traveled outside the restoration zone (two deaths). Meningeal worm has accounted for less than 1 percent (two deaths) of the overall mortality during the first two years. Nevertheless, there is evidence that this parasite may have a greater impact on the survivorship of juvenile elk born in Kentucky. Of the 34 calves born in Kentucky during 1998, at least 5 (15 percent) died as a result of meningeal worm. Given that calves born in Kentucky during 1998 were not radio-collared, the actual rate of brainworm deaths may be higher. Thus Kentucky-born calves should become the focus of intensive monitoring to determine the effects of meningeal worm on Kentucky's reintroduced elk population.

Reproductive rates of Kentucky elk were 20 to 30 percent lower than rates observed in the source populations (Kimball and Wolfe 1979; Taber et al. 1982). Forty-five adult cows translocated during the first year produced 29 calves (61 percent) during the 1998 calving season. This same transloca-tion cohort consisted of 52 adult cows in 1999, of which 25 calves were born (48 percent). The low calving rate observed in 1998 was not surprising in that it followed a low survivorship that year. The low calving rate observed in 1999 was the result of Allee effects (factors that inhibit population growth, such as difficulty finding mates, which result once a population drops below some critical density) created largely by the long-distance movements of 22 adult females (42 percent) and their subsequent reproduc-tive isolation (Larkin et al. in press). An aerial survey conducted in January 2000 suggests that cows translocated in the second year calved at a 52 per-cent rate. A recent elk restoration effort in Ontario also observed lower than expected reproduction during calving seasons immediately following translocation (Joe Hamr, Cambrian College, pers. comm.). Reductions may be due to abortions triggered by stress as well as injuries from trapping and translocation. Thus a 20 to 30 percent reduction in calving may be expected in the season immediately following translocation.

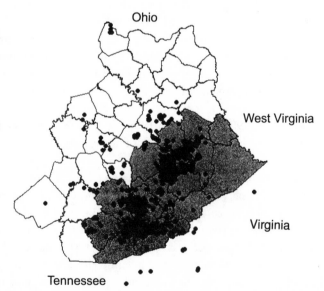

Figure 5.2. Map of the Kentucky Elk Restoration Zone and the >13,000 VHF radio-telemetry locations (dots) recorded during the first two years of the project.

Kentucky elk exhibit little postrelease wandering. Some 419 of 484 animals (87 percent) released during the first three years remained within 20 km of their release site. Only 25 of the 484 radio-instrumented elk (5 percent) released during the first three years of the project left the designated restoration zone (Figure 5.2). Despite extensive prerelease planning, we failed to recognize the potential for elk to cross state boundaries. More important, we did not involve neighboring states in the development of an interstate elk movement protocol. It was not until October 1998 (ten months after release), when an elk entered Virginia, that representatives from West Virginia, Virginia, Tennessee, and Kentucky met to discuss such occurrences. Although all three neighboring states are considering elk reintroduction, their solutions to interstate movements ranged from immediate removal of elk (Virginia), removal of only problem elk (Tennessee), to no need for action (West Virginia). After two and a half years, seven animals (1.5 percent) have crossed state boundaries (four in Virginia; three in Tennessee).

Murphy's Law in Action

Public support, adequate funding, biological feasibility, and experienced personnel are not enough to assure the success of a large-scale restoration. Further, all the well-intended planning and training are insufficient to fend off a galaxy of surprises and oversights. Because restoration efforts such as ours involve large numbers of animals, depend on a diverse array of people and agencies, and cover long distances between capture and release, Murphy's Law is an ever-present specter. Unpredictable events during the initial years of the Kentucky Elk Restoration Program were related to weather, equipment, animal behavior, and human fallibility.

Whereas cold weather and snow enhance elk accessibility in the western United States, capture success can be dictated by day-to-day weather patterns. Helicopters used as platforms for Arizona net-gunning were grounded for days by an unexpected blizzard—which postponed trapping until personnel, disease testing, state veterinarians, and researchers could be rescheduled. Ultimately this delay was responsible for a 50 percent reduction in the number of elk translocated from Arizona during the 2000 translocation campaign. In Utah, mild winter weather rendered baited traps ineffective—which resulted in our failure to meet the 200-elk objective in each of the first two years. A shipment of 51 elk from Utah was stranded for three days on the side of Interstate 80 during a 1998 Nebraska blizzard—which resulted in the death of 26 of the animals (more than 50 percent) from capture myopathy within four weeks after release. Narrow, icy roads prevented truck access to a release site designated for an experiment examining the effects of translocated elk on resident white-tailed deer (*Odocoileus virginianus*) and coyotes (*Canis latrans*)—which resulted in a lost opportunity to measure immediate behavioral responses. Aerial telemetry flights are routinely canceled and rescheduled due to rain, snow, fog, and wind. Not only does this compromise research findings by reducing the sample size, but it has prevented us from discovering the cause of death for many of our study animals due to advanced stages of decomposition.

Inadequate corrals at some capture sites prolonged handling times, increased stress to elk, allowed escapes, and caused injuries that prevented translocation. Although most animals survived their bouts with inadequate facilities, the associated stress likely contributed to the first year's mortalities. Postrelease monitoring was also impaired by equipment failures. The study of detailed elk movements through the landscape was compromised when 7 of the 13 GPS collars failed within three months after release. Furthermore, 40 percent of the VHF radio-collars placed on male calves failed before they matured. Not only did this prevent data collection, but it prematurely ended a male reproductive success study.

Despite Kentucky's vast elk restoration zone, some elk moved beyond its boundaries. (The size of the zone is itself a complicated factor in locating radio signals of widely dispersed individuals.) An adult bull dispersed 170 km from the release site, and seven other elk crossed into neighboring states. Unless retrieved, these animals are lost to the restored population. The retrieval or removal of elk in remote and rugged terrain requires the expenditure of unbudgeted funds and personnel. Although radio-collars make it easier to locate stray elk, their recapture is a distraction from routine monitoring. If initial attempts fail, elk tend to elude capture for months and may end up focusing negative media attention on the project. Out-of-state movements were a source of concern among neighboring wildlife agencies, as well, and resulted in three separate protocols for interstate elk issues. The postrelease travels of 22 female elk reduced the 1998 effective female population by 42 percent. Even though most of these cows remained within the restoration zone, a breeding season in the absence of males stalled population growth in a year already compromised by other factors.

Prerelease planning solved many problems related to the logistics of securing and moving elk to Kentucky. It also created a sense of control that disappeared once operations began. While our efforts may have advanced the state-of-the-art of long-distance elk restoration, unpredictable events will continue to compromise future efforts. Nonetheless, our experiences have

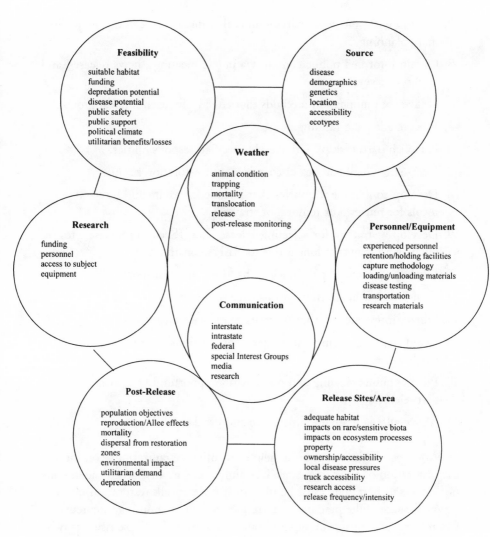

Figure 5.3. Network of contributing factors and considerations that influence the success of elk restoration.

revealed several guidelines. We offer these lessons here as the minimal components of successful elk restoration (Figure 5.3):

1. Conduct a thorough feasibility study to discover biological and cultural limitations.

2. Consult with state veterinarians and other disease experts during preliminary planning.

3. Obtain informed public consent via public meetings, news releases, and opinion surveys.

4. Utilize communication methods that effectively reach interest groups.

5. Procure adequate funding.

6. Establish partnerships with clearly specified responsibilities.

7. Designate a regional area of containment (restoration zone).

8. Develop containment policies that include intrastate and interstate protocols for retrieval or removal of stray elk.

9. Develop a strategy for selecting release sites. (Securing release sites on public land may take longer due to NEPA constraints.)

10. Involve experienced personnel for handling and transport.

11. Translocate wild elk from disease-free sources.

12. Obtain interstate transport permits to reduce transit time.

13. Avoid trapping during late winter or at locations experiencing unusually harsh winters.

14. Provide public viewing and participation opportunities starting with the first release.

15. Above all, maintain administrative and fiscal flexibility.

We hope our information is helpful to other organizations considering the restoration of large ungulates. Certainly there are myriad considerations not discussed here that must be examined once an elk restoration effort is deemed successful—predator reintroduction, sport hunting, depredation control, and elk impacts on other resources, for example. These issues may be even more contentious than the decision to restore elk and must be carefully examined in order to maintain public support, avoid ecological problems, and ensure the restored population's long-term viability.

Acknowledgments

We thank the Rocky Mountain Elk Foundation for financial and logistical support and Addington Enterprises and Orr Minerals for access to the study areas. C. Logsdon, D. Crank, and M. Wichrowski provided invaluable field assistance. L. Austin, M. Clemens, and J. Bingham provided expert piloting for elk tracking.

Literature Cited

Anderson, R. C., M. W. Lankester, and U. R. Strelive. 1966. Further experimental studies of *Pneumostrongylus tenuis* in cervids. *Canadian Journal of Zoology* 44:851–861.

Arnow, H. S. 1960. *Seedtime on the Cumberland.* New York: Macmillan.

Barbour, R. W., and W. H. Davis. 1974. *Mammals of Kentucky.* Lexington: University Press of Kentucky.

Beasom, S. L., W. Evans, and L. Temple. 1980. The drive net for capturing western big game. *Journal of Wildlife Management* 44:478–480.

Braun, E. L. 1950. *Deciduous Forests of Eastern North America.* New York: Hafner.

Bryant, L. D., and C. Maser. 1982. Classification and distribution. In J. W. Thomas and D. E. Toweill, eds., *Elk of North America.* Harrisburg, Pa.: Stackpole.

Cahalane, V. H. 1967. *Audubon Animals: The Quadrupeds of North America.* Maplewood, N.J.: Hammond.

Carpenter, J. W., H. E. Jordan, and B. C. Ward. 1973. Neurologic disease in wapiti naturally infected with meningeal worms. *Journal of Wildlife Disease* 9:148–153.

Diamond, J. M. 1984. "Normal" extinctions of isolated populations. In M. N. Nitecki, ed., *Extinctions.* Chicago: University of Chicago Press.

Eveland, J. F., J. L. George, N. B. Hunter, D. M. Forney, and R. L. Harrison. 1979. A preliminary evaluation of the ecology of the elk in Pennsylvania. In M. S. Boyce and L. D. Hayden-Wing, eds., *North American Elk: Ecology, Behavior, and Management.* Laramie: University of Wyoming Press.

Funkhouser, W. D. 1925. *Wild Life in Kentucky.* Frankfort: Kentucky Geological Survey.

Glenn, T. C., and D. R. Smith. 1993. Genetic variation and subspecific relationships of Michigan elk. *Journal of Mammology* 74:782–792.

Griffith, B., J. M. Scott, J. W. Carpenter, and C. Reid. 1989. Translocation as a species conservation tool: Status and strategy. *Science* 245:477–480.

Hill, J. D. 1976. *Climate of Kentucky.* Agricultural Experimental Station Progress Report. Lexington: University of Kentucky.

Kimball, J. F. Jr., and M. L. Wolfe. 1979. Continuing studies of the demographics of a northern Utah elk population. In M. S. Boyce and L. D. Hayden-Wing, eds., *North American Elk: Ecology, Behavior, and Management.* Laramie: University of Wyoming Press.

Kleiman, D. G. 1996. Reintroduction programs. In D. G. Kleiman, M. E. Allen, K. V. Thompson, and S. Lumpkin, eds., *Wild Mammals in Captivity: Principles and Techniques.* Chicago: University of Chicago Press.

Larkin, J. L., D. S. Maehr, J. J. Cox, M. W. Wickrowski, and R. D. Crank. In press. Factors affecting reproduction and population growth in a restored elk population. *Wildlife Biology.*

Lyon, L. J., and J. W. Thomas. 1987. Elk: Rocky Mountain majesty. In H. Kallman, ed., *Restoring America's Wildlife: 1937–1987.* Washington, D.C.: U.S. Fish and Wildlife Service.

Maehr, D. S., R. Grimes, and J. L. Larkin. 1999. Initiating elk restoration: The Kentucky case study. *Proceedings of the Annual Conference of Southeastern Fish and Wildlife Agencies* 53:350–363.

May, R. M. 1991. The role of ecological theory in planning a re-introduction of endangered species. *Symposium of the Zoological Society of London* 62:27–37.

McClung, J. A. 1986. *Sketches of Western Adventure.* Maysville, Ky.: L. Collins.

McCullough, J. F. 1969. *The Tule Elk: Its History, Behavior, and Ecology.* Berkeley: University of California Press.

McDonald, H. P., and R. L. Blevins. 1965. *Reconnaissance Soil Survey of Fourteen Counties in Eastern Kentucky.* Washington, D.C.: USDA.

McFarlan, A. C. 1943. *Geology of Kentucky.* Baltimore: Waverly.

Mech, L. D. 1983. *Handbook of Animal Radio-Tracking.* Minneapolis: University of Minnesota Press.

Murie, O. J. 1951. *The Elk of North America.* Harrisburg, Pa.: Stackpole.

Overstreet, J. C. 1984. *Robinson Forest Inventory: 1980–1982.* Lexington: University of Kentucky, College of Agriculture, Department of Forestry.

Phillips, J. 1997. *Technical Proposal for Free-Ranging Elk in Kentucky.* Frankfort: Kentucky Department of Fisheries and Wildlife Resources.

Pond, D. B., and B. W. O'Gara. 1994. Chemical immobilization of large mammals. In T. A. Bookhout, ed., *Research and Management Techniques for Wildlife and Habitats.* Bethesda, Md.: Wildlife Society.

Raskevitz, R. F., A. A. Kocan, and J. H. Shaw. 1991. Gastropod availability and habitat utilization by wapiti and white-tailed deer sympatric on range enzootic for meningeal worm. *Journal of Wildlife Disease* 27:92–101.

Samuel, W. M., M. J. Pybus, D. A. Welch, and C. J. Wilke. 1992. Elk as a potential host for meningeal worm: Implications for translocation. *Journal of Wildlife Management* 56:629–639.

Schemnitz, S. D. 1994. Capturing and handling wild animals. In T. A. Bookhout, ed., *Research and Management Techniques for Wildlife and Habitats.* Bethesda, Md.: Wildlife Society.

Seton, E. T. 1927. *Lives of Game Animals.* Vol. 3, pt. 1. Garden City: Doubleday.

Severinghaus, C. W., and R. W. Darrow. 1976. Failure of elk to survive in the Adirondacks. *New York Fish and Game Journal* 23:98–99.

Shoemaker, H. W. 1939. Vanished game. In A. Ely, H. E. Anthony, and R. M. Carpenter, eds., *North American Big Game*. New York: Scribner.

Simberloff, D., D. Doak, M. Groom, S. Trombulak, A. Dobson, S. Gatewood, M. E. Soulé, M. Gilpin, C. Martinez del Rio, and L. Mills. 1999. Regional and continental restoration. In M. E. Soulé and J. Terborgh, eds., *Continental Conservation*. Washington, D.C.: Island Press.

Stephenson, J. H. 1935. Michigan has elk. *Michigan Conservation* 5:3.

Taber, R. D., K. Raedeke, and D. A. McCaughran. 1982. Population characteristics. In J. W. Thomas and D. E. Toweill, eds., *Elk of North America*. Harrisburg, Pa.: Stackpole.

Wathen, G., L. Marcum, A. Peterson, D. Scott, and B. Layton. 1996. Elk reintroduction in Tennessee: An evaluation of its potential (draft). N.p.: Tennessee Wildlife Resources.

Wharton, M. E., and R. W. Barbour. 1991. *Bluegrass Land and Life: Land Character, Plants, and Animals of the Inner Bluegrass Region of Kentucky*. Lexington: University Press of Kentucky.

Witmer, G. 1990. Reintroduction of elk in the United States. *Journal of the Pennsylvania Academy of Science* 64:131–135.

Case 1. Restoration of White-Tailed Deer in Kentucky: From Absence to Overabundance

JONATHAN W. GASSETT

Kentucky has long been known for its abundance of wildlife. Indeed, it was commonly referred to as "the happy hunting ground" by Native Americans and early settlers. But the rapid influx of people in the early 1800s, combined with the perception that game was inexhaustible, soon caused declines in all species that were sought for food and fur. Unregulated market and subsistence hunting, coupled with degradation of suitable habitat, left many species of wildlife in dire straits. Elk (*Cervus elaphus*), gray wolf (*Canis lupus*), mountain lion (*Puma concolor*), and black bear (*Ursus americanus*) eventually disappeared. Although restrictions on harvest were implemented as early as the mid-1700s, it took generations of regulation experiments, effective law enforcement, and reintroduction before even a portion of Kentucky's large mammals rebounded.

Remnants of white-tailed deer (*Odocoileus virginianus*) occurred in pockets of inaccessible habitat, but the species had largely been depleted from the region by the early 1900s (Blackard 1971). In Kentucky, fewer than 2000 deer remained—primarily on the Jones-Keeney Refuge in Caldwell County and at the old Kentucky Woodlands Refuge. What followed was the most extensive wildlife restoration attempt in the commonwealth. From these remnant populations deer were trapped and transplanted to suitable but vacant range. Soon these areas became sources for more transplants (Blackard 1971). The growth of the deer population from around 1000 animals in the 1920s to around 690,000 in 2000 is testament to effective management and the ability of the white-tailed deer to respond to colonization opportunities.

Early Restoration

In 1927, the deer herd in Kentucky reached its all-time low of less than 1000; at this point there were populations in only 4 of the state's 120 counties (Barick 1951; Blackard 1971). By 1945 the herd had grown to about 2000—due primarily to increased protection and less so to immigration from adjacent states. Most of these animals inhabited the Kentucky Woodlands National Wildlife Refuge (now part of the U.S. Forest Service's Land-Between-the-Lakes). Kentucky's first introductions occurred here in the 1920s and consisted of 30 white-tailed deer from Wisconsin (McIntosh 1962). When the refuge closed in 1935 due to political problems, poachers soon reduced the herd to less than 100 (McIntosh 1962). Franklin D. Roosevelt designated the area as a national wildlife refuge in August 1938, whereupon law enforcement and habitat management by the Bureau of Biological Survey turned it into the source for restoring deer throughout the state.

Refuge Establishment

From 1946 to 1952 an influx of Pittman-Robertson funds fueled the establishment of 12 state wildlife refuges that were closed to hunting (Strode 1951). Previously only two refuges had been established (Hardy 1953). By 1952, refuges at Pennyrile State Forest, Green River WMA, Beaver Creek WMA, Three Forks Wildlife Restoration Area, Robinson Forest Refuge, Pine Mountain, Lewis County, Kentucky Ridge, Ford (Redbird Unit), Dewey Lake, Bernheim State Forest, and Mammoth Cave National Park had become part of the deer restoration system and were the recipients of all translocated deer (Strode 1951). As the deer herd expanded from these

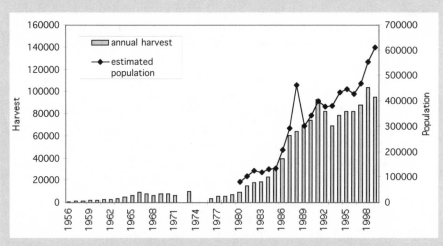

A half century of white-tailed deer population growth supports an increasing annual hunter harvest.

refuges, public support for the project swelled, the need for new refuges declined, and the Kentucky Department of Fish and Wildlife Resources adopted a statewide approach to deer restoration that focused on increasing the number of managed hunting areas (Barber 1963).

Habitat Improvement

The objectives of habitat work at the beginning of deer restoration focused the creation of forest openings and supplemental food plantings primarily on National Forest land and some private property. Other habitat improvements included the establishment of waterholes and mineral licks. These efforts were exceedingly costly, however, and were deemphasized so that resources could be directed at continued trapping and relocation (Barber 1963). Similarly, the planting of food plots was reduced. In their place, deer habitat was enhanced by promoting small clear-cuts on private and public land that were virtually cost-free to the agencies and landowners (Barber 1963).

Trapping and Restocking

Although 80 deer from Wisconsin and a few more from Oklahoma and Tennessee were released in Kentucky between 1939 and 1953, the vast majority (99 percent) originated from native Kentucky herds (Hardy 1953; McDonald and Miller 1993). Counties that appeared unsuccessful after the first translocation of 46 to 75 deer were restocked with an additional 50 animals (Nelson 1967). After stocking, each county remained closed to hunting for at least five years (Barber 1963). Kentucky's first legal modern firearm deer hunt was a 1956 buck-only season in 26 counties that culminated in a harvest of 750 bucks. (See the accompanying figure.)

Modern Restoration and Current Status

Even with the reopening of annual hunting seasons, deer herds continued to increase throughout Kentucky. Growth was spotty, and it was soon apparent that white-tailed deer were faring better in some areas than in others. Initially the blame for slow growth was placed on free-ranging dogs, particularly in east Kentucky (Nelson 1967). But it became evident that poaching was the more serious problem (Smith 1974). Increased efforts by law enforcement helped to curtail the poaching problem to a point where herd growth was more uniform throughout the state.

As refuge populations grew, surplus deer were moved to counties that showed little or no growth. Mammoth Cave stood out as the primary source for restocked deer throughout the 1960s and 1970s (Nelson 1967; McDonald and Miller 1993). From 1961 to 1974, some 50 to 75 deer were stocked per county to spread deer across the state more uniformly. Throughout the restocking period 3454 bucks and 6010 does were trapped and relocated in Kentucky. By 1976, deer had recovered sufficiently that mandatory harvest check stations were established (3476 deer were registered) to provide additional protection and a means of monitoring local and regional herds.

While mandatory harvest checks and restocking succeeded in the western and central regions of the commonwealth, they were not so successful in the mountainous Cumberland Plateau of eastern Kentucky. Although disturbance by free-ranging dogs and relatively unproductive habitat may have retarded recovery in this impoverished part of the state, poaching by subsistence hunters was the primary factor. Deer numbers remained low until 1984, when the Kentucky Department of Fish and Wildlife Resources (KDFWR) initiated a 15-year program to move 500 animals to each county in which population growth was stagnant or slow. Current estimates (KDFWR, unpublished data) suggest that these efforts have resulted in stable or increasing populations of at least 1000 deer per county.

The Kentucky deer population currently numbers around 690,000 animals. Hunter success closely tracks the growth of the herd (see the accompanying figure). Hunting is allowed in all 120 counties and, as growth continues, seasons are liberalized. The hunting season, in some fashion, extends from mid-September through mid-January. Current regulations provide for unlimited antlerless harvest in problem counties and for an extended firearms season for antlerless deer. To maintain the quality of bucks to which Kentucky hunters have become accustomed, however, harvest of antlered bucks is still restricted to one per hunter per year.

The value of the herd to hunters continues to increase due to herd quality and reasonable license fees. Over the past five years, the quality of Kentucky's deer herd, as measured by number of record book bucks taken, ranks among the best in the country. Apart from its value to hunters, Kentucky's deer herd provides other intrinsic values—for example, tourism related to the herd (hunting and viewing) contributes approximately $231 million to Kentucky's annual economy (Southwick Associates 1997).

The Future of Deer in Kentucky

White-tailed deer are undoubtedly Kentucky's first wildlife success story and, as such, have provided the foundation for the subsequent repatriation of wild turkey (*Meleagris gallopavo*), river otter (*Lontra canadensis*), peregrine falcon (*Falco peregrinus*), and, most recently, elk. While these other species are still recolonizing areas of Kentucky, the restoration of white-tailed deer is largely complete. As deer numbers continue to increase, they are becoming super-abundant to the point of exceeding social carrying capacity—the number that people will tolerate—in many areas. Because deer overabundance results in nuisance complaints, automobile collisions, and agricultural damage, today's deer management is less concerned with restoration than with population control—a job that becomes harder as the human population becomes more urban. This problem is further compounded by a declining hunter population: the front line of deer control. The difficulties in controlling the deer population, combined with the recent repatriation of elk to the eastern part of the state, will force wildlife biologists to become even more vigilant and innovative when managing these populations.

Literature Cited

Barber, H. L. 1963. Kentucky deer restoration: A report on the progress and present status of deer management in Kentucky. *Happy Hunting Ground* 19(3):2.

Barick, F. B. 1951. Deer restoration in the southeastern United States. *Proceedings of the Annual Conference of the Southeastern Association of Fish and Wildlife Agencies* 5:342–367.

Blackard, J. J. 1971. Restoration of the white-tailed deer in the southeastern United States. M.S. thesis, Louisiana State University, Baton Rouge.

Hardy, F. C. 1953. Kentucky's deer population growing. *Happy Hunting Ground* 9(6):10.

McDonald, J. S., and K. V. Miller. 1993. *A History of White-Tailed Deer Restocking in the United States, 1878 to 1992,* Research Publication 93-1. Watkinsville, Ga.: Quality Deer Management Association.

McIntosh, D. 1962. "Between the Rivers Area" wealthy in historic events. *Happy Hunting Ground* 18(1):30.

Nelson, L. K. 1967. Speaking of wildlife—through the eyes of a biologist. *Happy Hunting Ground* 23(6):25.

Smith, M. 1974. Some straight talk about deer. *Happy Hunting Ground* 30(6):7.

Southwick Associates. 1997. *The Economic Importance of Hunting: Economic Data on Hunting Throughout the Entire United States.* Report for the International Association of Fish and Wildlife Agencies. Federal Aid in Wildlife Restoration Grant Agreement 14-48-98210-97-G047. Arlington, Va.

Strode, D. H. 1951. Restoration of big game is paying off. *Happy Hunting Ground* 7(3):16.

Chapter 6

Outcomes of Hard and Soft Releases of Reintroduced Wolves in Central Idaho and the Greater Yellowstone Area

STEVEN H. FRITTS, CURTIS M. MACK, DOUGLAS W. SMITH, KERRY M. MURPHY, MICHAEL K. PHILLIPS, MICHAEL D. JIMENEZ, EDWARD E. BANGS, JOSEPH A. FONTAINE, CARTER C. NIEMEYER, WAYNE G. BREWSTER, AND TIMMOTHY J. KAMINSKI

Prior to European settlement, gray wolves (*Canis lupus*) were present over most of North America and virtually all of what would become the contiguous 48 states, including the northern Rocky Mountain region (Young 1944). By about 1930 they were eradicated from the American West (Lopez 1978; McIntyre 1995; Hampton 1997). Restoration of wolves to portions of the northern Rocky Mountains, particularly Yellowstone National Park (YNP), was advocated for several decades (Leopold 1945; Weaver 1978; Klinghammer 1979) until the U.S. Fish and Wildlife Service (USFWS) reintroduced a "nonessential experimental population" to central Idaho and YNP (Figure 6.1).

Gray wolves have been relocated under a variety of circumstances (Merriam 1964; Mech 1966; Henshaw and Stephenson 1974; Weise et al. 1975; Henshaw et al. 1979; Fritts et al. 1984, 1985; Bangs et al. 1995, 1998; Klein 1995), but no large-scale restoration attempt using wild wolves had occurred prior to this program. Although two fundamental strategies are available—

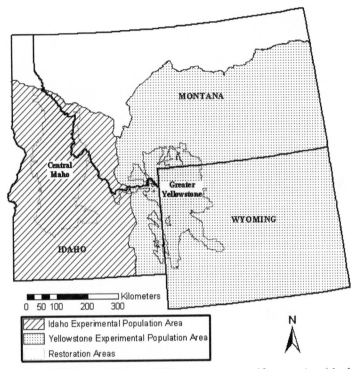

Figure 6.1. Central Idaho and Greater Yellowstone gray wolf restoration (shaded) and experimental population areas.

"hard" release and "soft" release—both were unproven for establishing a wild population. Hard release is an immediate and direct release into the new environment; soft release is a delayed release from a temporary enclosure (Fritts 1993). The degree of stimulation by humans at the time of release is a major variable that distinguishes these approaches. Soft release involves construction of acclimation pens and temporary husbandry. The two approaches differ as well in their demands of personnel, facilities, and cost. Hard releases have been the most common technique for wildlife reintroduction throughout North America, though with variable success (Griffith et al. 1989).

The objective of the authors and the agencies we represented was to establish populations of wolves in central Idaho and the Greater Yellowstone area as quickly as practical. In doing so we tested both methods, hard and soft release, to refine and optimize subsequent releases and to gain information that will benefit future wolf reintroductions. In this chapter we examine wolf reintroduction design and the outcomes of the hard and soft releases in central Idaho and YNP.

Restoration Areas

The central Idaho area is about 53,613 km^2 of rugged mountains (99 percent federal ownership), with almost 16,200 km^2 of designated wilderness and a human density of 1 person/km^2. Wild ungulates in the area include mule deer (*Odocoileus hemionus*), white-tailed deer (*O. virginianus*), elk (*Cervus elaphus*), mountain goat (*Oreamnos americanus*), bighorn sheep (*Ovis canadensis*), and moose (*Alces alces*). Mountain lion (*Puma concolor*), black bear (*Ursus americanus*), and coyote (*Canis latrans*) are sympatric. Virtually all motorized activity in wilderness areas is prohibited.

The Greater Yellowstone area covers 64,750 km^2 (76 percent federal ownership). YNP at the center is 9000 km^2 and is surrounded by six national forests. Average human density is about 2 persons/km^2. Wild prey include elk, mule and white-tailed deer, moose, bighorn sheep, bison (*Bison bison*), pronghorn (*Antilocapra americana*), and mountain goats. Black bear, grizzly bear (*Ursus arctos*), mountain lion, and coyote are sympatric.

The cores of both the central Idaho and Greater Yellowstone areas do not support livestock and are some of the most remote lands in the United States outside of Alaska. Low human densities, public landownership, and large size make them refugia for large carnivores. Hunting and livestock production do not occur inside YNP, and the number of people in the park fluctuates seasonally. Lamar Valley in the northeastern part of YNP was selected for initial releases because of the high density of ungulates, presence of an access road, and records of wolf activity before wolves were eradicated (Weaver 1978).

Approach

In early 1993 we surveyed 53 biologists who had experience with wild or captive wolves and asked them how a reintroduction might be carried out. Although most recommended some form of soft release, the span of opinions in the 31 responses, together with the scarcity of reintroduction experience, made it clear that whatever approach we selected would follow an adaptive management strategy. Any perception of insurmountable problems or inadequate planning was likely to bring pressure to terminate the project by those already opposed to it (Fischer 1995). Hard release of individual wolves was planned for Idaho; soft release of small family groups was used in YNP. Wild captured wolves were translocated from Canada in areas with habitat similar to restoration areas (Bangs and Fritts 1996; Fritts et al. 1997) and were free of rabies, tuberculosis, and brucellosis (see Case 2 in this volume), and contained few livestock.

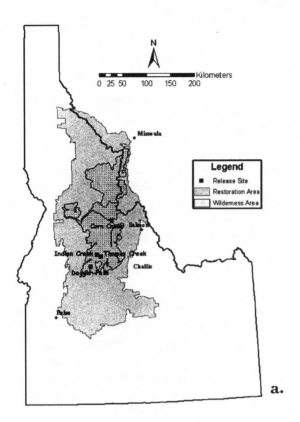

a.

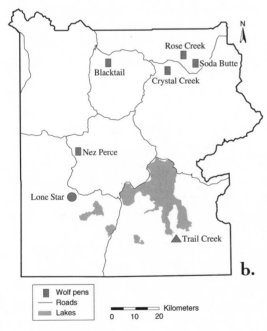

b.

Figure 6.2. *a.* Locations of hard-release sites in central Idaho in 1995 and 1996. *b.* Locations of acclimation pens where wolves were soft-released in Yellowstone National Park in 1995 and 1996. The dot represents a temporary enclosure from which the Lone Star pair was released. The triangle represents a pen used for the second release of the Soda Butte pack. After denning on private land north of the park, the wolves were captured, returned to the park, acclimated, and released again.

Release sites in central Idaho offered aerial or ground access and had year-round populations of elk and deer (Figure 6.2a). Wolves brought to Idaho generally were nonbreeding adults and yearlings from the same Canadian packs that provided wolves for YNP. We did not view the relatedness of some wolves as an advantage or disadvantage so long as they were from several packs.

Three pens were constructed in YNP in summer 1994 using a 3-m-high, 9-gauge, chain-link fence with an inward overhang and a 1.2-m ground apron (Figure 6.2b). These pens were 0.4 to 0.8 ha in size and were built at least 8 km apart (Phillips and Smith 1996). Two additional pens were constructed for 1996 releases. All sites but one were accessible by roads.

1995 RELEASES. In January 1995, 29 wolves were captured by helicopter darting, net-gunning, and live-snaring near Hinton, Alberta, and transported to central Idaho (1000 km) and YNP (1100 km) via cargo aircraft and ground transport (Fritts et al. 1997). All wolves were radio-collared, marked with plastic numbered tags in both ears, and also marked with PIT tags (passive integrated transponders) (Bangs and Fritts 1996; Fritts et al. 1996, 1997). All wolves were treated identically during transport regardless of release method.

Fifteen wolves from seven Canadian packs were sent to the Idaho release sites in January 1995 and immediately released. Four wolves spent nearly 90 hours in their crates due to bad weather and a delay caused by an American Farm Bureau legal challenge to the program. The first four Idaho wolves were released at Corn Creek along the Main Salmon River. The eleven that followed were set free at Thomas Creek and Indian Creek along the Middle Fork of the Salmon River (Figure 6.2a). No food was provided at the release sites.

Fourteen wolves in three groups (from four Canadian packs) were shipped to YNP in January 1995. The first eight spent approximately 40 hours in their crates due to the Farm Bureau's legal challenge. YNP groups consisted of three, five, and six wolves and were held at the Rose Creek, Soda Butte, and Crystal Creek pens, respectively (Table 6.1; Figure 6.2b).

Wolf husbandry is described by Phillips and Smith (1996, 1997) and by Johnson (Case 2 in this volume). In most instances wolves adapted to the pen within one week, and no significant conflict between wolves was known to occur. Some wolves damaged their teeth biting the fence when humans approached the pens. Tooth damage occurred primarily among adult wolves of both sexes, but there was no indication that killing ability or survival in the

Table 6.1. Basic Information on Packs of Wild Wolves Soft-Released in Yellowstone National Park in 1995 and 1996

Pack	Number of individuals	Natural pack?[a]	Bred in captivity before release?	Remained together after release?
1995: Crystal Creek	6	natural	no	yes
Rose Creek	3	artificial	yes	yes[b]
Soda Butte	5	natural	yes	yes
1996: Nez Perce	6	natural	yes	no
Lone Star	2	artificial	yes	yes
Chief Joseph	4	artificial	no	yes[c]
Druid Peak	5	artificial	no	yes[d]

[a] "Natural pack" means that all members were part of the same wild pack when captured in Canada. "Artificial" means that pack members were from different wild packs but were penned together and released together in Yellowstone.
[b] A pup separated from the breeding pair after six days.
[c] A pup separated from the group the first day after release.
[d] After several weeks with the pack, the alpha female began making long forays in May and began traveling alone in mid-August.

wild was affected. The wolves ate the food provided, howled, interacted, and bred in captivity. Acclimation time averaged 67.9 days in 1995, and the wolves took an additional 2 to 10 days (a mean of 5.7 days) to exit pens after they had been opened by opening the gate or removing a section of fence.

1996 RELEASES. Wolves were captured in January by helicopter-darting in the vicinity of Fort St. John, British Columbia. Altogether, 37 wolves from eight packs were transported to central Idaho (1400 km) and YNP (1550 km) (Bangs and Fritts 1996; Fritts et al. 1996). Twenty wolves were taken by snowmobile to a single release site at Dagger Falls along the headwaters of the Middle Fork of the Salmon River in Idaho (Figure 6.2a). This group included three 9-month-old pups, two of which were litter mates that were released with an adult female from the same wild pack.

Deliveries of eleven and six wolves were made to YNP on 23 January and 27 January. Seventeen wolves from four packs were placed in four pens in YNP during January 1996. Group sizes were two, four, five, and six wolves (Table 6.1). The Nez Perce pack included the breeding male and female from a wild pack; the other three groups (Lone Star pair, Druid Peak pack, Chief Joseph pack) included breeding female wolves and breeding age male wolves from different packs.

The Druid Peak pack and the Nez Perce pack were allowed to exit the pens in which they were acclimated: one in northern YNP (the old Rose Creek pen) and the other in the Firehole River–Nez Perce Creek area (Figure 6.2b). We chose not to release the Chief Joseph pack and the Lone Star pair from their pens inside the territories of packs released in 1995. After being acclimated in the Crystal Creek pen, the Chief Joseph pack was transported to the recently vacated Nez Perce pen, 53 km to the southwest, and then allowed to exit after recovering from the anesthesia. The Lone Star pair was immobilized and left to awaken in an incomplete temporary enclosure 61 km to the southwest (Figure 6.2b). In 1996 the acclimation time was 72 days, and the wolves took an additional 0 to 12 days (mean 4.2) to exit the pens.

Monitoring

Most radio-tracking in Idaho was done in a Cessna 206 fixed-wing aircraft. All locations were recorded with GPS units. Funding limitations restricted the frequency of flights to about one or two per month. Wolves were located from the ground as well during routine fieldwork. Wolves in YNP were located by aircraft as often as weather permitted for two weeks, then twice per week until midsummer, and then once per week thereafter. Some packs were observable from vantage points in the park.

Data Analysis

We examined initial and ultimate travel direction from release sites and distance moved from release sites by 25 June, 15 September, at home range establishment, and at the "ultimate" location. "Initial direction" was the first known location away from the release site (or from a point within 2 km of the release site) that was followed by a pattern of movement away from the site (Fritts et al. 1984). "Ultimate direction" was the compass bearing from the release site to the site of mortality, recapture (management action), last known location, or the approximate center of an established home range (Fritts et al. 1984). Ultimate distance was the straight-line distance between those points. Each wolf was treated as a separate and independent observation in the statistical analysis even though many appeared to be traveling together.

We examined ultimate distance with analysis of variance using three factors as independent variables at two levels: year of reintroduction (1995, 1996), restoration area (Idaho, YNP), and sex. We did not use an age variable in the ANOVA model because the number of known yearlings in Idaho was small and visual comparison of adults and yearlings suggested no difference between the groups. Moreover, a *t*-test revealed no difference in ultimate distance between YNP adults and yearlings.

Variance in ultimate distance between wolves at different factor levels (1995 vs. 1996), for example, was unequal due to outliers. To meet the ANOVA assumption of homogeneity of variances among groups (assessed using a Levin's test), the square root of ultimate distance (SQRTD) was used as the dependent variable. ANOVA was not performed on the June and September measures because those metrics were too similar to the ultimate distance to warrant separate statistical analysis. We compared distances from release sites to centers of territories established in the two restoration areas by using a Mann-Whitney *U*-test. Variances for the two groups were equalized using a $\log_{10}$ transformation.

Directions from release sites were evaluated using statistics for circular distributions (Batschelet 1981; White and Garrott 1990). We calculated mean angles, mean bearings, and angular deviation (*s*) for the initial and ultimate azimuths of wolf travel (Batschelet 1981). We evaluated the extent that directions deviated from a uniform circular distribution by calculating *r*—a measure of concentration of azimuths scaled from 0 (bearings dispersed) to 1 (highly concentrated)—and compared them with critical values (Batschelet 1981). If wolf movements were nonrandom, we further tested the hypotheses that initial and ultimate directions of wolf movements were toward the capture site in Canada using a *V*-test (Batschelet 1981; White and Garrott 1990). We used direction to a point halfway between the two capture sites in Alberta and British Columbia for those analyses.

We tested for associations between directions (both initial and ultimate) and year, restoration area, and sex by using log-linear analyses. This initial step was used to determine differences in the direction of wolf movements at different levels of these three factors. Where movements were significantly ($p < 0.05$) associated with a factor (direction varied by restoration area, for instance), we calculated circular statistics separately for each level of the respective factor (Idaho separately from YNP, for instance). For that analysis, initial travel directions were grouped into two categories: north (271–90°) and south (91–270°).

We used two-tailed Fisher's exact tests to validate the log-linear analysis. Expected cell counts used in calculating chi-square statistics for log-linear analysis should be five or greater, but ours were usually less than five. We applied Fisher's exact tests to all possible 2 × 2 contingency tables generated from combinations of restoration area, sex, and year to identify any two-way or three-way interactions that remained undetected in log-linear analysis.

Due to a small sample of pups released in Idaho, we did not use wolf age as a variable in the analysis. However, we were able to test for differences in travel direction by age (adults versus pups; north versus south) for YNP releases using a two-tailed Fisher's exact test prior to the log-linear analysis. Pups were captured in Canada at eight or nine months of age and released before their first birthday. These individuals were either pups (Idaho) or year-lings (YNP) when their initial directions were measured. Typically, wolves captured as pups in Canada were yearlings when their ultimate distances were measured.

We also compared survival and mortality causes, sociality (persistence of packs released in YNP and ability to form packs in Idaho), home range estab-lishment, reproduction, and livestock killed. Information on depredations was collected in coordination with USDA/APHIS, Wildlife Services. When wolves were relocated to acclimation pens because they killed or threatened livestock, ultimate distance and ultimate direction were measured using the site of recapture.

Results

We analyzed two release protocols in different areas, both of which ultimately succeeded. Comparing behavior of wolves in the two areas was not straight-forward, especially in view of the well-documented individualistic behavior of wolves (Fritts et al. 1984; Mech 1987). Nonetheless, there were certain commonalities among wolves and restoration areas and also some distinct differences. Although wolves that were hard-released into central Idaho and soft-released into YNP established populations, behavioral processes for achieving population status were dissimilar in many respects, corresponding to the challenges the wolves had to overcome.

Direction and Distance Moved

Fifty of the 65 wolves (77 percent) initially moved northward (271°–90°). Initial direction was not known for one Idaho wolf. Mean travel direction for all wolves was 36° and was concentrated about 50° east of the homeward direction (Figure 6.3).

In YNP, there were no significant differences between adults and pups in initial travel direction (p = 0.45). Because the initial movements of three Idaho pups did not appear to differ from those of Idaho adults, we pooled wolves by age in subsequent tests. Log-linear analysis indicated that direc-tions varied by sex (partial association chi square: 4.1; p = 0.043), but not by year, restoration area, or three-way interactions (p > 0.15). Analysis of 2 × 2 tables using Fisher's exact tests supported this result with one exception:

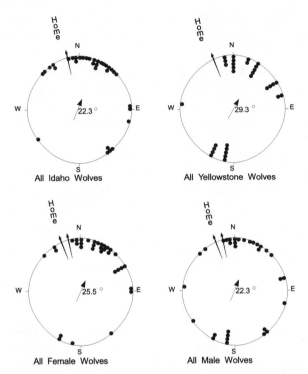

Figure 6.3. Direction of initial movement from release sites by wolves released in central Idaho and Yellowstone National Park in 1995 and 1996. Different release sites were used in each restoration area; release sites were standardized for measurement of directions and distances. Arrows indicate mean azimuths (travel direction). Differences between sexes were significant, whereas there were no significant differences by restoration area.

Idaho wolves initially moved northward more than YNP wolves in 1995, although this association between restoration area and direction was weak (p = 0.08). Thus we calculated separate sets of circular statistics for males and females, pooling restoration areas and years (Table 6.2).

Initial direction moved by females in Idaho and YNP was nonrandom (p < 0.001) and directional (p < 0.0001) toward home in Canada (Figure 6.3). Direction taken by males was random (p = 0.084) with a weak trend (p = 0.047) toward the capture site (Figure 6.3). This finding applied to males during both years and at both restoration areas, except that Idaho males tended to move slightly more northerly than YNP males in 1995 (Table 6.2a).

Wolves released in Idaho traveled extensively within a few weeks of

Table 6.2. Analysis of Circular Statistics

A. INITIAL AZIMUTH (INITIAL DIRECTION TRAVELED
FROM RELEASE SITE) OF ALL RELEASED WOLVES

	Mean bearing	Angular deviation	r (p value)	V test u (p value)
Females	25.5°	53.5°	0.65 (<0.001)	3.81 (<0.0001)
Males	22.3°	91.9°	0.28 (0.084)	1.68 (0.047)

B. ULTIMATE AZIMUTH (DIRECTION FROM
RELEASE SITES) OF ALL RELEASED WOLVES

	Mean bearing	Angular deviation	r (p value)	V test u (p value)
Idaho	29.6°	73.1°	0.443 (0.001)	2.73 (0.0032)
Yellowstone	168.7°	94.5°	0.257 (0.132)	no test

release. Most moved a short distance immediately after release and then remained in one area for one to three weeks before exploring widely. There was considerable variation in the timing of movements away from release sites. Male B-2 remained near the site for several weeks, for example, while female B-10 moved dozens of kilometers immediately after release. Female B-13 moved 88 km east of her release site in nine days before being illegally shot. By June, Idaho wolves averaged 84 km from release sites in 1995 (range: 35–223 km) and 79 km from their release sites in 1996 (range: 6–236 km).

For the first two weeks after release, most YNP wolves stayed near their pens and remained grouped. All packs but one tended to linger near their pens before traveling more widely. By June 1995, some 12 free-ranging Yellowstone wolves averaged 22 km (range: 3–34 km) from their release sites. Movement from release sites was higher in 1996 (average in June: 55 km; range: 22–125 km) because the breeding female and three female siblings from the Nez Perce pack (the only intact pack brought to Yellowstone in 1996) traveled 48 km the first day. They continued traveling to the northeast for four days, averaging 53 km/day before temporarily restricting movements to an area near Red Lodge, Montana. From there, the adult female continued northward, leaving the younger wolves behind.

Postrelease behavior and movements of the 1995 YNP wolves were categorized into three periods: exploration near release pens (2–14 days); wide-ranging exploration (3–35 days); and return to the area around the pen and home range establishment (more than 35–40 days). During week 3 of 1995, all three groups made exploratory moves to the northeast and north that extended for 80, 60, and 32 km. Movements of three of the four 1996-

released groups were similar, with the exception of the Nez Perce pack. Five of the seven groups that survived the early release period established territories in the vicinity of their acclimation pens, including one (the Rose Creek pack) that was recaptured and returned to the park (Phillips and Smith 1996, 1997). The Lone Star pack (a pair) had no chance to establish nearby because of the female's death and the male's dispersal.

As in late June, measures of distances from release sites at later periods were consistently greater and more variable for Idaho wolves than for YNP wolves. By mid-September of the release year Idaho wolves averaged 88 km (range: 17–224 km) from their release sites compared to 31 km (range: 3–120 km) for YNP wolves. For ultimate distance, averages were 98 km (range: 9–282 km) for Idaho wolves and 59 km (range: 2–166 km) for YNP wolves. Following initial northerly movements, Idaho wolves moved randomly (sometimes in a zigzag fashion through heavily dissected drainages) for the first three months. For several individuals, movements encompassed much of the central Idaho restoration area. Although patterns of movements by single wolves appeared to be random and unpredictable in direction and purpose, they often re-used travel routes to and from familiar areas separated by great distances. Single wolves moved long distances in short time periods to areas used by other wolves (related or not). Wolves visited familiar areas repeatedly. Those visits were often brief (a matter of days) and separated by extensive movements between visits.

The movements of female B-10 were the most extensive: a minimum of 800 km accumulated over 14 radiolocations in five months. Male B-2 moved only 200 km over a similar period. Only one wolf completely left the Idaho experimental population area: male B-14 traveled throughout central Idaho for almost three years before joining a nonreintroduced resident female and her two pups 258 km away in western Montana. A separate analysis, using all radio-fixes, showed that wolves drifted farther from their release sites in the first five or six months of freedom (Mack and Laudon 1998).

Movements of unpaired wolves in Idaho were similar to the wide-ranging movements of lone wolves elsewhere in North America (Fritts and Mech 1981; Mech 1987). The area used by male B-14 covered at least 18,130 km^2 from 1995 to 1998 (Mack and Laudon 1998). Most long-range movements took wolves well north of release sites but still within the experimental population area. After four to six months, the extensive and unpredictable movements became more localized as pairs formed. The change in pattern after pairing was typically abrupt, as is true for newly formed pairs elsewhere (Rothman and Mech 1979; Fritts and Mech 1981). By November 1996,

Idaho wolves were distributed over 47,170 km² (Mack and Laudon 1998) whereas Yellowstone wolves were found within 12,690 km².

We found no differences in ultimate distance between YNP pups and adults (p = 0.83), so their data were pooled in the ANOVA. That analysis indicated there were significant differences (p = 0.01) in ultimate distances between restoration areas, but no differences between sexes and years (p > 0.24; Figure 6.4). A significant area-year interaction (p = 0.042) indicated that the differences in restoration area were contingent on the year. There were significant differences in ultimate distance by reintroduction area (p = 0.004) in 1995 and no difference by sex (p = 0.62). There were no significant differences in ultimate distance between areas and sexes in 1996 (p > 0.61). Removing the outlier Nez Perce pack from these analyses resulted in a highly significant difference between Idaho and YNP wolves (p = 0.0021) (Figure 6.4).

There were no significant differences between adults and pups in ultimate

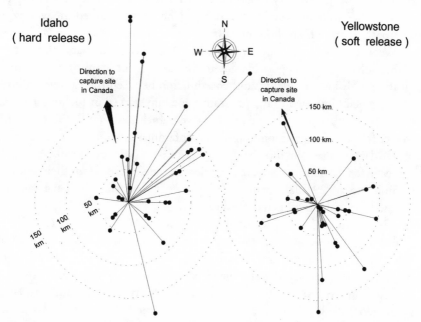

Figure 6.4. Ultimate direction and distance for wolves released in central Idaho and Yellowstone National Park in 1995 and 1996. Multiple release sites were used in each area; release sites are standardized for measurement of directions and distances.

direction of Yellowstone wolves ($p = 1.0$) and no differences in Idaho adults and pups based on a subjective inspection of the data. Log-linear analysis indicated that ultimate direction varied by restoration area (partial association chi square $= 14.1$; $p = 0.0002$), but not by sex, year, or three-way interactions ($p > 0.34$). Analysis of 2×2 tables using Fisher's exact tests supported this result. Therefore we calculated circular statistics separately for the two sites, pooling data by sexes and years. The results indicate that ultimate directions for Idaho wolves were nonrandom ($p < 0.001$) and significantly ($p < 0.0032$) toward the capture site in Canada (Table 6.2b; Figure 6.4). Ultimate directions for YNP wolves were random ($p > 0.132$). (Thus, no V test was made for movement toward the capture site). Based on angular deviation, directions of YNP wolves were more variable than those of Idaho wolves.

Survival and Mortality

All but two wolves survived handling, holding, transport, and reintroduction. One was killed by an errant dart during capture in Alberta and the other was euthanized for mandatory rabies tests after biting the thumb of a biologist. Because they were not confined in pens, Idaho wolves entered the wild with less tooth damage than those in YNP.

We were able to account for the fates of all YNP wolves and all but two Idaho wolves. This was possible due to the intensity of monitoring, communication with local residents, and coordination between the various government agencies involved in the program. One pup starved in Idaho; the other two associated with an adult and survived. Including YNP, the other 65 wolves found sufficient food despite the unfamiliar environment. At least 31 of 35 Idaho wolves (88.6 percent) and 25 of 31 YNP wolves (80.6 percent) survived for 12 months after their release. Three years after their release, a minimum of 29 Idaho wolves (82 percent) were still alive, whereas 13 Yellowstone wolves (42 percent) were alive.

There were 21 known deaths after three years: 4 in Idaho and 17 in YNP (Table 6.3). Illegal shooting and depredation control were more common in the greater Yellowstone area. Fourteen of the 21 deaths (67 percent) were human-related, mostly due to illegal shooting and government livestock depredation control. Differences in accessibility by humans probably accounted for the difference in illegal shooting between the two areas. Because wolves were released in remote areas where contact and conflicts with humans were less likely, humans caused a lower proportion of all mortalities in both areas than was the case with naturally recolonizing wolves in Montana (Bangs et al. 1998).

Intraspecific aggression among established packs increased mortality in

Table 6.3. Causes of Mortality Among Reintroduced
Wolves During the First Three Years

Cause of mortality	Central Idaho	Yellowstone
Illegally shot	1	7
Depredation control	1	4
Vehicle collision	0	1
Other wolves	0	2
Accident	0	2
Starvation	1	0
Cougar	1	0
Other natural cause	0	1

YNP. The alpha male from the Crystal Creek pack and a yearling and two-year-old female with a litter from the Rose Creek pack were killed by the Druid Peak pack. In addition, the Druid Peak pack likely killed a litter of the Crystal Creek pack.

Sociality

In Idaho there were nine instances in which several pack members (two to six) were released from the same Canadian pack. We documented subsequent associations between former pack members in five of these nine related groups. Interactions were intermittent, lasting only a few days, occurred more often within nine months of release, and rarely occurred between more than two former pack members at one time. No Idaho wolf paired with a former pack member. The most frequent and long-term interaction was between an adult female and two pups that remained in contact through October of the release year. Later, two of these three wolves were known to have pair-bonded with wolves that originated from other packs.

Wolves in Idaho formed pairs in as little as 16 days in 1995. By the end of September 1995, three pairs had exhibited localized movements. The first pairing of Idaho wolves in 1996 occurred 63 days after release, and this pair established a territory less than 60 days later. Eight pairs formed in 1996 included two pairings of animals released in 1995. Pairs produced litters in their first breeding season after pairing in 82 percent of the cases. The number of pairs and packs producing pups increased to ten in 1998 (Figure 6.5). By fall 1998, all the surviving females of breeding age had paired and formed packs. Idaho packs increased from 11 in November 1996 to 14 in 1998. Pairs in Idaho were permanent unless one member died.

After release, six of seven YNP groups stayed together and five survived

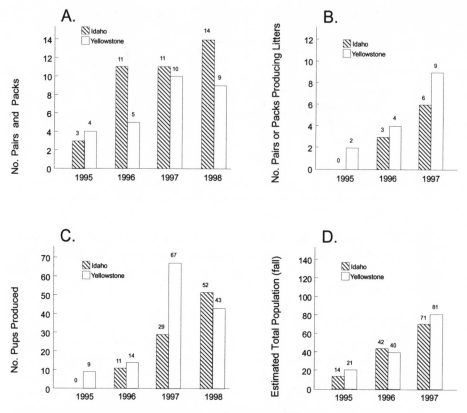

Figure 6.5. Performance of the newly established wolf populations in central Idaho and the Greater Yellowstone area through 1998: *a.* Number of pairs and packs in the populations. *b.* Number of pairs or packs producing pups. *c.* Number of pups produced. *d.* Estimated total population of wolves in each restoration area.

to establish actual or potential breeding units. One of our greatest concerns was that penned wolves, especially those from different packs, would separate and disperse after release. Nonetheless, the four groups that were made up of wolves from different packs remained together after release and two of them bred in captivity (Phillips and Smith 1996). Interestingly, the only group in which the alpha pair separated was the naturally occurring Nez Perce pack. In that instance, the alpha female and two younger females immediately departed the pen, leaving the males to exit later. Two subadults left other packs within a week after release. All other separations occurred after several months. Although we considered this normal behavior, it could have been hastened by confinement with conspecifics. Reintroduced wolves formed three new pairs in 1996, and all packs reproduced in 1997. By the end of 1998, 9 of 11 packs in the Greater Yellowstone area were the result of reintroduced wolves; the other two packs involved the offspring of reintroduced wolves. Six other reintroduced wolves dispersed and failed to pair because of human-caused mortality.

Home Range Establishment

Both hard- and soft-released wolves settled into home ranges, but those that were soft-released established home ranges sooner. In Idaho, home range establishment coincided with pair formation; home ranges were established 27 to 256 km (mean: 92 km) from release sites. YNP packs that stayed together took about a month to settle into home ranges that became centered 4 to 23 km from release pens (mean: 14.5 km; five pack territories included their release pens). Median distances of home ranges from release sites were greater in Idaho than in YNP (Mann-Whitney $T = 0$; $p = 0.0003$). In YNP, individuals from the original groups paired with individuals from others to form additional territories in and near YNP, some of which were more distant from release sites.

The hard-release protocol in Idaho evidently contributed to a more widespread pack distribution. Yet, the fact that some Yellowstone wolves were recaptured and returned to the park also played a role. Entire or potential packs were confined to acclimation pens on three occasions after settling in unsuitable areas or threatening livestock. Each established a home range near their second release site; thus the second acclimation refocused home range selection. The only home range shifts occurred as a result of aggression by the Druid Peak pack toward the Crystal Creek and Rose Creek packs; the Crystal Creek pack was completely ousted from the Lamar Valley.

Reproduction

Of the initial reintroduced wolves, at least 22 Idaho wolves (63 percent) gave birth to or fathered pups through 1998, and two more did so by 2000. Reproduction and recruitment in Idaho were delayed compared to YNP because of the type of release. Releases there occurred only three to six weeks before the breeding season. None of the three original pairs reproduced in 1995, but all three produced pups in 1996. Twenty-four of the 31 YNP wolves (77 percent) gave birth to or fathered pups, nine of which resulted from breeding in acclimation pens. One of the four groups released in 1996 successfully produced pups. Altogether, four of seven groups that were acclimated in Yellowstone bred in captivity and produced litters. Yellowstone litter production grew from two in 1995 to nine in 1997. Three packs had multiple litters in

1997 and two did so in 1998. Multiple litters probably resulted from the soft-release protocol whereby a dominant male was combined with several unrelated females—thus relaxing the inbreeding avoidance characteristic of wolves in packs (Smith et al. 1997). Including the introduced wolves, Idaho and YNP populations grew at annual rates of 101 percent and 75 percent from 1995 to 1998, respectively. In this calculation the releases in 1995 and 1996 mask the effect of delayed start of reproduction in Idaho (Figure 6.5).

Depredation on Livestock

Five Idaho wolves (14 percent) were involved in depredations, although a total of 11 (31 percent) may have been associated with packs that killed livestock. From 1995 to 1998, Idaho wolves killed 15 cattle and 64 sheep (Bangs et al. 1998; Mack and Laudon 1998). In response, six wolves were translocated (two were translocated twice), and two were killed in control actions (Bangs et al. 1998). Ranchers were compensated more than $13,800 for livestock losses. Of the 31 wolves released in YNP, 6 (19 percent) were involved in depredations or were involved, implicated, or associated with packs that killed 8 cattle and 20 sheep during 1995–1998. Government agents killed four wolves. Fourteen reintroduced wolves and their pups were translocated to reduce conflicts with livestock and provide those wolves greater security (Bangs and Fritts 1996). Compensation to ranchers totaled $13,000. Hard-released wolves in Idaho killed a few more livestock than soft-released wolves despite a higher abundance of livestock near YNP (Bangs and Fritts 1996). Hard-released wolves might have been more prone to encounter livestock as a result of their wider-ranging tendencies. We saw no indication that reintroduced wolves were more inclined to kill livestock than naturally occurring wolves.

Conclusions

Both hard- and soft-release techniques established breeding wolf populations. Although our original plan was to release wolves for five years, pack establishment and reproduction allowed us to stop at the end of year 2 by which time 66 wolves had been released. Perhaps the most important components of success were the wolves' ability to endure the rigors of reintroduction and the suitability of release habitat. We learned that combining wolves from different packs did not hamper—and may even have improved—the soft-release procedure by promoting reproduction. All but one soft-released pack remained together, indicating that temporary captivity does not compromise pack cohesiveness. Further, the level of contact with humans needed to maintain captive wolves probably did not increase their tolerance of people. As has been found elsewhere, hard releases seem to guarantee that wolves will not remain together, (Fritts et al. 1984, 1985; Bangs et al. 1995).

Most wolves in this and other studies appeared to move toward home after release even though translocated up to 1000 or even 1550 km (Fritts 1993). Nonetheless, homeward movement was not a significant factor in the outcome of the program because the restoration areas were very large. Such homing tendencies have been exhibited by translocated cougars (Belden and Hagedorn 1993; Ruth et al. 1998), black bears (Rogers 1988), other wolves (Henshaw and Stephenson 1974; Weise et al. 1975; Fritts et al. 1984), and a variety of mammals (Bovet 1992). The concentration of travel directions some 50° east of the homeward direction, apparent in both sexes and both restoration areas, had no ready explanation.

Why females initially moved homeward more than males is unknown. The pattern did not extend to home range establishment. The ultimate directions of hard-released wolves tended toward their capture sites, however, whereas soft-released wolves were distributed more randomly. We believe the shorter distances moved by YNP wolves resulted from the soft-release procedure. During acclimation, YNP wolves apparently lost some motivation to return home and were less likely to scatter from the release site. Evidently soft-released packs are likely to return to the vicinity of their release sites after a period of exploration. Thus soft releases give managers an element of control not possible in hard releases. Soft releases and occasional management actions to return wolves to captivity encouraged their establishment in the most desirable areas of the Greater Yellowstone area. Wolves released in Idaho generally moved well north of the center of the release area. The disintegration of the YNP Nez Perce pack was unexpected but may have been related to noise from snowmobiles. No other pack was subject to that disturbance. Based on this case, we suspect that the degree of stimulation by humans while wolves are penned and at the time of release could well affect the outcome. The release of the Nez Perce pack was far from a failure, though, as the divided pack became dispersers that ultimately produced seven litters.

The extremely rugged terrain of Idaho with its complex networks of ridges and drainages might have helped attenuate homeward movement and general exploratory activity. The ability of wolves to navigate through central Idaho suggests cognitive mapping that is tied to landscape features (Dyer 1998). Fritts et al. (1984) reported that some translocated wolves in Minnesota returned to the vicinity of previous locations and then moved away in different directions, apparently using those locations as reference points. Cases of wolves leaving territories, traveling great distances, and then returning again are known (Fritts and Mech 1981; Messier 1985; Mech 1987; Fuller 1989). Merrill and Mech (2000) have documented the long-distance travels of four wolves that returned to territories after traveling up to 494 km away.

High survival (more than 80 percent the first year) contributed to early population establishment in both release areas and is better than translocations of canids elsewhere (Fritts et al. 1985; Carbyn et al. 1994; Bangs et al. 1998). Survival among the YNP wolves might have been higher if release pens had been more widely spaced. Future reintroductions might reduce intraspecific conflict by spacing release pens much greater than 8 km apart. No deaths from intraspecific aggression occurred among hard-released wolves through 1998.

Because of some breeding in captivity, reproduction occurred sooner with the soft-release technique. This enabled population growth to begin earlier in the Yellowstone area. But because of the ability of wide-ranging wolves in Idaho to find one another, pair-bond, and reproduce, the recruitment within that population was delayed only a year. By 1998, pup production had caught up to the YNP level. This result demonstrates the demographic advantage of having more breeding units early in a wolf population's history.

Both populations have continued to grow since 1998, and both appear well established today. By December 2000 there were about 192 wolves in central Idaho and 177 in the Greater Yellowstone area. Clearly, release technique was not the only factor that determined restoration success. It appears that if landscape conditions, prey availability, wolf restoration stock, and early release management are suitable (Phillips and Smith 1997; Mack and Laudon 1998; Smith 1998; Bangs et al. 1998), the choice of hard versus soft release seems to matter little. Nonetheless, hard releases may be advantageous if the size of the area can accommodate wolves wandering without encountering people or killing livestock. The technique is relatively inexpensive as well, and involves less husbandry. If the size of the area is restricted, however, then a soft release should be used to limit postrelease movements. Because few areas are as extensive as central Idaho, soft releases are likely to be preferred in future wolf restoration efforts.

Acknowledgments

Numerous people contributed to the success of this endeavor. For a comprehensive list, see the acknowledgments in Bangs and Fritts (1996), *Wildlife Society Bulletin* 24 (1996):780; Fritts et al. (1997); and Bangs et al. (1998). We would especially like to mention the contribution of J. Varley, B. Ruesink, R. Heberger, T. Koch, K. Laudon, E. Garton, K. Lawrence, M. Collinge, and D.

Guernsey. D. Jennings, J. Crantz, and S. Stevenson prepared maps and figures. J. Larkin and M. Depoy-Smith of the University of Kentucky assisted with data analysis. We thank the field crews who captured wolves in Canada and those who worked in Idaho and the Greater Yellowstone area. Without their extraordinary dedication and skills this effort would not have been possible.

Literature Cited

Bangs, E. E., S. H. Fritts, D. R. Harms, J. A. Fontaine, M. D. Jimenez, W. G. Brewster, and C. C. Niemeyer. 1995. Control of endangered gray wolves in Montana. In L. N. Carbyn, S. H. Fritts, and D. R. Seip, eds., *Ecology and Conservation of Wolves in a Changing World.* Edmonton: Canadian Circumpolar Institute.

Bangs, E. E., and S. H. Fritts. 1996. Reintroducing the gray wolf to central Idaho and Yellowstone National Park. *Wildlife Society Bulletin* 24:402–413.

Bangs, E. E., S. H. Fritts, J. A. Fontaine, D. W. Smith, K. M. Murphy, C. M. Mack, and C. C. Niemeyer. 1998. Status of gray wolf restoration in Montana, Idaho, and Wyoming. *Wildlife Society Bulletin* 26:785–798.

Batschelet, E. 1981. *Circular Statistics in Biology.* New York: Academic Press.

Belden, R. C., and B. W. Hagedorn. 1993. Feasibility of translocating panthers into northern Florida. *Journal of Wildlife Management* 57:388–397.

Bovet, J. 1992. Mammals. In F. Papi, ed., *Animal Homing.* London: Chapman and Hall.

Carbyn, L. N., H. J. Armbruster, and C. Mamo. 1994. The swift fox reintroduction program in Canada from 1983 to 1992. In M. L. Bowles and C. J. Whelan, eds., *Restoration of Endangered Species.* Cambridge: Cambridge University Press.

Dyer, F. C. 1998. Cognitive ecology of navigation. In R. Dukas, ed., *Cognitive Ecology.* Chicago: University of Chicago Press.

Fischer, H. 1995. *Wolf Wars.* Helena: Falcon Press.

Fritts, S. H. 1993. Reintroductions and translocations of wolves in North America. In R. S. Cook, ed., *Ecological Issues on Reintroducing Wolves into Yellowstone National Park.* Scientific Monograph NPS/NRYELL/NRSM-93/22. Washington, D.C.: National Park Service.

Fritts, S. H., and L. D. Mech. 1981. Dynamics, movements, and feeding ecology of a newly protected wolf population in northwestern Minnesota. *Wildlife Monographs* 80:1–79.

Fritts, S. H., W. J. Paul, and L. D. Mech. 1984. Movements of translocated wolves in Minnesota. *Journal of Wildlife Management* 48:709–721.

———. 1985. Can relocated wolves survive? *Wildlife Society Bulletin* 13:459–463.

Fritts, S. H., E. E. Bangs, J. A. Fontaine, and M. R. Johnson. 1996. Capture, handling, and transport of wolves for the Yellowstone and central Idaho reintroduction. In *Wolves of America Conference Proceedings.* Washington, D.C.: Defenders of Wildlife.

Fritts, S. H., E. E. Bangs, J. A. Fontaine, M. R. Johnson, M. K. Phillips, E. D. Koch, and J. R. Gunson. 1997. Planning and implementing a reintroduction of wolves to Yellowstone National Park and central Idaho. *Restoration Ecology* 5:7–27.

Fuller, T. K. 1989. Population dynamics of wolves in north-central Minnesota. *Wildlife Monographs* 105:1–41.

Griffith, B., J. M. Scott, J. W. Carpenter, and C. Reed. 1989. Translocation as a species conservation tool: Status and strategy. *Science* 245:477–479.

Hampton, B. 1997. *The Great American Wolf.* New York: Holt.

Henshaw, R. E., and R. O. Stephenson. 1974. Homing in the gray wolf, *Canis lupus. Journal of Mammalogy* 55:234–237.

Henshaw, R. E., R. Lockwood, R. Shideler, and R. O. Stephenson. 1979. Reintroduction of wolves into the wild (workshop). In E. Klinghammer, ed., *Behavior and Ecology of Wolves.* New York: Garland STMP Press.

Klein, D. R. 1995. The introduction, increase, and demise of wolves on Coronation Island, Alaska. In L. N. Carbyn, S. H. Fritts, and D. R. Seip, eds., *Ecology and Conservation of Wolves in a Changing World.* Edmonton: Canadian Circumpolar Institute.

Klinghammer, E., ed. 1979. *The Behavior and Ecology of Wolves.* New York: Garland STMP Press.

Leopold, A. 1945. Review of the Wolves of North America. *Journal of Forestry* 43:98.

Lopez, B. H. 1978. *Of Wolves and Men.* New York: Scribner.

Mack, C. M., and K. Laudon. 1998. *Idaho Wolf Recovery Program: Recovery and Management of Gray Wolves in Idaho.* Progress report 1995–1998. Lapwai, Idaho: Nez Perce Tribe, Department of Wildlife Management.

McIntrye, R. 1995. *War Against the Wolf: America's Campaign to Exterminate the Wolf.* Stillwater, Minn.: Voyageur Press.

Mech, L. D. 1966. *The Wolves of Isle Royale.* Fauna Series, no. 7. Washington, D.C.: National Park Service.

———. 1987. Age, season, direction, distance, and social aspects of wolf dispersal from a Minnesota pack. In B. D. Chepko-Sade and Z. T. Halpin, eds., *Dispersal Patterns.* Chicago: University of Chicago Press.

Merriam, H. R. 1964. The wolves of Coronation Island. *Proceedings of the Alaska Science Conference* 15:27–32.

Merrill, S. B., and L. D. Mech. 2000. Details of extensive movements by Minnesota wolves (*Canis lupus*). *American Midland Naturalist* 144:428–433.

Messier, F. 1985. Solitary living and extra-territorial movements of wolves in relation to social status and prey abundance. *Canadian Journal of Zoology* 63:239–245.

Phillips, M. K., and D. W. Smith. 1996. *The Wolves of Yellowstone*. Stillwater, Minn.: Voyageur Press.

———. 1997. *Yellowstone Wolf Project: Biennial Report 1995–1996*. YCR-NR-97-4. Yellowstone National Park: Yellowstone Center for Resources.

Rogers, L. L. 1988. Homing tendencies of large mammals: A review. In L. Nielsen and R. D. Brown, eds., *Translocation of Wild Animals*. Milwaukee: Wisconsin Humane Society.

Rothman, R. J., and L. D. Mech. 1979. Scent marking in lone wolves and newly formed pairs. *Animal Behavior* 27:750–760.

Ruth, T. K., K. A. Logan, L. L. Sweanor, M. G. Hornocker, and L. J. Temple. 1998. Evaluating cougar translocation in New Mexico. *Journal of Wildlife Management* 62:1264–1275.

Smith, D., T. J. Meier, E. Geffen, L. D. Mech, L. G. Adams, J. W. Burch, and R. K. Wayne. 1997. Is incest common in gray wolf packs? *Behavioral Ecology* 8:384–391.

Smith, D. W. 1998. *Yellowstone Wolf Project: Annual Report 1997*. YCR-NR-98-2. Yellowstone National Park: Yellowstone Center for Resources.

Weaver, J. 1978. *The Wolves of Yellowstone*. Resource Report 14. Washington, D.C.: U.S. Department of Interior, National Park Service.

Weise, T. F., W. L. Robinson, R. A. Hook, and L. D. Mech. 1975. *An Experimental Translocation of the Eastern Timber Wolf*. Audubon Conservation Report 5. New York: National Audubon Society.

White, G. C., and R. A. Garrott. 1990. *Analysis of Wildlife Radio-Tracking Data*. San Diego: Academic Press.

Young, S. P. 1944. The wolves of North America. I: Their history, life habits, economic status, and control. In S. P. Young and E. A. Goldman, *The Wolves of North America*. Washington, D.C.: American Wildlife Institute.

Chapter 7

Health Aspects of
Large Mammal Restoration

JOSEPH K. GAYDOS AND JOSEPH L. CORN

Infectious and noninfectious diseases have caused major harm to populations of free-ranging wildlife (Bellrose 1951; Scott 1988; Heuschele 1991; Fairbrother et al. 1996). Consequently, no large-mammal restoration project should proceed without consideration of disease risks. Diseases may result in reduced fitness of the restoration target, impacts on wildlife and livestock in the release area, and impacts on human health. Failure to recognize and avoid a disease process can annul the tremendous personal, financial, and political efforts that drive all wildlife restoration projects.

In discussing health issues in wildlife restoration, most large mammal restoration efforts may be classified as externally supplemented or internally managed. Externally supplemented restoration involves species that may be endangered or even extirpated in a portion of their range. In these situations conspecific populations in other geographic areas provide a source for translocation. The cheetah (*Acinonyx jubatus*), wapiti (*Cervus elaphus*), gray wolf (*Canis lupus*), and grizzly bear (*Ursus arctos horribilis*) are examples. Internally managed projects, by contrast, involve populations that are small in size, do not have external genetic linkages, and thus cannot involve the translocation of individuals from nearby populations. Examples include the Hawaiian monk seal (*Monachus schauinslandi*), mountain gorilla (*Gorilla gorilla beringei*), and Sumatran rhinoceros (*Diceros sumatrensis*). Although disease risks must be considered for both categories of restoration projects, mitigation will vary between the two.

In this chapter we discuss the health aspects of externally supplemented and internally managed restorations. Examples are used to demonstrate how disease problems can thwart successful restoration and how proactive effort can help prevent problems.

Externally Supplemented Efforts

Externally supplemented restoration involves the translocation of genetically similar individuals in order to establish a new population or augment an existing one. The health of the population, as well as the individual, must be considered prior to and throughout any translocation effort. Although population health concerns should always take precedence, the individual's health is important, too, because the translocation of individuals in poor health can undermine the effort's success.

Population Level Health

Population-level health concerns in wildlife translocation include the potential for introducing a new disease agent or parasite into a release area, aggravation of an existing disease problem by introduction of additional pathogens, and exposure of translocated wildlife to diseases already present in the release area. Although all these scenarios are of concern, the potential to introduce a disease into a release site via translocation is the most significant (Cunningham 1996). Prevention of such introductions is much easier than trying to eradicate or control a disease after it is established.

Large mammals are host to numerous organisms—bacteria, rickettsias, viruses, fungi, helminths, protozoa, and arthropods—and should be viewed as "biological packages" (Davidson and Nettles 1992). It is impossible to separate an animal from these organisms, and occasionally disease agents can be inadvertently moved with translocated animals. A classic example is the introduction of plains bison (*Bison bison*) infected with both bovine brucellosis and bovine tuberculosis into Wood Buffalo National Park, Canada (Fuller 1966; Environmental Assessment Panel 1990; chapter 9 in this volume). Efforts to control these diseases are hampered by political and biological factors alike. Bovine brucellosis and bovine tuberculosis can be transmitted to domestic livestock and humans. In 1985, Canada's domestic cattle herd was declared free of bovine brucellosis, and Agriculture Canada soon expects to declare the country's cattle free of bovine tuberculosis. Thus the bison of Wood Buffalo National Park will be one of the last remaining animal herds infected with either disease in Canada (Environmental Assessment Panel 1990). Effective treatment or immunization methods do not exist for either disease, and test-and-slaughter programs are unlikely to succeed in eradicating them (Environmental Assessment Panel 1990). Other examples include the private-sector movement of raccoons from Florida into the Appalachian mountain region and its link to the current raccoon rabies epizootic in the mid-Atlantic and northeastern United States (Nettles and Martin 1979; Krebs et al. 1999), the introduction of liver flukes to Italy via translocation of elk from North America (Bassi 1893, cited in Kistner 1982), and the detection of *Elaphostrongylus cervi* in quarantined red deer (*C. e. elaphus*) being imported from New Zealand into Canada (Gajadhar and Yates 1994).

Davidson and Nettles (1992) have developed an assessment model that provides a conceptual approach for the identification of disease risks in the translocation of free-ranging wildlife. This model permits evaluation of two potential translocation scenarios: introduction and establishment of an exotic disease in the release area and intensification of a preexisting enzootic disease situation. The model includes a two-tiered process: first delineate the potential for a pathogen to persist at a release site; then assess the pathological consequences of the pathogen in wildlife, domestic animals, and humans. The prototype was used in an assessment of helminths found in translocated raccoons. Each species of helminth was rated for its potential to become established in the release area and for its pathological potential (Schaffer et al. 1981). Although use of this model has been based on the necropsy of animals during ongoing translocation programs, a proactive approach would be more useful in preventing the introduction of unwanted pathogens. There are problems associated with a proactive application, however, including the lack of diagnostic and laboratory support for wildlife disease investigations and the lack of data on the geographic distribution, host susceptibilities, and pathological capabilities of pathogens in wildlife (Davidson and Nettles 1992).

Also of concern is the potential for exposure of translocated wildlife to pathogens already present at the release area—especially if the translocated animals are from a population without previous exposure to a certain pathogen. Neurological disease resulting from infection by *Parelaphostrongylus tenuis* is a concern when elk, caribou (*Rangifer sp.*), and moose (*Alces alces*) are translocated from western to eastern North America (Anderson and Prestwood 1981). Neurological disease and mortality caused by this parasite have been documented in populations of elk established in Arkansas (SCWDS, unpublished data) and eastern Kentucky (chapter 5 in this volume) and have been implicated in a failed elk reintroduction in New York (Severinghaus and Darrow 1976). Another example is the death of lechwe (*Kobus leche*) due to heartwater after animals were translocated into an area where the rickettsia *Cowdria ruminantium* and a tick vector were present (Pandey 1991 cited in Woodford and Rossiter 1993). Within two months of the translocation, 56 lechwe had died—rendering the translocation attempt unsuccessful (Woodford and Rossiter 1993). The risk assessment model offers a useful conceptual framework for addressing such enzootic diseases. Such an assessment would identify the pathogens known to be enzootic in a potential release area as well as the pathological consequences of exposure of translocated wildlife to such pathogens before animals were captured and moved.

A comprehensive disease risk assessment provides the information needed to evaluate the potential for introducing pathogens and exposing animals to them. With this information, a risk reduction protocol can facilitate translocation, assuming that the assessment does not contraindicate the program. Disease risk reduction methods in wildlife translocation include surveillance for disease agents in wildlife, monitoring of populations that may serve as sources of transplant stock, and prophylactic treatment. The method depends on the diseases of concern. Surveillance, for example, should be designed to detect disease agents relative to geographic area or population. Surveillance also can be used for a population selected to serve as a source of animals for future translocations. This process has been described in detail for the translocation of elk (Corn and Nettles 2001). Similarly, bighorn sheep translocated in Wyoming all come from a single population that is monitored for *Brucella* sp. and *Psoroptes* spp. (Thorne et al. 1992). Populations of bighorn that serve as sources of animals for translocation in California are monitored as well, and animals are translocated only from herds that are free of psoroptic mange and other pathogens of concern (Thorne et al. 1992). Arizona and Idaho test the individual animals to be translocated (Thorne et al. 1992). Prophylactic treatment may also be included in risk reduction. Some situations dictate treatment with acaricides to eliminate ticks. In addition, vaccination might be used to immunize translocated wildlife against diseases that are present in the release area. Such prophylactic treatment could protect released animals but not their progeny.

It is impossible to identify all potential disease agents and parasites or to precisely predict their impact on restoration. Such an assessment will involve qualitative judgments based on the information available and, in turn, will determine the methods to reduce disease risk.

Individual Animal Health

The success of a translocation depends on the survival and successful reproduction of the founder stock. Trauma, capture myopathy, or infectious diseases may be associated with the stress of capture, quarantine, transport, and release. Ultimately these factors can reduce the size of the founder population. Care must be taken to plan and execute the translocation in a manner that will minimize the potential for trauma and stress at all points of the capture, quarantine, and translocation.

Selecting the right capture technique is the first step in minimizing stress and trauma. This is a multifactorial decision that should be species-specific and appropriate for the local terrain, climate, and season. In some situations, it may be safer to capture large mammals without using chemical immobilization. Extensive evaluation of capture methods of free-ranging bighorn sheep (*Ovis canadensis*) has revealed that use of a net gun has considerable advantages over the use of ground nets and, moreover, that chemical immobilization should be used only when all other alternatives are contraindicated (Kock et al. 1987). In other situations, chemical immobilization alone or in combination with other capture techniques will help reduce capture-associated problems and minimize human risk—especially when handling large carnivores and large, horned ungulates (Haviernick et al. 1998). If chemical immobilization is necessary, texts such as Kreeger (1999), species-specific literature, and professional guidance should be used to assure appropriate drug dosages and administration techniques.

Selection of transport means and the quarantine facility is the next step in minimizing individual animal health problems. These phases of translocation result in temporary crowding, promote disease transmission, increase stress, and can lead to recrudescence of latent infections (Woodford and Rossiter 1993). For example, a wild, clinically normal black rhinoceros (*Diceros bicornis*) that harbored the protozoa *Trypanosoma brucei* developed clinical trypanosomiasis and died 25 days after capture despite treatment attempts (Clausen 1981). As with capture technique, the means of animal transport and the nature of the quarantine facility will depend on the species and number of animals. When large ungulates such as elk are transported in semi-tractor cattle trailers, for example, the floors should be covered with hay, sharp corners should be padded, and animals should have sufficient personal space to avoid trampling. Regardless of the technique, transport and quarantine should be accomplished in as little time as distance and disease testing permit. A new concept for reducing injuries and stress associated with transport and quarantine of many ungulates is the use of long-acting neuroleptics (Ebedes 1992). These single-dose, slow-release drugs can relieve anxiety and stress for a week or more with minimal drowsiness and sedation. As drug pharmacology improves and undesirable side effects such as anorexia are diminished, the use of these agents may become widespread.

Internally Managed Efforts

If a population is genetically unique and limited in size, translocation is not a restoration option. In these single, localized populations of endangered or threatened species, a single disease event has the power to impact or eliminate the entire population. The introduction of mosquitoes (*Culex quinquefasciatus*) and avian malaria (*Plasmodium relictum*) to the Hawaiian Islands, for example, is believed to have played a major role in the decline and extinction of native Hawaiian honeycreepers (*Drepanidinae*) (Atkinson et al. 2000). Similarly, it is speculated that bluetongue virus infection may have caused the disappearance of the desert bighorn from the Trans-Pecos area of Texas (Robinson et al. 1967). Thus disease risk assessment and intervention for internally managed restoration projects must focus on prevention of catastrophic disease and treatment of individuals when population viability may depend on the performance of key individuals.

Biologists overseeing internally managed restoration projects must determine which disease agents present a risk to the population. This may be difficult to do as these populations often are understudied from a wildlife health perspective and the potential disease threats may be unknown. A list of potential disease risks should include noninfectious causes such as toxins and environmental contaminants, infectious diseases to which the animals are known to be susceptible, and potential infectious diseases that are known to infect closely related domestic animals or sympatric wild animals. Diseases with severe pathological consequences must be evaluated for potential avenues of intervention. Development of a disease-free population is not a realistic goal of any wildlife health program. The goal is prevention or control of diseases that may impact the population.

A protocol should be developed that delineates which interventions are to be taken as well as how and when they will be made. Intervention should encompass traditional methods of disease control or treatment, such as vaccination and deworming, and habitat management that minimizes infection risk. Increases in population size are a common goal of large-mammal restoration projects. When populations increase beyond carrying capacity, however, stress on individual animals may be increased and disease processes may be intensified. The syndrome of malnutrition and parasitism in white-tailed deer, for example, is a well-studied density-dependent phenomenon whereby white-tailed deer herds that exceed the local carrying capacity exhibit increased pathogenic and nonpathogenic internal parasites (Eve and Kellogg 1977; Davidson and Doster 1997).

Additionally, methods for evaluating the program must be predetermined. These should include identification of the parameters to be measured, methods of data collection, methods for assessing the data collected, and the time frame involved (Wobeser 1994). Comparing the ability of a disease intervention program to achieve its stated objectives to the cost of administering the program (Karesh 1993) is a good technique for developing a cost-effectiveness analysis. Such analyses will benefit individual projects by ensuring that resources are used efficiently. Ultimately, a disease surveillance program should be developed that includes seriologic surveys and complete examinations of dead animals by qualified diagnosticians. Such a program would allow for the continual assessment of new disease threats and assist in evaluating the effects of intervention.

The use of vaccines exemplifies the difficulties encountered when trying to intensively manage disease in an internally managed restoration project. Vaccines have long been a pillar of disease prevention in domestic animals. Very few vaccines, however, have been tested in wild animals (Bittle 1993). Vaccines that have been determined to be safe and effective in preventing disease in domestic animals may induce fatal disease in wild animals (Brown and Scott 1960; Bush et al. 1976; Carpenter et al. 1976). Moreover, the protective immunity conferred by a single vaccination is rarely lifelong. This means that repeated vaccinations would be necessary in most situations. Depending on the average length of immunity conferred by a vaccine, the population density, the size of the susceptible population, the method of vaccine delivery, and the cost of administration, populationwide vaccination programs may be prohibitively expensive and completely impractical.

Case Studies

In light of increasing human populations and shrinking habitat, internally managed restoration may become more widespread. Three case studies are presented here to demonstrate the complexities involved in disease risk assessment and disease management in such restoration efforts. Additionally, they offer insight into successful and unsuccessful attempts to mitigate health concerns in single, localized populations.

The Florida Panther

The endangered Florida panther (*Puma concolor coryi*) exists in a small population in southern Florida's Big Cypress and Everglades physiographic regions (USFWS 1987). Roelke et al. (1991) describe the development of a disease assessment and management plan for what began as an internally managed restoration project. At the start of disease assays in 1984, researchers compiled a comprehensive list of infectious disease agents that might affect the panther. This list included infectious agents that had been documented in wild or captive Florida panthers and closely related species, agents documented in more distantly related species within the same order, and agents that could be transmitted to Florida panthers from sympatric carnivores and prey. Samples taken from living and dead panthers, as well as those obtained through serendipity (Butt et al. 1991), helped identify potential problems including 61 environmental contaminants, tumors, anomalies, parasites, and infectious disease agents (Forrester 1992).

Based on virulence and potential for control, three viruses (rabies virus, feline panleukopenia virus, feline calicivirus) and the hookworm, *Ancyclostoma pleuridentatum*, were selected for intensive management. For control of the viruses, a killed rabies vaccine and a killed multivalent vaccine were chosen based on prior demonstration of efficacy and safety in clinical trails using nondomestic felids. An injectable anthelmintic was chosen for hookworm control based on lack of adverse reactions in the previous six years of field use in the Florida panther. Panthers were vaccinated and dewormed on an annual or biannual basis depending on the capture interval (Roelke and Glass 1992).

Although it is difficult to demonstrate the population benefits of medical intervention, one example of a panther that developed rabies suggests that vaccination against this disease may have prevented this animal and others from succumbing to rabies. Prior to a 1987 National Park Service (NPS) decision to discontinue vaccinations, all captured panthers routinely received rabies vaccinations. Subsequent to this decision, one unvaccinated collared panther later died of rabies (Roelke et al. 1991). Although the NPS reversed its decision against vaccination, this animal probably would have survived natural exposure to the virus had it been vaccinated.

A retrospective analysis of Florida panther mortality revealed that highway collisions were responsible for 46.9 percent of documented deaths (Maehr et al. 1991). Trauma often is a major wildlife mortality factor and in some situations must be treated with the same importance as infectious diseases. In this situation, roadkills occurred throughout southwestern Florida, but were concentrated especially on two highways. Wildlife underpasses (a form of habitat management) have reduced the number of panthers killed on at least one of these major highways (Maehr 1997).

The Mountain Gorilla

The Mountain Gorilla Veterinary Project (MGVP) is a full-time program developed in central Africa in 1986 to treat mountain gorillas accidentally captured in snares set by subsistence hunters (Hastings et al. 1991; Foster 1993). Its goals are to provide full-time veterinary care, evaluate long-term health-care needs, provide veterinary training to gorilla researchers, and coordinate veterinary research applicable to mountain gorilla management (Foster 1992, 1993). A scientific advisory committee developed a protocol to deal with human-caused injury, life-threatening conditions, and circumstances that may threaten the health of the population (Macfie 1992). The MGVP has worked closely with local and national officials. And despite concerns about administering medical care that is superior to what is available to the local human population, the program has become a source of local pride and economic assistance (Foster 1993).

The MGVP has achieved its goal of coordinating gorilla veterinary research applicable to mountain gorilla management (Mudakikwa and Sleeman 1997; Sleeman et al. 1998; Graczyk et al. 1999; Nizeyi et al. 1999). It has also achieved its goals of providing full-time veterinary care for the mountain gorillas and providing veterinary training to gorilla researchers in the area. As a result of the MGVP, numerous snares have been removed from gorillas (Macfie 1992), and in 1988 more than 65 gorillas were vaccinated against measles in the face of a potential morbillivirus-like respiratory outbreak (Hastings et al. 1991). Population impacts resulting from veterinary intervention have been more difficult to quantify due to the MGVP's conservative guidelines preventing the collection of blood samples for the sole purpose of research or baseline data. And due to the endangered status of the species, treatment and nontreatment groups have not been monitored and compared. Nonetheless, the MGVP's protocol demonstrates the potential for such programs to operate even in remote, economically challenged regions of the world.

Colorado Bighorn Sheep

A pneumonia complex, including verminous pneumonia caused by the lungworm *Protostrongylus stilesi*, was documented as the proximate cause of lamb mortality and population decline of bighorn sheep in the Sand Creek drainage of the Sangre de Cristo Mountains in Colorado (Woodard et al. 1974). Infectious lungworm larvae acquired by adult ewes can remain dormant until the ewe becomes pregnant, at which time the larvae pass through the placenta and infect the fetuses. Lambs are born with parasites that mature into adult lungworms within 30 to 45 days. These parasites can overwhelm the lungs and increase susceptibility to bacterial pneumonia. Schmidt et al. (1979) thought that changes in land use and human encroachment were restricting the bighorn populations to landscape islands. The high densities of these isolated populations, they speculated, led to constant and prolific parasite shedding and reinfection near bedding grounds and subsequent poor lamb recruitment. The circumstances surrounding management of this population did not permit natural control of lungworm infection via dispersion of sheep among islands. As a result, extensive field tests were used to develop a treatment program to increase lamb survival. Because increased lamb survival was the goal of the lungworm control program, evaluation of the benefit of intervention was based on comparing lamb survival in treated versus nontreated ewes. Ewes treated with Cambendazole delivered in ensiled apple pomace in early spring had higher lamb survival (Schmidt et al. 1979).

Conclusions

Potential disease risks should be determined during the planning stages of a wildlife restoration program. Whether a restoration program involves translocating wildlife or bolstering a unique wildlife population, the potential for the occurrence of a disease agent should be determined and the pathological consequences assessed. Risk reduction protocols or health programs can be developed for the population being restored based on the specific disease agents or parasites of concern. Clear goals and methods for evaluating results should be developed concomitantly with risk assessments and mitigation efforts. The application of such clear planning and diligent implementation has resulted in several modern wildlife success stories—stories that might well have ended less satisfactorily without considering the health aspects of restoring populations.

Acknowledgments

This work was supported through sponsorship from the fish and wildlife agencies of Alabama, Arkansas, Florida, Georgia, Kansas, Kentucky, Louisiana, Maryland, Mississippi, Missouri, North Carolina, Puerto Rico, South Carolina, Tennessee, Virginia, and West Virginia. Funds were provided by the Federal Aid to Wildlife Restoration Act (50. Stat. 917) and through Grant Agreement 14-45-0009-94-906, National Biological Service, U.S. Department of the Interior. Support also was received through Cooperative Agreement 94-9-6-13-0032-CA, Veterinary Services, Animal and Plant Health Inspection Service, U.S. Department of Agriculture.

Literature Cited

Anderson, R. C., and A. K. Prestwood. 1981. Lungworms. In W. R. Davidson, F. A. Hayes, V. F. Nettles, and F. E. Kellog, eds., *Diseases and Parasites of White-Tailed Deer*. Miscellaneous Publication 7. Tallahassee, Fla.: Tall Timbers Research Station.

Atkinson, C. T., R. J. Dusek, K. L. Woods, and W. M. Iko. 2000. Pathogenicity of avian malaria in experimentally-infected Hawaii Amakihi. *Journal of Wildlife Diseases* 36:197–204.

Bassi, R. 1893. *Distomum magnum* (bassi) in Italia ed in America. *Il Moderno Zooiatro* 4:269–270.

Bellrose, F. C. 1951. Effects of ingested lead shot upon waterfowl populations. *Proceedings of the Transactions of the North American Wildlife Conference* 16:125–133.

Bittle, J. L. 1993. Use of vaccines in exotic animals. *Journal of Zoo and Wildlife Medicine* 24:352–356.

Brown, R. H., and G. R. Scott. 1960. Vaccination of game with lapinised rinderpest virus. *Veterinary Record* 72:1232–1233.

Bush, M., R. J. Montali, D. Brownstein, A. E. James, and M. J. G. Appel. 1976. Vaccine-induced canine distemper in a lesser panda. *Journal of the American Veterinary Medical Association* 169:959–960.

Butt, M. T., D. Bowman, M. C. Barr, and M. E. Roelke. 1991. Iatrogenic transmission of *Cytauxzoon felis* from a Florida Panther (*Felis concolor coryi*) to a domestic cat. *Journal of Wildlife Diseases* 27:342–347.

Carpenter, J. W., M. J. G. Appel, R. C. Erickson, and M. N. Novilla. 1976. Fatal vaccine-induced canine distemper virus infection in black-footed ferrets. *Journal of the American Veterinary Medical Association* 169:961–964.

Clausen, B. 1981. Survey for trypanosomes in black rhinoceros (*Diceros bicornis*). *Journal of Wildlife Diseases* 17:581–586.

Corn, J. L., and V. F. Nettles. 2001. Health protocol for translocation of free-ranging elk. *Journal of Wildlife Diseases* 37(3):413–426.

Cunningham, A. A. 1996. Disease risks of wildlife translocations. *Conservation Biology* 10:349–353.

Davidson, W. R., and V. F. Nettles. 1992. Relocation of wildlife: Identifying and evaluating disease risks. *Transactions of the North American Wildlife and Natural Resources Conference* 57:466–473.

Davidson, W. R., and G. L. Doster. 1997. Health characteristics and white-tailed deer population density in the Southeastern United States. In W. J. McShea, H. B. Underwood, and J. H. Rappole, eds., *The Science of Overabundance*. Washington, D.C.: Smithsonian Institution Press.

Ebedes, H. 1992. Long-acting neuroleptics in wildlife. In H. Ebedes, ed., *The Use of Tranquillizers in Wildlife*. Department of Agricultural Development. Pretoria: Sinoville Printers.

Environmental Assessment Panel. 1990. *Northern Diseased Bison*. Quebec: Federal Environmental Assessment Review Office.

Eve, J. H., and F. E. Kellogg. 1977. Management implications of abomasal parasites in southeastern white-tailed deer. *Journal of Wildlife Management* 41:169–177.

Fairbrother, A., L. N. Locke, and G. L. Hoff. 1996. *Noninfectious Diseases of Wildlife*, 2nd ed. Ames: Iowa State University Press.

Forrester, D. J. 1992. *Parasites and Diseases of Wild Mammals in Florida*. Gainesville: University Press of Florida.

Foster, J. W. 1992. Mountain gorilla conservation: A study in human values. *Journal of the American Veterinary Medical Association* 200:629–633.

———. 1993. Health plan for the mountain gorillas of Rwanda. In M. Fowler, ed., *Zoo and Wild Animal Medicine: Current Therapy*. Vol. 3. Philadelphia: Saunders.

Fuller, W. A. 1966. The biology and management of bison of Wood Buffalo National Park. *Wildlife Management Bulletin Series* 1(16).

Gajadhar, A., and W. D. Yates. 1994. Diagnosis of *Elaphostrongylus cervi* infection in New Zealand red deer (*Cervus elaphus*) quarantined in Canada, and experimental determination of a new extended prepatent period. *Canadian Veterinary Journal* 35:433–437.

Graczyk, T. K., L. J. Lowenstine, and M. R. Cranfield. 1999. *Capillaria hepatica* (Nematoda) infections in human-habituated mountain gorillas (*Gorilla gorilla beringei*) of the Parc National de Volcans, Rwanda. *Journal of Parasitology* 85:1168–1170.

Hastings, B. E., D. Kenny, L. J. Lowenstine, and J. W. Foster. 1991. Mountain gorillas and measles: Ontogeny of a wildlife vaccination program. In *Proceedings of the American Association of Zoo Veterinarians*.

Haviernick, M., S. D. Cote, and M. Festa-Bianchet. 1998. Immobilization of mountain goats wih xylazine and reversal with idazoxan. *Journal of Wildlife Diseases* 34:342–247.

Heuschele, W. P. 1991. Impacts of major infectious diseases on free-living wildlife. In *Proceedings of the American Association of Zoo Veterinarians*.

Karesh, W. B. 1993. Cost evaluation of infectious disease monitoring and screening programs for wildlife translocation and reintroduction. *Journal of Zoo and Wildlife Medicine* 24:291–295.

Kistner, T. P. 1982. Diseases and parasites. In J. W. Thomas and D. E. Toweill, eds., *Elk of North America* Harrisburg, Pa.: Stackpole Books.

Kock, M. D., D. A. Jessup, R. K. Clark, C. E. Franti, and R. A. Weaver. 1987. Capture methods in five subspecies of free-ranging bighorn sheep: An evaluation of drop-net, drive-net, chemical immobilization and the net-gun. *Journal of Wildlife Diseases* 23:634–640.

Krebs, J. W., J. S. Smith, C. E. Rupprecht, and J. E. Childs. 1999. Rabies surveillance in the United States during 1998. *Journal of the American Veterinary Medical Association* 215:1786–1798.

Kreeger, T. J. 1999. *Handbook of Wildlife Chemical Immobilization*. Fort Collins: Wildlife Pharmaceuticals.

Macfie, E. J. 1992. An update on the current medical management program for Rwanda's mountain gorillas: Veterinarians as population managers, and the effects of war. In *Proceedings of the Joint Meeting of the American Association of Zoo Veterinarians and the American Association of Wildlife Veterinarians.*

Maehr, D. S. 1997. *The Florida Panther: Life and Death of a Vanishing Carnivore.* Washington, D.C.: Island Press.

Maehr, D. S., E. D. Land, and M. E. Roelke. 1991. Mortality patterns of panthers in southwest Florida. *Proceedings of the Annual Conference of the Southeastern Association of Fish and Wildlife Agencies* 45:201–207.

Mudakikwa, A. B., and J. M. Sleeman. 1997. Analysis of urine from free-ranging mountain gorillas (*Gorilla gorilla beringei*) for normal physiological values. In *Proceedings of the American Association of Zoo Veterinarians.*

Nettles, V. F., and W. M. Martin. 1979. General physical parameters and health characteristics of translocated raccoons. *Proceedings of the Annual Conference of the Southeastern Association of Fish and Wildlife Agencies* 32:71–74.

Nizeyi, J. B., R. Mwebe, A. Nanteza, M. R. Cranfield, G. R. N. N. Kalema, and T. K. Graczyk. 1999. *Cryptosporidium* sp. and *Giardia* sp. infections in mountain gorillas (*Gorilla gorilla beringei*) of the Bwindi impenetrable national park, Uganda. *Journal of Parasitology* 85:1084–1088.

Pandey, G. S. 1991. Heartwater (*Cowdria ruminantium*) with special reference to its occurrence in Zambian wildlife. *Centre for Tropical Veterinary Medicine Newsletter* 52:6.

Robinson, R. M., T. L. Hailey, C. W. Livingston, and J. W. Thomas. 1967. Bluetongue in the desert bighorn sheep. *Journal of Wildlife Management* 31:165–168.

Roelke, M. E., D. J. Forrester, E. R. Jacobson, and G. V. Kollias. 1991. Rationale for surveillance and prevention of infectious and parasitic disease transmission among free-ranging and captive Florida panthers (*Felis concolor coryi*). In *Proceedings of the Annual Conference of the American Association of Zoo Veterinarians.*

Roelke, M. E., and C. M. Glass. 1992. Strategies for the management of the endangered Florida panther (*Felis concolor coryi*) in an ever shrinking habitat. In *Proceedings of the Joint Meeting of the American Association of Zoo Veterinarians and the American Association of Wildlife Veterinarians.*

Schaffer, G. D., W. R. Davidson, V. F. Nettles. 1981. Helminth parasites of translocated raccoons (*Procyon lotor*) in the southeastern United States. *Journal of Wildlife Diseases* 17:217–227.

Schmidt, R. L., C. P. Hibler, T. R. Spraker, and W. H. Rutherford. 1979. An evaluation of drug treatment for lungworm in bighorn sheep. *Journal of Wildlife Management* 43:461–467.

Scott, M. E. 1988. The impact of infection and disease on animal populations: Implications for conservation biology. *Conservation Biology* 2:40–56.

Severinghaus, C. W., and R. W. Darrow. 1976. Failure of elk to survive in the Adirondacks. *New York Fish and Game Journal* 23:98–99.

Sleeman, J. M., K. Cameron, A. B. Mudakikwa, S. Anderson, J. E. Cooper, B. Hastinds, J. W. Foster, E. J. Macfie, and H. M. Richardson. 1998. Field anesthesia of free ranging mountain gorillas (*Gorilla gorilla beringei*) from the Virunga volcano region, Central Africa. In *Proceedings of the American Association of Zoo Veterinarians.*

Thorne, E. T., M. W. Miller, D. A. Jessup, and D. L. Hunter. 1992. Disease as a consideration in translocating and reintroducing wild animals: Western state wildlife management agency perspectives. In *Proceedings of the Joint Meeting of the American Association of Zoo Veterinarians and the American Association of Wildlife Veterinarians.*

U.S. Fish and Wildlife Service (USFWS). 1987. Florida panther (*Felis concolor coryi*) recovery plan. Prepared by the Florida Panther Interagency Committee for the U.S. Fish and Wildlife Service, Atlanta.

Wobeser, G. A. 1994. *Investigation and Management of Disease in Wild Animals.* New York: Plenum Press.

Woodard, T. N., R. J. Gutierrez, and W. H. Rutherford. 1974. Bighorn lamb production, survival, and mortality in south-central Colorado. *Journal of Wildlife Management* 38:771–774.

Woodford, M. H., and P. B. Rossiter. 1993. Disease risks associated with wildlife translocation projects. *Revue scientifique et technique, Office International des Epizooties* 12:265–270.

Case 2. Health Aspects of Gray Wolf Restoration

MARK R. JOHNSON

While livestock losses and real or imagined threats to humans may be problematic in large carnivore restoration, handling protocols and disease testing are no less important. The recent wolf (*Canis lupus*) reintroductions in Idaho and Yellowstone National Park (see chapter 6 in this volume) presented challenging public relations and animal husbandry dilemmas. Before government agencies could confidently proceed with the translocation of 66 wolves from Canada, state-of-the-art handling procedures were needed to reduce the risk of disease transmission and enhance individual wolf survival. The protocol that emerged from these translocations may be of use to others considering the restoration of large carnivores.

During 1995 and 1996, wolves were captured by helicopter darting in January (Bangs and Fritts 1996; Fritts et al. 1997), transported to a central processing facility, examined, treated, and held individually before transport to the United States. Each wolf was then processed a second time, placed into an aluminum crate, and transported to either central Idaho or Yellowstone National Park (see chapter 6). The protocol (Johnson 1994b) evaluated individual health, protected wolves from disease during and after translocation, minimized the potential for wolves to introduce new diseases or parasites into the release area, and provided opportunities to gather data and samples for collaborative research (Forbes and Boyd 1997).

Principal Diseases of Concern

Rabies (Johnson 1995), tuberculosis, and brucellosis (Johnson 1992) were the primary untreatable diseases of concern. Therefore, donor populations

Processing Protocol for Chemically Immobilized Wolves

First chemical immobilization (initial capture)
Drug dose: Telazol 500 mg per wolf

INTRAMUSCULAR
1. Mark animals with eartags and microchips (Trovan, Burnsville, MN).
2. Conduct physical exam for signs of infectious disease, clinical illness, and injuries. Monitor temperature, pulse, respiration every 15 minutes. Monitor each wolf with pulse oximeter (Nellcor, Pleasanton, CA).
3. Collect the following samples:
   a. Four 10-ml SST blood tubes for serology and serum chemistry.
   b. Three 7-ml EDTA blood tubes for genetics.
   c. Two 7-ml EDTA blood tubes for hematology and blood parasites.
   d. One 1-oz fecal sample for internal parasites.
   e. Two 2-oz fecal samples for genetics, *Echinococcus*, and cortisol studies.
   f. Collect ectoparasites if present and skin scrapings if lesions are present.
4. Prophylactic medications:
   a. Vaccinate for canine parvovirus, canine disemper, infectious canine hepatitis, leptospirosis, and parainfluenza (DHLPP, Ft. Dodge, IA) and rabies (Imrab, Rhone Merieux, Athens, GA).
   b. Administer Ivermectin (Merck, Rahway, NJ) at 0.2 mg/kg, subcutaneous.
   c. Administer Praziquantel (Droncit, Haver-Lokart, Shawnee, KS) at 0.2 ml/2.3 kg, intramuscular.
   d. Spray with Para S pyrethrin spray (Vetkem, Dallas, TX).
   e. Give procaine penicillin G and benzathine (Durapen, Vetkem, Dallas, TX) at 1 cc/6.8 kg, intramuscular.
   f. Administer 500 cc lactated Ringers solution subcutaneous in several places.

Preparation for translocation
Drug dose: Telazol 500 mg per wolf

INTRAMUSCULAR
1. Conduct physical exam and monitor temperature, pulse, and respiration.
2. Collect the following samples:
   a. Two 10-ml SST tubes for serum chemistry, postvaccination serology, and banking.
   b. Two 7-ml EDTA tubes for hematology.
   c. Two 1-oz fecal samples for internal parasites and banking.
3. Repeat prophylactic medications except rabies.
4. Brush animal to remove plant debris from bedding material.
5. Apply radio-collar.
6. Complete International Health Certificate (if no signs of infection are present) and rabies vaccination form.

were selected from areas free of these diseases (Johnson 1994a). In addition, each wolf was vaccinated for rabies (see the table) and serologically tested for brucellosis and rabies before transport. *Echinococcus* spp. was of concern because of its zoonotic potential (hydatid disease). Each wolf was treated twice: once at first capture and again before translocation. Because canine parvovirus and canine distemper posed a potential threat to translocated wolves (Gese et al. 1997; Johnson et al. 1994), each wolf was vaccinated. Wolves were also treated to prevent the transmission of internal and external parasites such as hookworms (*Ancyclostoma caninum*) and roundworm (*Toxocara canis*) (Kreeger et al. 1990), sucking lice (*Anoplura* spp.), *Sarcoptes scabiei*, and chewing lice (*Trichodectes canis*).

Processing Protocol

The details of processing each anesthetized wolf are described in the table. Significant attention was given to minimizing stress and maximizing animal health. During capture, transport, and processing, a cloth headcover was placed on each anesthetized wolf and, whenever possible, sounds (including talking) were minimized. Routine monitoring included physical examination and regular checking of temperature, pulse, respiration, and pulse oximetry. All vital signs were documented in chronological order to anticipate physiological trends (Johnson 1994c).

Temporary Confinement

Each wolf was kept in a chain-link holding kennel (2 m high by 2 m wide by 4 m long) for one to ten days until enough wolves were gathered for transport into the United States. Ambient temperatures averaged 35°C, so kennel floors were lined with 60 cm of straw, straw-bale dens were provided, and kennels were covered with cloth tarps to reduce noise and chewing on chain-link fencing. Road-killed elk and moose, as well as ice, were provided. (Most wolves ignored these offerings).

Transport

Wolves were transported awake in individual crates designed to meet International Air Transporter Association (www.iata.org) specifications. A veterinarian escorted each shipment of wolves and inspected wolves at least every four hours. Most wolves spent less than 24 hours in crates (Bangs and Fritts 1996). Wolves held longer than 12 hours were given ice blocks for water.

Release

Acclimation pens doubled as quarantine facilities for wolves released into Yellowstone Park. Wolves were released immediately into central Idaho and were not quarantined prior to release.

Release Criteria

Wolves were free of clinical and infectious disease and were serologically negative for brucellosis before translocation. Because no wolves tested positive
for rabies or brucellosis, surplus animals were returned to their packs.

Postrelease Evaluation

It is difficult to evaluate the success of these protocols for preventing transmission of disease into the receiving environment without testing a broad
spectrum of sympatric wildlife species following the release of wolves. Extensive literature review and peer review supported a protocol that alleviated
opposition to the wolf reintroduction program based on fear of disease.
Moreover, capture-related mortalities of wolves were low and no infectious
outbreaks occurred.

Literature Cited

Bangs, E. E., and S. H. Fritts. 1996. Reintroducing the gray wolf to central Idaho and
Yellowstone National Park. *Wildlife Society Bulletin* 24:402–413.

Forbes, S. H. and D. K. Boyd. 1997. Genetic structure and migration in native and
reintroduced Rocky Mountain wolf populations. *Conservation Biology*
11:1226–1234.

Fritts, S. H., E. E. Bangs, J. A. Fontaine, M. R. Johnson, M. K. Phillips, E. D. Koch,
and J. R. Gunson. 1997. Planning and implementing a reintroduction of wolves to
Yellowstone National Park and central Idaho. *Restoration Ecology* 5(1):7–27.

Gese, E. M., R. D. Schultz, M. R. Johnson, E. S. Williams. R. L. Crabtree, and R.
L. Ruff. 1997. Serological survey for diseases in free-ranging coyotes (*Canis
latrans*) in Yellowstone National Park, Wyoming. *Journal of Wildlife Diseases*
33:47–56.

Johnson, M. R. 1992. The disease ecology of brucellosis and tuberculosis in potential
relationship to Yellowstone wolf populations. In *Wolves for Yellowstone? A Report to
the United States Congress.* Vol. 4: Research and Analysis. Yellowstone National
Park: National Park Service.

———. 1994a. Addressing disease concerns in source wolf populations. Wolf Reintroduction Working Paper 1. Unpublished report on file. Yellowstone National
Park: National Park Service,

———. 1994b. Protocol for addressing disease aspects for gray wolf reintroduction.
Wolf Reintroduction Working Paper 2. Unpublished report on file. Yellowstone
National Park: National Park Service.

———. 1994c. Gray wolf translocation manual. Wolf Reintroduction Working Paper
3. Unpublished report on file. Yellowstone National Park: National Park Service.

———. 1995. Rabies in wolves and its potential role in a Yellowstone wolf population. In L. N. Carbyn, S. H. Fritts, and D. R. Seip, eds., *Ecology and Conservation of Wolves in a Changing World.* Edmonton: Proceedings of the Second North American Symposium on Wolves.

Johnson, M. R., D. K. Boyd, and D. H. Pletscher. 1994. Serologic investigations of canine parvovirus and canine distemper in relation to wolf (*Canis lupus*) pup mortalities. *Journal of Wildlife Diseases* 30:270–273.

Kreeger, T. J., U. S. Seal, M. Callahan, and M. Beckel. 1990. Treatment and prevention with ivermectin of dirofilariasis and ancyclostomiasis in captive gray wolves (*Canis lupus*). *Journal of Zoo and Wildlife Medicine* 21:310–317.

Chapter 8

Restoring the Mexican Gray Wolf to the Mountains of the Southwest

WENDY M. BROWN AND DAVID R. PARSONS

Mexican wolf (*Canis lupus baileyi*) recovery is one of three gray wolf restoration efforts in the United States conducted by the U.S. Fish and Wildlife Service (USFWS). The Endangered Species Act (ESA) of 1973 (16 USC 1531-1544) saved the Mexican wolf from almost certain extinction. All known Mexican wolves descend from seven founders that were captured from the wild and bred in captivity (Parsons 1996). Two challenges faced by Mexican wolf recovery include the use of naive, captive-reared wolves for release stock and sociopolitical opposition to restoring predators. Following years of planning and public involvement, reintroduction of Mexican wolves began in March 1998 (Parsons 1998). In this chapter we describe the history and ecology of the Mexican wolf and its narrow escape from extinction. Details and early results of ongoing efforts to restore this extirpated predator to the American Southwest are presented. We hope this information will be of interest and value to others pursuing carnivore restoration efforts.

History and Ecology

Although Mexican wolves once were abundant in the Southwest (Young and Goldman 1944), three centuries of conflict with domestic livestock preceded today's restoration challenge (Brown 1983). Wolves persisted longer in the Southwest than elsewhere in the western United States due to source populations in Mexico (Young and Goldman 1944; Leopold 1959; McBride 1980; Brown 1983).

Wolf control by the Predatory Animal and Rodent Control branch of the U.S. Biological Survey (now the USFWS) in the early 1900s was strongly supported by noted Southwestern naturalists and "wolfers" such as Vernon Bailey, J. Stokely Ligon, Stanley P. Young, and Aldo Leopold. The federal government reported more than 900 wolves killed in New Mexico and Arizona from 1915 to 1925, and it is believed that an even greater number were killed from 1890 to 1915 (Young and Goldman 1944; Brown 1983). Leopold (1930, 1936, 1937) later advocated conservation and restoration of predators, but this conversion came too late to stop the wolf's decline. The Mexican wolf was extirpated from the United States in the mid-1900s (Brown 1983; Bednarz 1988). McBride (1980) estimated that no more than 50 breeding pairs survived in Mexico in 1978. Mexico's use of Compound 1080 in the 1950s resulted in a rapid decline in wolf populations (Brown 1983) and recent surveys in Mexico have failed to confirm their presence (J. A. Carrera and J. F. Gonzales, Protección de la Fauna Mexicana A.C., Saltillo, Coahuila, unpublished data).

In April 1990, several environmental organizations filed suit against the Departments of the Interior and Defense alleging failure to implement provisions of the ESA (*Wolf Action Group, et al. vs. United States,* Civil Action CIV-90-0390-HB, U.S. District Court, New Mexico). The USFWS hired a Mexican wolf recovery coordinator in October 1990 and began to develop reintroduction plans. The litigants signed a settlement agreement in 1993, after the USFWS agreed to implement the Mexican Wolf Recovery Plan and the National Environmental Policy Act (NEPA) process for reintroduction of Mexican wolves to the wild. This step ultimately led to the development of an environmental impact statement (EIS) for reintroduction of the Mexican wolf to the Southwest (USFWS 1996).

Although there is confusion and disagreement over North American gray wolf taxonomy (Brewster and Fritts 1995), morphometric and molecular genetic data suggest that the Mexican wolf is a distinct subspecies (see Hall and Kelson 1959; Bogan and Mehlhop 1983; Wayne et al. 1992; Nowak 1995; Hedrick et al. 1997). García-Moreno et al. (1996:384) have concluded that Mexican wolves were the "most [genetically] distinct grouping of North American wolves supporting their distinction as an endangered subspecies."

The Mexican wolf is the southernmost subspecies of gray wolf in North America (Nowak 1995). But because wolves can disperse long distances (Fritts 1983; Gese and Mech 1991; Boyd and Pletcher 1999), the boundaries between subspecies are zones of genetic intergradation (Young and

Goldman 1944; Mech 1970; Brewster and Fritts 1995). Based on a review of taxonomy and dispersal, the Mexican Wolf Recovery Team has described its historic range as portions of central and northern Mexico, western Texas, southern New Mexico, and southeastern and central Arizona (Parsons 1996) (Figure 8.1).

Natural history information on the Mexican wolf is sketchy and suspect (Brown 1983; Gipson 1995; Gipson et al. 1998). Weights of Mexican wolves range from 25 to 45 kg (Young and Goldman 1944; Leopold 1959; McBride 1980). Adults measure from 140 to 170 cm in total length, and 72 to 80 cm in shoulder height. The pelt is a mix of gray, black, brown, and rust hues, with light-colored underparts. McBride (1980) has reported a mean litter size of 4.5 from eight dens in Mexico. The size of 155 litters of Mexican wolves born in captivity ranged from 1 to 9 with a mean size of 4.2 (D. P. Siminski, pers.

Figure 8.1. Probable historical range of the Mexican gray wolf.

comm.). Captive females come into estrus from February through March; gestation takes 63 days.

Historically, Mexican wolves preyed primarily on white-tailed deer (*Odocoileus virginianus*), mule deer (*O. hemionus*), and collared peccary (*Tayassu tajacu*) (Leopold 1959; Brown 1983). Elk (*Cervus elaphus*), pronghorn (*Antilocapra americana*), bighorn sheep (*Ovis canadensis*), beaver (*Castor canadensis*), rabbits (*Sylvilagus* spp.), hares (*Lepus* spp.), and small mammals may have provided alternative prey (Bailey 1931; Leopold 1959; Brown 1983). In the late 1800s, livestock numbers increased and native ungulate populations declined because of habitat degradation and unregulated hunting, which led to increased livestock depredation by Mexican wolves (Brown 1983; USFWS 1987; Dunlap 1988).

Most Mexican wolf collections have come from pine (*Pinus* spp.), oak (*Quercus* spp.), and pinyon (*Pinus edulis*)–juniper (*Juniperus* spp.) woodlands and from adjacent grasslands more than 1372 m in elevation (Brown 1983). Wolves apparently avoided desert scrub and semidesert grasslands, which provided limited cover, water, and prey (Young and Goldman 1944; Brown 1983).

Removing top predators may induce overabundance of large herbivores and mesopredators (Terborgh et al. 1999; chapter 3 in this volume)—ecological changes that may cause a cascade of effects that disrupt processes such as herbivory, predation, competition, and behavioral exclusion resulting in ecosystem simplification, loss of ecosystem stability, and loss of biological diversity (Terborgh et al. 1999). This top-down control may explain Leopold's (1937) observation of the absence of coyotes and lack of deer irruptions in the Sierra Madre in Chihuahua, Mexico, where viable wolf populations remained.

Background

The Mexican Wolf Recovery Plan was approved by the director of the USFWS and Mexico's director general of the Dirección General de la Fauna Silvestre in 1982 (USFWS 1982). The plan's objective is clear: "To conserve and ensure the survival of *C. l. baileyi* by maintaining a captive breeding program and re-establishing a viable, self-sustaining population of at least 100 Mexican wolves in the middle to high elevations of a 5,000 mi² [13,000 km²] area within the Mexican wolf's historic range" (USFWS 1982:23). Given the natural and human-caused isolation of suitable wolf habitat, geographically separate reintroductions will likely be required to ensure long-term conservation of the subspecies. The resulting metapopulation may require habitat linkages or active demographic and genetic management to ensure adequate gene flow among subpopulations and genetic variation within them (Lande and Barrowclough 1987; Dobson et al. 1999).

Captive Breeding

Following the listing of the Mexican wolf as endangered in 1976 (41 *Federal Register* 17736), five wild wolves were captured in Durango and Chihuahua, Mexico (McBride 1980). Three of them became the founders of a captive breeding program for future reintroduction (later termed the McBride lineage) (Parsons and Nicholopoulos 1995; Parsons 1996, 1998).

Based on DNA studies, two additional lineages of captive Mexican wolves—the Ghost Ranch lineage in the United States and the Aragón lineage in Mexico—were certified for inclusion in the official breeding program in July 1995. As of July 2000, 203 Mexican wolves were in 45 zoos and wildlife sanctuaries in the United States and Mexico (Siminski 2000). This population is the result of captive breeding within and among the three officially accepted lineages.

Captive population management minimizes inbreeding and maximizes retention of the genetic variation of the seven founders (AZA 1994). With only seven founders inbreeding cannot be avoided, and population management has resulted in a mean inbreeding coefficient of 0.2134 (Siminski 2000). No evidence of inbreeding depression has been detected to date (Hedrick et al. 1997; Kalinowski et al. 1999).

The Reintroduction Plan

The interior secretary approved reintroduction of Mexican wolves to establish a wild population of no fewer than 100 wolves in March 1997. The plan authorized the release of genetically expendable, captive-raised Mexican wolves into the Apache National Forest in eastern Arizona (USFWS 1996; Parsons 1998). Wolves are expected to recolonize the Blue Range wolf recovery area, which includes all of the Apache National Forest in Arizona and the Gila National Forest in New Mexico (Figure 8.2). The White Sands wolf recovery area may be used as an additional reintroduction area if the 100-wolf objective cannot be met in the Blue Range area. The Blue Range area is multiple-use national forest land; the White Sands area is closed to most human uses except limited hunting and military training.

The plan calls for three to five annual releases of about three family groups of wolves. Population growth will result from natural reproduction to achieve a self-sustaining wild population of at least 100 wolves in eight to ten years (USFWS 1996). The USFWS and its partners use "adaptive manage-

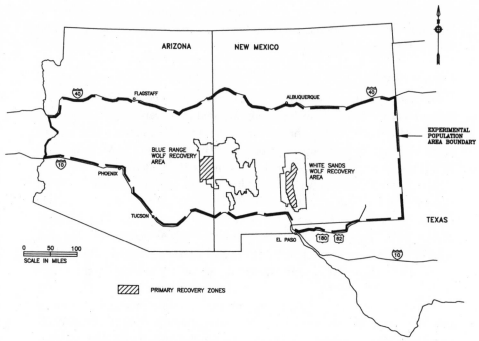

Figure 8.2. Recovery areas and experimental population boundary of reintroduced Mexican gray wolves.

ment" to adjust the number of animals released, release methods, and other aspects of implementation as needed.

Released wolves and their progeny were designated in a special rule (63 *Federal Register* 1752) as a "nonessential, experimental population" under provisions of Section 10(j) of the ESA. It allows the interior secretary to authorize the release of an "experimental population" of a threatened or endangered species where it has been extirpated and if reintroduction will further the conservation of the species. Wolves that leave the restoration zone will be captured and returned (Figure 8.2). Furthermore, the rule allows limited taking of Mexican wolves to reduce conflict with human activities. The level of take allowed by the rule is not expected to preclude recovery of the Mexican wolf—a required finding under Section 10(j). The USFWS thinks the management flexibility in the "nonessential, experimental" designation was critical to achieving the tolerance of the project's opponents (Phillips et al. 1995; Bangs et al. 1998; Parsons 1998).

The USFWS established cooperative agreements from 1998 to 2000 with the Arizona Game and Fish Department, USDA Wildlife Services Division,

New Mexico Department of Game and Fish, and the White Mountain Apache Tribe. Biologists from these agencies constitute the field management team that implements the Mexican Wolf Interagency Management Plan (USFWS 1998). An Interagency Management Advisory Group—composed of representatives from federal and state agencies as well as county and tribal governments—meets periodically to review reintroduction progress and address emerging issues. Defenders of Wildlife, a nongovernmental conservation organization, reimburses livestock owners for verified losses of livestock killed by Mexican wolves. They also provide other subsidies, such as veterinary expenses for injured animals and fencing to reduce depredation.

Impact Assessment and Outreach

Public "scoping" meetings were held from 1991 through 1994 to identify social and biological concerns and to formulate alternative plans for restoring Mexican wolves (Groebner et al. 1995; USFWS 1996). Fourteen open-house meetings and three hearings were held in 1995 to receive comments on a draft EIS (USFWS 1996, 1997a). Some 18,000 comments resulted in several changes that were incorporated in the final EIS (USFWS 1996). Four additional public meetings and hearings were conducted in 1996 to obtain comments on a special rule that would designate reintroduced Mexican wolves as a nonessential, experimental population. A site-specific environmental assessment and public review determined that constructing release pens and holding wolves in them prior to their release would not cause significant impacts (USFWS 1997b). In 2000, prior to exercising its authority under the special rule to translocate wolves from the primary wolf recovery zone in Arizona (the release area) to the Gila National Forest in the secondary wolf recovery zone (the population expansion area), the USFWS conducted another environmental assessment (USFWS 2000). The two public hearings on wolf translocation plans were extremely contentious—demonstrating the persistence of strong opinions both for and against the restoration of wolves in the Southwest. After reviewing more than 9000 comments, the USFWS determined that the process of translocating wolves to the Gila National Forest would not result in significant impacts (USFWS 2000).

Wolf Releases

Candidates for release were moved to acclimation facilities where contact between wolves and humans was minimized and carcasses of road-killed native prey (mostly deer and elk) supplemented their routine diet of processed canine food. Genetically and socially compatible breeding pairs were established and evaluated for physical, reproductive, and behavioral suitability for release. Generally wolves selected for release were well represented genetically by other members of the captive population and were not more than four years old. Some pairs produced pups in captivity before release, and their pups and occasionally yearlings were included in the release group.

Wolves selected for release were radio-collared and moved to 0.13-ha chain-link pens at release locations (Parsons 1998). There they were held for about two months through the breeding season and released either before parturition or before pups were 12 weeks old to encourage site fidelity. We developed a portable, heavy-gauge nylon mesh pen reinforced with light-weight electric wire powered by solar cells for use at more remote release sites (USFWS, unpublished data). Wolves were transported to remote areas in specially designed panniers carried by mules and held in these "soft mesh" pens for up to 65 days before they were released or escaped.

Wolves were monitored daily from the ground and twice weekly from the air. Monitoring was most intensive during the initial weeks after release to determine when wolves began hunting. If we found wolves near roads or other areas of human activity, they were chased on foot, horseback, or all-terrain vehicles. We also used rubber bullets, cracker shells, and slingshots to encourage a flight response to humans.

Wolves were supplementally fed for two to four months depending on time of year, availability of vulnerable prey, and whether pups were present. Supplemental feeding was gradually discontinued when wolves began killing prey. Suspected wolf depredation on livestock was investigated by USDA Wildlife Services specialists using standard techniques (Roy and Dorrance 1976; Paul and Gipson 1994).

Results

Fifty released Mexican wolves experienced many different fates. Most readily adapted to life in the wild; some did not. Results are summarized below.

Wolf Releases

From March 1998 through August 2000, at least 60 free-ranging wolves in eight family groups inhabited the Blue Range wolf recovery area. These included 19 wolves released as adults (over 2 years old), 15 as yearlings (between one and two years old), 16 as pups (under one year old), and at least 10 that were born in the wild. (Table 8.1) Released adults were paired for over a year in captivity and seven had previously produced offspring. Six of eight

Table 8.1. Status and Fates of Reintroduced Mexican Gray Wolves in Arizona and New Mexico: March 1998–August 2000

	Releases	Wild births[a]	Recaptures[b]	Re-releases	Deaths[c]	Missing	Remaining
1998	13	1	3	—	5	2 (1)	4
1999	22 (12)	6	9 (6)	—	2	3 (3)	18 (9)
2000	15 (4)	3	18	10	1	4	23 (7)
Total	50 (16)	10	30 (6)	10	8	9 (4)	

Note: Numbers in parentheses indicate pups of the year.
[a] Indicates pups observed; total numbers unknown.
[b] Returned to captive management facility. Four individuals were released and recaptured more than once.
[c] Indicates known deaths in the wild.

pairs were released with yearlings from prior years. The smallest released group was a breeding pair that had not previously reproduced; the largest included two adults, two yearlings, and four pups. In 2000, nine wolves in two packs were recaptured and translocated into the secondary recovery zone in New Mexico.

If one member of a breeding pair was killed or removed, we attempted to provide the remaining animal with a new mate. We accomplished this with two males during the first year by recapturing the males and placing them with females in on-site acclimation pens. By August 2000, at least 23 free-ranging wolves in six family groups, including nine adults, seven yearlings, and seven pups, inhabited the Blue Range wolf recovery area. The status of four yearling wolves was unknown.

Reproduction

In 1998, wolf groups were released about one month prior to whelping and were supplementally fed about 50 percent of total needs for no more than two months. In 1999, three of four pairs were allowed to whelp in acclimation pens and all groups were supplementally fed about 100 percent of nutritional needs until pups were eight weeks of age. Denning or reproduction was confirmed for all but one released pairs that were together during a breeding season. No pups survived in 1998, but 11 of 18 known pups in four litters born in 1999 survived to 1 January 2000. Reproduction was confirmed for four of six pairs in 2000, but the number of pups was unknown (Table 8.1). Of 26 known pups, 10 were wild-born and 16 were born in acclimation pens. One pair has surviving offspring from both the 1999 and 2000 breeding seasons.

Recaptures and Mortality

Of 50 wolves released, 26 (52 percent) were recaptured due to livestock depredation (15 wolves in two large family groups, including 6 pups); dispersal outside the designated recovery area (7 wolves); nuisance behavior such as frequenting human settlements (4 wolves); population management such as replacement of a lost mate (3 wolves); and trapping injury (1 wolf). Of 7 wolves recaptured for leaving the recovery area, 6 were yearlings that were likely dispersing. Ten of the recaptured wolves were returned to the wild (4 were subsequently recaptured), 6 were permanently removed from the wild, 9 were awaiting rerelease, and 5 (all pups) died in captivity. Eight wolves have died in the wild: five were shot, two were struck by vehicles, and one was killed by a mountain lion. The fate of nine wolves was unknown (Table 8.1).

Predation

The earliest confirmed kill made by a wolf pack occurred 23 days after release. Of 41 native ungulates killed or scavenged by wolves, 35 (85 percent) were elk, 3 were mule deer, and 3 were white-tailed deer. Of 30 elk for which age could be determined, 16 (53 percent) were calves and 14 (47 percent) were adults. Of 21 elk for which sex could be determined, 14 (67 percent) were female and 7 (33 percent) were male.

Livestock Depredation

Nine livestock depredations including five adult cattle, one calf, one sheep, and non-fatal injuries to one calf and one miniature horse colt were confirmed. All but two depredations (the miniature horse and the sheep) were caused by two wolf packs living in lower-elevation areas where cattle grazing occurred year-round and where deer were the primary native prey. By August 2000, the Defenders of Wildlife Wolf Compensation Trust had paid $6,018 to ranchers for confirmed livestock losses or injuries caused by Mexican wolves (Defenders of Wildlife 2000).

Outreach

Biweekly updates were posted on a Mexican wolf website (http://mexican-wolf.fws.gov/), as were detailed project newsletters. In rural communities, updates were posted in libraries, post offices, and agency offices. Project newsletters, which included a citizen commentary column, were mailed to more than 2000 recipients. Press releases were issued on important events related to the reintroduction effort. Project staff gave presentations to key stakeholders such as livestock producers, hunters, guides, and environmental organizations. Field staff living in local communities responded to local concerns and interact with livestock producers, hunters, and others interested in the project. A citizen advisory group was established in one community. A toll-free Mexican wolf hotline was available for citizens to report wolf sightings or wolf-related incidents such as livestock depredation. The USFWS sponsored a workshop in 1998 to solicit stakeholders' opinions regarding priorities for funding Mexican wolf research (Thompson et al. 2000) and funded a food habits research project that received a high-priority rating.

Discussion

Mexican wolf recovery is progressing despite many challenges, ranging from the inexperience of captive-raised wolves to litigation by opposing interests. Innovative approaches and management flexibility are critical to success. Below, we discuss challenges and successes of Mexican wolf restoration.

Biological and Ecological Aspects

Limited genetic diversity will have unknown consequences for Mexican wolf recovery. Although the captive population has been managed to reduce inbreeding depression, a population based on seven founders may be vulnerable to the effects of low genetic variability (Hedrick et al. 1997). Continued population growth should lessen the bottleneck effect.

Inexperienced animals also present a significant challenge to restoration. In North America, black-footed ferrets (*Mustela nigripes*) and red wolves (*Canis rufus*) are the only other carnivores to have been completely eliminated in the wild, bred in captivity, and returned to the wild. The effort to reintroduce red wolves to the Great Smoky Mountains National Park was aborted, primarily because of inadequate prey (B. Kelly, pers. comm.). Red wolf recovery in North Carolina continues but much more slowly than gray wolf restoration in the northern Rockies. Phillips et al. (1995) report that five years after initial reintroduction of captive red wolves in North Carolina, 36 had been released, 20 had been wild-born, and 20 (mostly wild-born animals) were free-ranging. In contrast, Bangs et al. (1998) report that approximately two and a half years after reintroduction of wild Canadian wolves to Yellowstone National Park, 31 had been released, seven packs had produced ten litters of pups, and 116 wolves were free-ranging. There are differences in habitat and scope among the three wolf recovery efforts. However, the key correlation between the red and Mexican wolf projects is reliance on captive stock. Early results of Mexican wolf recovery approximate those of red wolves.

Supplemental postrelease feeding appeared to enhance survival of all age classes and did not prevent wolves from learning to hunt. However, it may prolong wolves' tolerance of people. Holding pairs or family groups in acclimation pens appeared to enhance reproductive success and pup survival.

Although Mexican wolves established home ranges, learned to hunt native prey, and reproduced in the wild, intensive monitoring, supplemental feeding, and hazing were required for a few months following release. Frequent recaptures (52 percent) were not only costly but slowed the colonization process. Phillips et al. (1995) found that intensive management of reintroduced red wolves was required as well, reporting that 67 percent of red wolves released from 1987 to 1992 were recaptured at least once.

Although historical records suggest that their primary prey were deer (Leopold 1959; Brown 1983), reintroduced Mexican wolves have preyed primarily on elk. Reintroduced wolves in the Greater Yellowstone Ecosystem exhibit a similar tendency to kill elk (Bangs et al. 1998). Our experience suggests that successful pack establishment is enhanced when releases coincide with elk calving season. Elk calving in the Blue Range occurs from May to July and coincides more closely with wolf parturition (April–May) than does deer parturition (July–August). Livestock depredation has been the primary reason for removal of wolves in areas where deer are the primary prey and livestock graze year-round. Wolves released into areas where livestock were not present at least two months after release, and where elk were abundant, did not kill livestock.

Solitary yearlings tended to leave the designated recovery area and required additional resources to recapture them—even if they had not otherwise caused problems. Given that long-distance dispersal of young wolves is well documented (Ballard et al. 1983; Fritts 1983; Gese and Mech 1991), this will likely be an increasing management burden for the project as wolf numbers increase. In order to accommodate natural dispersal and promote natural population growth, the boundaries of the recovery area must grow or be eliminated. This step would require additional analysis and documentation under NEPA and revisions of the recovery plan and experimental population rule. Until this happens, natural colonization of otherwise suitable wolf habitat will be prevented.

Mortality was problematic in 1998 when almost half of the released animals were shot. Wolf wariness appears to increase after release, however, so those that survive the first year become less vulnerable. No shootings have been documented since November 1998. We attribute this to learning in the wolves and to hunter education and law enforcement.

Sociopolitical Aspects

Public support for Mexican wolf recovery is generally strong and broad-based. Prerelease surveys conducted between 1988 and 1995 showed that a majority of residents in Arizona and New Mexico support wolf recovery (Biggs 1988; Johnson 1990; Duda and Young 1995). Nonetheless, 58 percent of the residents surveyed in Greenlee County, Arizona (most of which is in the reintroduction area) opposed wolf reintroduction owing to fears of livestock depredation and human danger (Schoenecker and Shaw 1997). A postrelease survey in Greenlee County showed that such anxiety had lessened (K. Schoenecker, pers. comm.).

During the planning period, many public officials opposed Mexican wolf reintroduction—including the governors of New Mexico and Arizona, members of the New Mexico Game and Fish Commission (appointed by the governor), and the director of the New Mexico Department of Game and Fish (hired by the commission). Originally the Arizona Game and Fish Department and its governor-appointed commission favored an experimental release at White Sands Missile Range in New Mexico. The decision to release captive wolves only in the Arizona portion of the Blue Range wolf recovery area stemmed from a USFWS policy that precluded wolf reintroduction if it was opposed by the state's wildlife management agency or the land management agency with jurisdiction in a proposed wolf recovery area. Organized public opposition took the form of an unsuccessful 1998 lawsuit by the New Mexico Cattle Growers' Association and eight related groups to stop the reintroduction project completely (*New Mexico Cattle Grower's Association, et al. vs. United States*, Civil Action CIV-98-0367-HB/LFG).

Some local residents have responded to wolf recovery as an economic opportunity. Defenders of Wildlife has assisted ranchers in marketing "Wolf Country Beef," which promotes wolf recovery through the sale of beef at a premium price. Basically ranchers ask a higher price for meat produced on land that also supports wolves. Another ranch in the recovery area has initiated its own "predator-friendly" marketing scheme and offers a wolf-oriented bed and breakfast service.

Summary and Conclusions

Reintroduced Mexican wolves face the social and political challenges presented by centuries of antipredator prejudice, misinformation, and real conflicts with livestock. Mexican wolf recovery is further complicated by the project's dependence on captive-reared animals and a multiple-use landscape, which includes cattle grazing and lacks a large core refuge such as Yellowstone National Park that is free of livestock and hunting. Even so, Mexican wolves may fare better than red wolves because they have been reintroduced in a landscape with fewer people

After two and a half years, some wolves have learned to hunt native prey, avoid killing livestock, and reproduce in the wild. Clearly this subspecies has benefited from intensive reintroduction planning, implementation, public outreach, and the experiences of previous wolf restoration efforts. The results of those efforts suggest that once a few pairs of wild-born animals are established, Mexican wolf recovery will accelerate dramatically. As the population grows with the production of wild-born wolves, managers must prepare for increasing interactions with livestock, humans, and their pets.

Although the NEPA process provided a forum for public involvement and validation of the government-sponsored Mexican wolf restoration process, the need for education and public outreach will continue long after the population is established and self-sustaining. For long-term success, Mexican wolf recovery must continue to involve all interest groups as well as the scientific community. Most conservation biologists agree that full recovery will require more than one population of 100 wolves. The fates of Mexican wolves in the Blue Range recovery area will provide a critical baseline against which additional efforts are judged.

Acknowledgments

The U.S. Fish and Wildlife Service funds Mexican wolf recovery, with assistance from the Arizona Game and Fish Department and the New Mexico Department of Game and Fish. The USDA Forest Service and USDA Wildlife Services are critical cooperators. Defenders of Wildlife funds and implements the livestock compensation program. Captive population management is funded by zoos and sanctuaries and the American Zoo and Aquarium Association, which provides expertise in small-population management. V. Asher, D. Boyd, and S. Naftal led the field crew in Arizona. A. Armistead led wolf/livestock conflict management efforts. We thank all other past and current members of the interagency field team, P. Frame, P. Morey, T. Peltier, D. Stark, and A. Watts, as well as the many dedicated volunteers, for data collection, monitoring and managing wolves. We are especially grateful to N. Smith for planning and executing wilderness releases, notably for designing wolf panniers and packing wolves on good mules. V. Asher, N. Smith, and P. Frame designed soft-mesh pens for wilderness releases. T. Johnson contributed significantly to project planning, and D. Groebner provided important support and technical expertise to the recovery effort. P. Jenkins coordinated the preparation of the environmental impact statement. C. Buchanan and P. Siminski managed the captive population and provided important data for this chapter. A. Armistead, C. Buchanan, R. Drewien, D. Groebner, and B. Kelly reviewed the manuscript. We are indebted to the countless other organizations and individuals we could not name here, but without whom Mexican wolf recovery could not take place.

Literature Cited

American Zoo and Aquarium Association (AZA). 1994. *The 1994 Mexican Wolf SSP Master Plan.* Compiled by D. P. Siminski, Arizona-Sonora Desert Museum, Tucson.

Bailey, V. 1931. *Mammals of New Mexico.* North American Fauna 53. Washington, D.C.: U.S. Department of Agriculture, Bureau of Biological Survey.

Ballard, W. B., R. Farnell, and R. O. Stephenson. 1983. Ecology of an exploited wolf population in southcentral Alaska. *Wildlife Monographs* 98:1–54.

Bangs, E. E., S. H. Fritts, J. A. Fontaine, D. W. Smith, K. M. Murphy, C. M. Mack, and C. C. Niemeyer. Status of gray wolf restoration in Montana, Idaho, and Wyoming. *Wildlife Society Bulletin* 26:785–798.

Bednarz, J. A. 1988. *The Mexican Wolf: Biology, History, and Prospects for Reestablishment in New Mexico.* Endangered Species Report 18. Albuquerque: U.S. Fish and Wildlife Service.

Biggs, J. R. 1988. Reintroduction of the Mexican wolf into New Mexico—an attitude survey. M.S. thesis, New Mexico State University, Las Cruces.

Bogan, M. A., and P. Mehlhop. 1983. Systematic relationships of gray wolves (*Canis lupus*) in southwestern North America. Occasional Paper 1. Albuquerque: Museum of Southwestern Biology.

Boyd, D. K., and D. H. Pletscher. 1999. Characteristics of dispersal in a colonizing wolf population in the Central Rocky Mountains. *Journal of Wildlife Management* 63:1094–1108.

Brewster, W. G., and S. H. Fritts. 1995. Taxonomy and genetics of the gray wolf in western North America: A review. In L. N. Carbyn, S. H. Fritts, and D. R. Seip, eds., *Ecology and Conservation of Wolves in a Changing World.* Edmonton: Canadian Circumpolar Institute.

Brown, D. E. 1983. *The Wolf in the Southwest: The Making of an Endangered Species.* Tucson: University of Arizona Press.

Defenders of Wildlife. 2000. *Wolf Compensation Trust Payments.* Washington, D.C.: Defenders of Wildlife.

Dobson, A., K. Ralls, M. Foster, M. E. Soulé, D Simberloff, D. Doak, J. A. Estes, L. S. Mills, D. Mattson, R. Dirzo, H. Arita, S. Ryan, E. A. Norse, R. F. Noss, and D. Johns. 1999. Corridors: Reconnecting fragmented landscapes. In M. E. Soulé and J. Terborgh, eds., *Continental Conservation: Scientific Foundations of Regional Reserve Design Networks.* Washington, D.C.: Island Press.

Duda, M. D., and K. C. Young. 1995. New Mexico residents' opinions toward Mexican wolf reintroduction. Contract report for League of Women Voters of New Mexico. Harrisonburg, Va.: Responsive Management.

Dunlap, T. R. 1988. *Saving America's Wildlife.* Princeton: Princeton University Press.

Fritts, S. H. 1983. Record dispersal of a wolf from Minnesota. *Journal of Mammalogy* 64:166–167.

García-Moreno, J., M. D. Matocq, M. S. Roy, E. Geffen, and R. K. Wayne. 1996. Relationships and genetic purity of the endangered Mexican wolf based on analysis of microsatellite loci. *Conservation Biology* 10:376–389.

Gese, E. M., and L. D. Mech. 1991. Dispersal of wolves (*Canis lupus*) in northeastern Minnesota, 1969–1989. *Canadian Journal of Zoology* 69:2946–2955.

Gipson, P. 1995. Wolves and wolf literature: A credibility gap. In R. McIntyre, ed., *War Against the Wolf: America's Campaign to Exterminate the Wolf.* Stillwater, Minn.: Voyageur Press.

Gipson, P., W. B. Ballard, and R. M. Nowak. 1998. Famous North American wolves and the credibility of early wildlife literature. *Wildlife Society Bulletin* 26:808–816.

Groebner D. J., A. L. Girmendonk, and T. B. Johnson. 1995. A proposed cooperative reintroduction plan for the Mexican wolf in Arizona. Technical Report 56. Phoenix: Arizona Game and Fish Department.

Hall, R. E., and K. R. Kelson. 1959. *The Mammals of North America.* New York: Ronald Press.

Hedrick, P., P. S. Miller, E. Geffen, and R. Wayne. 1997. Genetic evaluation of the three Mexican wolf captive lineages. *Zoo Biology* 16:47–69.

Johnson, T. B. 1990. Preliminary results of a public opinion survey of Arizona residents and interest groups about the Mexican wolf. Phoenix: Arizona Game and Fish Department.

Kalinowski, S. T., P. W. Hedrick, and P. S. Miller. 1999. No evidence for inbreeding depression in Mexican and red wolves. *Conservation Biology* 13:1371–1377.

Lande, R., and G. F. Barrowclough. 1987. Effective population size, genetic variation, and their use in population management. In M. E. Soulé, ed., *Viable Populations for Conservation.* New York: Cambridge University Press.

Leopold, A. 1930. Game management in the national forests. *American Forests* 36:412–414.

————. 1936. Threatened species: A proposal to the Wildlife Conference for an inventory of the needs of near-extinct birds and animals. *American Forests* 42:116–119.

————. 1937. Conservationist in Mexico. *American Forests* 43:118–120, 146.

Leopold, A. S. 1959. *Wildlife of Mexico: The Game Birds and Mammals.* Berkeley: University of California Press.

McBride, R. T. 1980. *The Mexican wolf (*Canis lupus baileyi*): A Historical Review and Observations on Its Status and Distribution.* Endangered Species Report 8. Albuquerque: U.S. Fish and Wildlife Service.

Mech, L. D. 1970. *The Wolf: The Ecology and Behavior of an Endangered Species.* Garden City, N.Y.: Natural History Press, Doubleday.

Nowak, R. M. 1995. Another look at wolf taxonomy. In L. N. Carbyn, S. H. Fritts, and D. R. Seip, eds., *Ecology and Conservation of Wolves in a Changing World.* Edmonton: Canadian Circumpolar Institute.

Parsons, D. R. 1996. Case study: The Mexican wolf. In E. A. Herrera and L. F. Huenneke, eds., New Mexico's natural heritage: Biological diversity in the Land of Enchantment. *New Mexico Journal of Science* 36:101–123.

————. 1998. "Green fire" returns to the Southwest: Reintroduction of the Mexican wolf. *Wildlife Society Bulletin* 26:799–807.

Parsons, D. R., and J. E. Nicholopoulos. 1995. The status of the Mexican gray wolf. In L. N. Carbyn, S. H. Fritts, and D. R. Seip, eds., *Ecology and Conservation of Wolves in a Changing World.* Edmonton: Canadian Circumpolar Institute.

Paul, W. J., and P. S. Gipson. 1994. Wolves. In *Prevention and Control of Wildlife Damage.* Lincoln: University of Nebraska Press, Cooperative Extension Service.

Phillips, M. K., R Smith, V. G. Henry, and C. Lucash. 1995. Red wolf reintroduction program. In L. N. Carbyn, S. H. Fritts, and D. R. Seip, eds., *Ecology and Conservation of Wolves in a Changing World.* Edmonton: Canadian Circumpolar Institute.

Roy, L. D., and M. J. Dorrance. 1976. *Methods of Investigating Predation of Domestic Livestock: A Manual for Investigating Officers.* Edmonton: Alberta Agriculture.

Schoenecker, K. A., and W. W. Shaw. 1997. Attitudes toward a proposed reintroduction of gray wolves in Arizona. *Human Dimensions of Wildlife* 2:42–55.

Siminski, D. P. 2000. *2000 Mexican Wolf SSP Annual Meeting and Reunion Binacional sobre el Lobo Mexicano.* Tucson: Arizona-Sonora Desert Museum.

Terborgh, J., J. Estes, P. Paquet, K. Ralls, D. Boyd-Heger, B. Miller, and R. Noss. 1999. The role of top carnivores in regulating terrestrial ecosystems. In M. E. Soulé and J. Terborgh, eds., *Continental Conservation: Scientific Foundations of Regional Reserve Design Networks.* Washington, D.C.: Island Press.

Thompson, B. C., J. S. Prior-Magee, M. L. Munson-McGee, W. Brown, D. Parsons, and L. Moore. 2000. Beyond release: Incorporating diverse publics in setting research priorities for the Mexican Wolf Recovery Program. *Transactions of the North American Wildlife and Natural Resources Conference* 65:278–291.

U.S. Fish and Wildlife Service (USFWS). 1982. Mexican wolf recovery plan. Albuquerque: U.S. Fish and Wildlife Service.

———. 1987. *Restoring America's Wildlife, 1937–1987: The First 50 Years of the Federal Aid in Wildlife Restoration (Pitman-Robertson) Act.* Washington, D.C.: U.S. Department of the Interior.

———. 1996. The reintroduction of the Mexican wolf within its historic range in the United States: Final environmental impact statement. Albuquerque: U.S. Fish and Wildlife Service.

———. 1997a. Notice of record of decision and statement of findings on the environmental impact statement on reintroduction of the Mexican gray wolf to its historic range in the southwestern United States. Albuquerque: U.S. Fish and Wildlife Service.

———. 1997b. Environmental assessment for the construction of Mexican wolf release pens in the Apache National Forest, Arizona. Albuquerque: U.S. Fish and Wildlife Service.

———. 1998. 1998 Mexican wolf interagency management plan. Albuquerque: U.S. Fish and Wildlife Service.

———. 2000. Environmental assessment for the translocation of Mexican wolves throughout the Blue Range wolf recovery area in Arizona and New Mexico. Albuquerque: U.S. Fish and Wildlife Service.

Wayne, R. K., N. Lehman, M. W. Allard, and R. L. Honeycutt. 1992. Mitochondrial DNA variability of the gray wolf: Genetic consequences of population decline and habitat fragmentation. *Conservation Biology* 6:559–569.

Young, S. P., and E. A. Goldman. 1944. *The Wolves of North America.* Washington, D.C.: American Wildlife Institute.

PART III

The Human Link

Many practicing wildlife biologists learned in Wildlife 101 that wildlife management is 90 percent people management. With a global population of more than 6 billion and growing, this notion is more true today than ever. The traditional approach to restoring large mammals is an agency-led effort where the public has little input into the process—let alone into the decision to restore a species or not. For the last four decades in the United States, however, the National Environmental Policy Act (NEPA) has gradually changed the way that agencies and the public interact. Today, as federal efforts proceed with the restoration of controversial carnivores (Roy et al. in chapter 10), managers can proceed with a confidence earned through an exhaustive interactive process that involves the public's voice at many levels. Successful projects will pull back the targets of NEPA review and Endangered Species Act recovery from the precipice of extinction (Schwartz in Case 3). Without strong public support, the best-designed restoration will remain on the planner's shelf. But even a project that has earned the green light through consensus can be dealt a lethal blow by a change in the political wind (Roy et al. in chapter 10).

While the bulk of the expertise may still reside within a natural resource agency, its decisions can no longer be made in a vacuum. It is becoming increasingly clear that successful large mammal restoration has no chance whatsoever without support from the public and consensus among strategic interest groups. When user groups are affected disproportionately—or their concerns are not considered at all—litigation is often the result. When management decisions are designed to correct the mistakes of past generations—or politics hold sway over common sense—the plan chosen is often the one that allows no change at all (Carbyn and Watson in chapter 9).

187

A recent change in large mammal restoration involves the creation of nonprofit organizations that have adopted conservation flagships such as elk and bighorn sheep. Income from dedicated members and special fund-raising events has transformed some of these organizations into powerful forces in land acquisition and wildlife management. And as umbrella species, many of these flagship mammals bring benefits to other key species and ecosystem processes. Another emerging trend in large mammal restoration is the role of people who possess both the economic and ethical capital to undertake significant conservation on their own. The unfolding story of desert bighorn sheep restoration (Krausman et al. in chapter 11) may serve as an international model for similar efforts.

Chapter 9

Translocation of Plains Bison to Wood Buffalo National Park: Economic and Conservation Implications

LUDWIG N. CARBYN AND DAVID WATSON

The most common purpose of translocation is to return a species to an area that was once part of its historical range but from which it has disappeared (IUCN 1998). Here we use the large-scale transfer of bison from southern to northern Canada as an example of an effort that has had long-lasting effects on the conservation of bison in and around North America's largest national park.

The translocation of more than 6000 plains bison (*Bison bison bison*) to an area with an established herd of about 1500 wood bison (*Bison bison athabascae*) in Wood Buffalo National Park (WBNP) occurred 75 years ago and is probably the largest megaherbivore translocation ever undertaken. Among the problems associated with the translocation is a high likelihood that introduced animals infected the resident herd with diseases and created a hybrid of two subspecies—issues that have evolved into chronic management problems. This chapter describes the consequences of the translocation with emphasis on both its long-term and short-term conservation and economic impacts.

Background

The total original range of wood bison in Canada covered an area of approximately 1,823,000 km² (WBRT 2000). The total area compromised by the presence of diseased bison is approximately 12 percent of the original range, or 218,500 km². Moreover, game ranching using bison has increased greatly since 1990. It is estimated that some 20,000 bison are raised on commercial ranches within the original wood bison range (WBRT 2000). It is difficult to prevent the spread of disease when infected and uninfected herds come into close contact. This applies both to domestic bison kept on ranches and to free-roaming bison released in the wood bison recovery program.

There has been considerable debate about the taxonomic status of "wood" bison in relation to "plains" bison. Rhoads (1897) presented the first formal taxonomic description of wood bison as a separate subspecies. Geist (1991) believes that North American subspecies were less distinct and that morphological differences were a reflection of environmental influences. Although studies are ongoing to determine the genetic basis of the subpopulations (C. Gates, pers. comm.), indications are that all "wood" bison have some introgression of prairie bison genes (Wilson and Strobeck 1998).

The park bison herds exist within a largely unmanipulated landscape where roads and seismic lines are minimal. Except for one short stretch in the northeastern corner, there currently are no major transport corridors through the park. There is an unpaved road linking the northern portion of the park with a central location at Peace Point on the Peace River. There is another link to the outside from the southwest that terminates at the park boundary. In 1999, commercial and political forces began an effort to link the southwestern road with an all-weather road along a 118-km right of way to the center of the park. Such a transport corridor, if approved, would result in bison deaths through vehicle collisions. It would also encourage increased movements of disease-free bison to undiseased areas (hard-packed snow along plowed roads may result in long-distance movements by bison) and influence wolf predation patterns (Carbyn 1997), making bison more vulnerable to killing by wolves.

Study Area and Methods

Wood Buffalo National Park covers 44,800 km² in Alberta and the Northwest Territories (Figure 9.1). The region is primarily subarctic wilderness with sparse human settlement. It is a remote park that receives fewer than 2000 visitors a year. Vegetation is boreal forest interspersed with muskeg, sand ridges, forest/sedge meadows, numerous shallow lakes, ponds, and meandering streams over karst topography. As part of the interior plains it is underlain by sedimentary rock with a few granite outcroppings bordering the Canadian Shield to the southeast. The Peace and Athabasca rivers have cre-

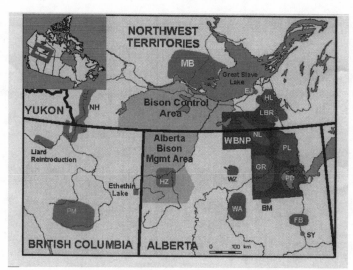

Figure 9.1. Location of Wood Buffalo National Park in relation to other free-roaming bison in northwestern North America (BM = Birch Mountain, FB = Firebag, GR = Garden River, HL = Hook Lake, LBR = Little Buffalo River, SY = Syncrude, PL = Pine Lake, WA = Wabasca, WZ = Wentzel Lake, EJ = Edjericon Ranch, MB = MacKenzie Bison, NL = Needle Lake, HZ = Hay Zama, NH = Nahanni, PM = Pink Mountain, PD = Peace-Athabasca Delta). Map reprinted with permission; J. Nishi, NWT Government (1996).

ated the 5000-km^2 Peace-Athabasca Delta, the prime habitat for park bison. The highest winter concentrations of bison are found along the sedge/forest interface near creeks and oxbows north of the Peace River and in the Peace-Athabasca Delta south of the Peace. Bison numbers have varied in both regions since records began in 1971 (Carbyn et al. 1998).

We used survey results from Parks Canada (D. Bergson, unpublished reports, 1994–1997) to determine the status of bison in Wood Buffalo National Park. Surveys were flown during late February and March in fixed-wing aircraft at an altitude of about 600 m and at a cruising speed of 180 km/hour. Transects were 2.5 km apart, extended across the width of the park, and were plotted on topographic maps. Information on sex and age (adults, calves, and yearlings) was obtained by herding bison with helicopters past ground observers and using pedestrian surveys where one to three observers approached bison at close range and noted age and sex composition of cow/calf herds (Carbyn et al. 1998). These data were used to estimate yearling survival and cow/calf ratios in early spring.

For information on program expenditures we relied on a combination of accessible government records, program final reports, and personal contact with people associated with various government research and conservation initiatives. It was difficult to get accurate figures on actual expenditures (including internal park staff costs and costs in kind) for the various research and management initiatives. Records of real costs therefore are higher but not available. We have been able to obtain minimum cost figures.

Herd History

Wood Buffalo National Park and the area around Great Slave Lake have been part of the northernmost North American bison range since before European colonization and may once have included parts of Alaska (R. Stephenson, pers. comm.). The first reports of bison in the area were made in 1772 by Samuel Hearne who, in the company of Dene hunters, killed bison along the east side of the Slave River (Figure 9.2). Other information from casual sightings and reports made by explorers, geologists, and government employees suggests that bison inhabited this area continually for more than 200 years (Carbyn et al. 1993).

Very little information on bison in the Wood Buffalo National Park region is available for the late eighteenth and most of the nineteenth centuries (Preble 1908). Toward the latter part of the nineteenth century, the most persistent records came from what would become Wood Buffalo National Park. Concerns were raised about declines in their numbers as esti-

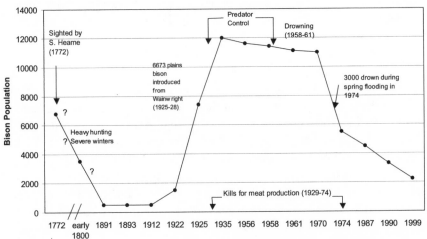

Figure 9.2. Bison population trends from 1772 to 1999 in Wood Buffalo National Park.

mates for the region dropped to as low as possibly 500 by the 1880s (Preble 1908 as reported by Jarvis in 1897). Low numbers of bison persisted at the northern edge of their range into the twentieth century. Although the numbers had increased by 1924, only a single viable population existed north of the Peace River and that was within Wood Buffalo National Park and in the adjacent Slave River Lowlands area.

Elsewhere in western North America, bison had been declining in the latter part of the nineteenth century. Only a few survivors remained in isolated pockets, and these were placed in captivity. From 1907 to 1912 the Canadian government purchased 716 plains bison from Michel Pablo, a rancher near Missoula, Montana. The "Pablo herd" originated from four bison captured by Native Americans in 1873, supplemented by 26 animals from Nebraska. The first shipment was delivered by railway to Elk Island National Park because their final destination, Buffalo National Park at Wainwright, Alberta, was not ready to receive them. Their numbers had increased to about 7000 by 1923 (Lothian 1976). Overcrowding at Wainwright (Graham 1924) led to the transfer of 6673 plains bison (4826 yearlings, 1515 two-year-olds, and 332 three-year-olds) to Wood Buffalo National Park. Those animals were shipped by railroad and crowded barges for a distance of about 1100 km, and as many as 50 percent may have died. In all "consignments," females were in the majority (Stevens 1954). The estimated duration of rail transport was two or three days. At Waterways, Alberta, the animals were transferred to barges, pushed by tugs down the Athabasca and Slave rivers for two or three days (under ideal weather conditions, which likely were not always present), and released into Wood Buffalo National Park. By 1926, introduced bison had crossed the Peace River and become residents of the Peace-Athabasca Delta. The park was expanded by approximately 17,400 km² to provide additional protection for the rapidly expanding herd.

The transfer of plains bison from Wainwright to wood bison range was opposed by conservationists who feared disease introduction and hybridization (Harper 1925; Howell 1925; Saunders 1925). Further, wood bison were increasing on their own by the time the plains bison were released (McHugh 1972). Nonetheless the interior minister claimed "it is not the intention to mix breeds in any way" and the release took place. Hybridization soon occurred, as feared, compelling the Harvard zoologist Thomas Barbour to observe that the transfer was "one of the most tragic examples of bureaucratic stupidity in all history" (McHugh 1972:306).

Roddy Fraser (pers. comm.), a resident of Fort Chipewyan, notes that those involved in the transfer reported that "many" of the bison died while in transit. As a boy he remembers peering through the boards on scows and seeing bison packed "tight as sardines." W. A. Fuller (pers. comm.) obtained firsthand accounts from bison handlers who saw "dead animals falling out of the cramped barges as the doors were opened." Further, Oldham (1947:1) stated that "there is a wide difference of opinion as to the actual numbers that survived the trip; also, no accurate information is available as to the number of animals that succumbed shortly after arrival due to injuries sustained enroute as well as those that died due to change of site."

Despite the losses, environmental conditions appeared ideal for the survivors. By 1928, the total bison population was estimated at 7500; by 1934 it numbered 10,000 to 12,000 (Soper 1941). Presumably the introduced bison quickly interbred with the established population. The infusion of plains bison genes (probably about a 4:1 ratio) diluted the local wood bison genotype and stopped what was a promising comeback by the indigenous northern bison. It has been suggested that the genetic composition of the bison in the delta may be largely that of wood bison while those north of the Peace River are genetically closer to plains bison (Van Zyll de Jong et al. 1995). Our own observations on the physical appearance of the bison confirm that bison throughout WBNP seem to exhibit phenotypic characteristics similar to those listed for wood bison (Geist and Karsten 1977; Geist 1991). Subsequent studies also have shown considerable movements of radio-collared bison throughout the park area (WBNP 1995).

Current Status

There were about 2300 bison in Wood Buffalo National Park in 1999. From 1990 to 1994 there was a low yearling recruitment and evidence of high wolf (*Canis lupus*) predation on calves (Carbyn et al. 1998). Bison numbers increased between 1994 and 1996 (to 2779) and then declined again after the floods in 1997 and 1998 (Figure 9.3). Cow/calf ratios in early spring averaged 30 calves per 100 cows over the 10-year period. Although pregnancy rates were not determined, they are likely to have been considerably higher as observations of calves at heel do not take account of losses due to early predation, stillbirths, abortions, and drowning. Survival of calves from spring to the yearling stage averaged 8 yearlings per 100 cows.

Though drowning is the largest "single point in time" mortality factor (Carbyn et al. 1993, 1998), wolf predation, coupled with reduced reproduction, has probably done more to limit bison population growth. Wolf predation maintains numbers at a lower level than would be the case without predation. Without predation, however, it is likely that density-dependent factors may become more important at some point.

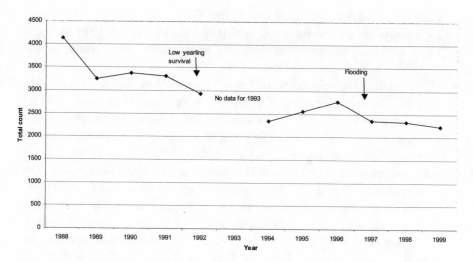

Figure 9.3. Population changes of bison in Wood Buffalo National Park from 1988 to 1999 as determined by aerial surveys.

Management Implications

The transfer of plains bison to wood bison range had a series of ecological and economic implications, not all of which are clearly understood. At the time of European settlement, bison in the Wood Buffalo National Park area existed in a more restricted region and were probably a single geographic/taxonomic entity that had developed some level of local adaptation. With the massive infusion of numbers in the 1920s, the ecological impact on range conditions must have been significant. Bison in northern latitudes feed heavily on *Carex atherodes* (slough sedge) and *Calamagrostis* spp. (reed grass) year-round (Reynolds and Peden 1987). The effects of seasonal use, productivity, change in plant composition, snow cover, and changing hydrological regimes on the availability and quality of forage on bison range in Wood Buffalo National Park are poorly understood, however.

Most research and management programs for park bison have centered on bovine diseases (tuberculosis and brucellosis). Tuberculosis was likely introduced to Wood Buffalo National Park in 1925 with the translocation of plains bison (Fuller 1966; Lothian 1976). Although the initial translocation protocol for Wainwright bison called for tuberculin testing of all animals, the procedure was discontinued because of the perception that only older animals were susceptible to disease. It was therefore assumed that if only younger animals were shipped, there would be virtually no chance of spreading the disease—a notion that was later proved false. Despite the presence of tuberculosis and wolves, bison increased until they were targeted in 1929 to supply meat to remote provincial settlements. The presence of tuberculosis was noted during early hunts (Mitchell 1976). Because of small-scale operations, all early kills were referred to as "hunts." When W. A. Fuller arrived in the area in 1947, the target number set for two communities (Salt Plains and Fort Chipewyan) was about 30 to 40 animals each (Fuller, pers. comm.). Fuller noted what may have been the first case of disease in 1946 when an old bull at Fort Smith was examined by a local medical doctor who suggested that it had tuberculosis. By the 1950s, the number of bison killed had increased. The discovery of tuberculosis led to slaughters designed to test for TB and remove reactors in an attempt to control the disease, which may have been causing a 2 percent annual mortality based on the advanced cases that were examined: they were considered to be at death's door. The management options discussed by park personnel from the 1950s to the early 1970s ranged from doing nothing to eradication of tuberculosis by enlisting the Canadian Air Force to kill every bison in and around the park (W. Stevens, pers. comm.). The latter option would have included a reintroduction with uninfected animals. No decisions have been made and the debate continues today.

Brucellosis, an infectious bacterial disease (*Brucella* spp.), was first confirmed in bison in 1956 (Fuller 1966). Its spread is facilitated by oral contact with contaminated placenta, blood, and aborted fetuses. Positive reactors to tests on bison in the park have varied from 3 to 62 percent (Mitchell 1976), with an average of 33 percent. Although brucellosis may or may not have been introduced into the park with plains bison, the presence of both diseases likely coincided with the plains bison introduction. Fuller (pers. comm.) determined from early testing results in the mid-1950s that infection rates of both diseases ranged from 35 to 39 percent.

Anthrax is caused by the bacterium *Bacillus antracis,* the spores of which are known to retain their viability in contaminated soil for many years. The first recorded outbreak of anthrax in the Wood Buffalo National Park vicinity occurred at Hook Lake, Northwest Territories, along the east side of the Slave River in 1962 (Novakowski et al. 1963); the first outbreak in Wood Buffalo National Park occurred in 1964. Between 1962 and 1971, a minimum of 1003 bison from the Hook Lake and Wood Buffalo National Park herds died of anthrax (Choquette et al. 1972). The most recent outbreak killed about 100 bison in 2000. How anthrax reached the bison in the park is not known. It may have been endemic to the park or it may have been introduced by outfitters' horses during the early 1960s. Anthrax is highly virulent in horses, however, so it is doubtful that the disease was introduced by this route. Another possible source is birds migrating from the United States, where the organism is believed to originate.

Anthrax has had a major impact on the bison populations in northern Alberta. Programs to vaccinate Wood Buffalo National Park bison against anthrax lasted from 1965 to 1977, during which time 27,943 bison were rounded up and treated at great expense. As many as 4000 or 5000 bison were herded about 60 km from the delta to the Lake One corral. Such a forced move resulted in corral-related injury, loss of calves by trampling, separation of calves from cows, goring by bulls, stress-related myopathies, and abortion (Hudson et al. 1976). The average recorded mortality for the animals vaccinated between 1964 and 1974 was 2 percent, with a one-year high in 1972 of about 7 percent (Millette and Sturko 1977). Wardens believe that only a small proportion of the stress-related deaths were in fact observed, so the actual mortality rate was probably higher than the recorded rate.

Red Meat Production

Bison harvests for meat began in 1929 (Mitchell 1976). Thus park bison were more than a component of Canada's natural heritage: they had become a resource for direct human consumption. During the 1950s and 1960s, the government administrators and society alike saw little conflict between bison harvest and bison preservation.

As bison harvest became commercialized, portable abattoirs (slaughterhouses) were replaced by fixed abattoirs to process up to 1000 bison per year. Between 1958 and 1963, the attempt to eliminate tuberculosis and brucellosis became an important consideration and bison were regularly tested at the abattoirs. Reactors were killed, and nonreactors were released. The last commercial slaughter occurred during Canada's centennial celebrations when bison meat was shipped to Montreal for Expo '67. Approximately 2700 bison were killed between 1950 and 1958; approximately 2200 were killed between 1959 and 1974 (Figure 9.2). The last major kill was approved and carried out in 1974 when 114 were killed.

The most recent evaluation of the disease problem in bison was undertaken by the 1988–1990 Federal Environmental Assessment Review Office (FEARO) panel. At that time the Canadian Department of Agriculture declared Canadian cattle free of brucellosis and promised that they soon would be free of tuberculosis. The sense of urgency about the possible infection of cattle via bison in Wood Buffalo National Park—and the economic consequences—led to the 1986 creation of an interjurisdictional committee to head off problems. The group offered four possible courses of action to stop disease transmission from bison to cattle. In 1988, the FEARO review panel was commissioned to examine these options. The option chosen by the FEARO panel called for the elimination of the herd and its replacement with disease-free and phenotypically pure wood bison (FEARO 1990).

The FEARO report triggered a lively debate among conservation organizations, Native Americans, animal welfare groups, national parks, and agricultural interests. In the end, herd replacement was not carried out even though Agriculture Canada offered to contribute the $20 million needed to do it. Action was delayed so that Native American interest in the bison herd could be better addressed. Consequently, the Northern Buffalo Management Board (NBMB) was established by the federal government in 1993. The Fort Smith Buffalo Users Group (1992) recommended the following guidelines for management of free-roaming buffalo:

1. Buffalo are sacred to aboriginal people. Any attempt to eradicate the population is a crime against Nature and the Creator of all life.

2. Humans must protect the bison and other wildlife from harm. Unnecessary killing or stress of bison must be avoided.

3. Aboriginal people who understand and are dependent on bison should be their primary stewards during research and management.

4. As co-managers, aboriginal people will continue to cooperate with governments in the care of bison as they did in the creation of Wood Buffalo National Park and will abide by and respect park laws.

5. The ecological integrity of Wood Buffalo National Park will be maintained by protecting the delicate balance among all components of the natural ecosystem.

6. The status of Wood Buffalo National Park as a World Heritage Site and its responsibility to the world in dealing with bison management must be acknowledged.

7. Wood Buffalo National Park bison are a single species regardless of phenotypic variation.

The NBMB report also recommended studies on bison status, population characteristics, human use, and traditional values.

Operating concurrently with the bison initiatives were concerns over the hydrology and ecosystem dynamics of the Peace-Athabasca Delta. Hydrological studies commenced prior to the FEARO hearings in the mid-1970s in response to changing water flow after the construction of the Bennett Dam in British Columbia. An increase in pulp and paper mill activities on

Table 9.1. Minimum Research and Management Expenditures in Wood Buffalo National Park and Vicinity: 1990–2000

Category	Amount (Can$)
Bison demography	6,803,000
Bison genetic analysis of subspecies	1,400,000
Buffer zone management and disease control	3,200,000
Endangered species management and recovery	4,474,900
Bison-related hydrology/range research	2,147,000
PAD ecology studies unrelated to bison	15,375,000
TOTAL	33,400,000

the rivers flowing into the delta prompted studies that focused on water quality, aquatic organisms (especially fish), and ecosystem health. This galaxy of concerns relating to bison and the delta ecosystem resulted in a proliferation of fragmented and compartmentalized projects with few spatial or temporal linkages among them. In all, more than $33 million was spent on these studies between 1990 and 2000 (Table 9.1).

Clearly the actions taken in the mid-1920s and the translocation of 6673 bison set the stage for the regional economics and bison management dilemma of the twenty-first century. The initial economic benefit of abundant bison in the region was red meat production from 1929 to the early 1970s. Control efforts for anthrax, though costly, were important to the economy of local settlements at Fort Smith, Northwest Territories, and Fort Chipewyan, Alberta. The most lasting economic impact continues to be disease monitoring. Tuberculosis and brucellosis are considered threats to humans, to nearby disease-free bison, to commercial bison, and to cattle. Unlike these two diseases, which require living hosts for disease transfer, anthrax does not.

Conservation Implications

The initial transfer of plains bison from Wainwright to Wood Buffalo National Park was conducted under the presumption that it would benefit the overcrowded Wainwright bison herd. Public opposition precluded slaughter as a means for population control, so translocation became a politically attractive alternative. Nonetheless, mixing of wood bison and plains bison was a threat to the "purity" of the small, remaining wood bison race. This was not considered to be a problem when bison translocation began. Nor was the transfer of diseased bison to a herd without infectious diseases a factor in the final decisions. Very early on Raup (1933) speculated that the "Wood Buffalo" were rapidly disappearing, but he suggested that some "purer" herds still existed in more remote areas. The disease issue eventually became a huge economic burden.

The Committee on the Status of Endangered Wildlife in Canada (COSEWIC) is responsible for producing the official Canadian Endangered Species list. Wood bison were recognized by COSEWIC as an endangered subspecies in 1978 but were downlisted to threatened in 1988 because their numbers had improved considerably. A national Wood Bison Recovery Team—operating under a federal/provincial/nongovernment authority known as the Committee for the Recovery of Nationally Endangered Wildlife (RENEW)—has been involved since 1990 in the planning of wood bison restoration in northern Canada and Alaska. One of the committee's principal tasks will be to determine how to continue recovery of the wood bison in the limited areas that are free of disease and hybrids.

Three decades of efforts to restore wood bison in the wild have created a herd of 2500—the number of animals that existed in 1925 (Gates et al. 1997). Further recovery of bison in their historic range in Canada has been viewed as problematic because of the presence of diseased hybrids in Wood Buffalo National Park. Gates et al. (1997) and others have suggested that disease eradication in the park would facilitate range expansion in the historic range by at least 12 percent. In 2000, the population was at 2818 animals in six free-roaming disease-free herds, 708 in four captive herds, and about 2900 in diseased, free-roaming herds—2.5 times the number of animals in 1992. The anticipated spread of diseases from infected Wood Buffalo National Park bison to those in the disease-free Mackenzie Bison Sanctuary has not occurred. Nonetheless, the possibility of disease transfer remains.

The pressure to eliminate diseased bison in WBNP and adjacent areas has grown in recent years as the number of both captive ranching operations and disease-free wild bison herds have increased. The recent Wood Bison Recovery Plan (April 2000 draft) states: "Elimination of bovine tuberculosis and brucellosis from bison herds in the vicinity of Wood Buffalo National Park would remove the largest single obstacle to the recovery of wood bison in Canada."

Supporters of herd replacement in accordance with FEARO's findings pointed in 1990 to three major issues:

- Action was needed immediately to prevent the spread of diseases to wood bison in the Mackenzie Bison Sanctuary. (Considering the events of the last ten years, the problem turned out to be exaggerated although the long-term implications are still not settled.)

- Action was needed to eliminate "hybrid" bison from the area so that wood bison recovery could proceed in the region as a whole. (Subsequent analyses have indicated that the genetic differences are very minor—in fact, bison in Wood Buffalo National Park are now considered "wood bison" and the population is evolutionarily a significant conservation unit.)

- Action was needed to reduce costs to agricultural interests for the monitoring of bovine diseases in cattle herds. (This is a serious ongoing issue.)

Those who opposed herd replacement argued:

- Herd replacement could not be proved to guarantee the elimination of the two named diseases in the region as a whole.

- There was not enough evidence to declare the "hybrid" bison as expendable to the gene pool of the northern bison populations.

- There was not enough evidence to suggest that it was disease in the Wood Buffalo National Park herd that caused its decline. Only through further study could there be a more accurate assessment on this issue.

Conclusion

Wood Buffalo National Park is unique due to its size and the existence of free-ranging bison herds. Although the transport of more than 6000 bison in the 1920s caused great initial mortality of the animals involved in the transplant, the "hybrid" population (mixing of local bison with introduced animals) grew rapidly. Wolf populations increased as a result and, together with other mortality factors, caused bison to decline from about 11,000 in the early 1970s to a current herd size of some 2700 animals. Debates over the management of the herd have intensified over the last decade. Normally transplants take place from occupied to vacant ranges from which animals spread to unoccupied areas. In this case, however, the transplant took place from occupied range to partially occupied range—primarily to reduce crowding at the source. There were both positive and negative ramifications of the transplant. The initial transfer of bison temporarily alleviated overcrowding in Wainwright National Park, and Wood Buffalo National Park was expanded by 17,400 km² to accommodate a larger bison herd. But mixing of bison with adaptations to two different biomes (boreal forest versus grasslands), as well as introduction of two bovine diseases, caused serious problems. Government and nongovernment agencies alike are still coming to grips with some of the negative consequences of disease transmission. The economic, conservation, and social impacts of actions taken 75 years ago are still with us.

Acknowledgments

A number of people went out of their way to provide us with information on costs of programs carried out over the years: I. van der Linden, C. Gates, T. Steele, F. Letchford, C. Strobeck, R. Wein, and D. Stewart. J. Nishi and G. Chisholm gave us the unpublished information. W. A. Fuller, C. Gates, B. Laishley, M. Raillard, and H. Reynolds reviewed the manuscript in earlier drafts. We are grateful to two anonymous reviewers and to V. Jespersen for typing the manuscript.

Literature Cited

Bison Research and Containment Program. 2000. Research Advisory Committee minutes, 1995–1999. Fort Smith: Parks Canada.

Carbyn, L. N. 1997. Unusual movement by bison, *Bison bison*, in response to wolf, *Canis lupus*, predation. *Canadian Field Naturalist* 91:461–462.

Carbyn, L. N., S. M. Oosenbrug, and D. Anions. 1993. *Wolves and Bison and the Dynamics Related to the Peace-Athabasca Delta in Canada's Wood Buffalo National Park*. Circumpolar Research Series, no. 4. Edmonton.

Carbyn, L. N., N. J. Lunn, and K. Timoney. 1998. Trends in the distribution and abundance of bison in Wood Buffalo National Park. *Wildlife Society Bulletin* 26:463–470.

Choquette, L. P. E., E. Broughton, A. Currier, J. G. Cousineau, and N. S. Novakowski. 1972. Parasites and diseases of bison in Canada. Pt. 3: Anthrax outbreak in the last decade in northern Canada and control measures. *Canadian Field Naturalist* 86:127–132.

Environment Canada. 2000. *Renew 98–99: Report on Endangered Species Activities*. Report 8. Ottawa: Canadian Wildlife Service.

Federal Environmental Assessment Review Office (FEARO). 1990. *Northern Disease Bison: Report of the Environmental Assessment Panel*. Ottawa: Ministry of Supply and Services Canada.

Fort Smith Buffalo Users Group. 1992. A co-management plan for Buffalo research and interim management. Unpublished report.

Fuller, W. A. 1952. *Report on the Buffalo Hunt, Wood Buffalo National Park*. Ottawa: Parks Canada.

———. 1966. The biology and management of bison of Wood Buffalo National Park. *Wildlife Management Bulletin* 1(16):1–52.

Gates, C., B. Etkin, and L. N. Carbyn. 1997. The diseased bison issue in northern Canada. In E. Thorne, M. Boyce, P. Nicoletti, and T. Kreeger, eds., *Proceedings of the National Symposium on Brucellosis in the Greater Yellowstone Area*. Cheyenne: Wyoming Game and Fish Department.

Geist, V. 1991. Phantom subspecies: The wood bison *Bison bison "athabascae"* (Rhoads) 1897 is not a valid taxon, but an ecotype. *Arctic* 44:283–300.

Geist, V., and P. Karsten. 1977. The wood bison (*Bison athabascae*) (Rhoads) in relation to hypotheses on the origin of the American bison (*Bison bison* Linnaeus). Zeitschrift für Säugetierkunde 42:119–127.

Graham, M. 1924. Finding range for Canada's buffalo. *Canadian Field Naturalist* 38:189.

Harper, F. 1925. Letter to the editor. *Canadian Field Naturalist* 39:45.

Howell, A. B. 1925. Letter to the editor. *Canadian Field Naturalist* 39:118.

Hudson, R. J., T. Tennessen, and A. Sturko. 1976. Behavioural and physiological reactions of bison to handling during an anthrax vaccination program in Wood Buffalo National Park. In J. G. Stelfox, ed., *Wood Buffalo National Park: Bison Research 1972–76*. 1976 annual report. Ottawa: CWS/Parks Canada.

International Union for the Conservation of Nature (IUCN). 1998. IUCN Guidelines for Re-introductions. Prepared by IUCN/SSC Re-introduction Specialist Group. Gland, Switz.: IUCN.

Lothian, W. F. 1976. *A History of Canada's National Parks.* Vol. 1. Ottawa: Parks Canada.

McHugh, T. 1972. *The Time of the Buffalo.* New York: Knopf.

Millette, F. J., and A. N. Sturko. 1977. *A Proposal to Review and Revise the Anthrax Control Program in Wood Buffalo National Park.* Ottawa: Parks Canada.

Mitchell, R. 1976. A review of bison management, Wood Buffalo National Park, 1922–1976. In J. G. Stelfox, ed., *Wood Buffalo National Park: Bison Research 1972–76*. 1976 annual report. Ottawa: CWS/Parks Canada.

Northern Buffalo Management Board. 1992. Northern Buffalo Management Program Report. Report submitted to the Government of Canada, Fort Smith, Northwest Territories.

Northern Rivers Ecosystem Initiative. 1999. *River News: Northern Rivers Ecosystem Initiative Newsletter,* vol. 1, no. 1. Edmonton: Environment Canada.

Novakowski, N. S., J. G. Cousineau, G. B. Kolenosky, G. S. Wilton, and L. P. E. Choquette. 1963. Parasites and diseases of bison in Canada. Pt. 2: Anthrax epizootic in the Northwest Territories. *Transactions of the North American Wildlife and Natural Resources Conference* 28:233–239.

Oldham, E. G. 1947. Buffalo counts in Wood Buffalo National Park, December 1946 and January 1947. Parks Canada Memo.

Peace Athabasca Technical Studies Group. 1996. Final report. Fort Chipewyan, Alberta: PADTS Communications.

Preble, E. A. 1908. A biological investigation of the Athabasca-Mackenzie Region. *North American Fauna* 27. Washington, D.C.: U.S. Department of Agriculture, Bureau of Biological Survey.

Raup, H. M. 1933. Range conditions in Wood Buffalo National Park of western

Canada with notes on the history of wood bison. *Bulletin 74* (Biological Service) 20:1–174.

Reynolds, H. W., and D. G. Peden. 1987. Vegetation, bison diets and snow cover. In *Bison Ecology in Relation to Agricultural Development in the Slave River Lowlands, Northwest Territories.* Occasional Paper 63. Edmonton: Canadian Wildlife Service.

Rhoads, S. N. 1897. Notes on the living and extinct species of North American bovidae. *Proceedings of the Academy of Natural Sciences of Philadelphia* 49:585–602.

Saunders, W. E. 1925. Correspondence to the editor. *Canadian Field Naturalist* 39:118.

Soper, D. 1941. History, range and home life of the northern bison. *Ecological Monographs* 2:347–412.

Stevens, W. E. 1954. Bison report September 1954. Parks Canada/CWS. Unpublished report. Wood Buffalo National Park, Fort Smith, Northwest Territories.

Van Zyll de Jong, S., C. Gates, H. Reynolds, and W. Olson. 1995. Phenotypic variation in remnant populations of North American bison. *Journal of Mammalogy* 76:391–405.

Wilson, G. A., and C. Strobeck. 1998. Microsatellite analysis of genetic variation in wood and plains bison. In L. Irby and J. Knight, eds., *International Symposium on Bison Ecology and Management in North America.* Bozeman: Montana State University.

Wood Buffalo National Park (WBNP). 1995. Bison movement and distribution study: Final report. Technical Report 94-08WB. Fort Smith, Northwest Territories.

Wood Buffalo Recovery Team (WBRT). 2000. *National Recovery Plan Proposal: Recovery of Nationally Endangered Wildlife (RENEW).* Ottawa: Canadian Wildlife Service.

Chapter 10

Restoration of Grizzly Bears to the Bitterroot Wilderness: The EIS Approach

JOHNNA ROY, CHRISTOPHER SERVHEEN,
WAYNE KASWORM, AND JOHN WALLER

The grizzly bear (*Ursus arctos horribilis*) inhabited most of western North America for thousands of years. Habitat loss and excessive killing by humans, however, reduced its numbers from 50,000 prior to European settlement to about 1000 bears today that occupy only 2 percent of their historic range in the lower 48 states (USFWS 1993). The grizzly was listed as threatened in the conterminous United States under the Endangered Species Act (ESA) in 1975. At that time they were known or thought to persist in six areas in the Northwest. As the designated lead agency, the U.S. Fish and Wildlife Service (USFWS) must develop and implement a recovery plan with the ultimate goal of removing the species from listed status. In this chapter we describe the evolution of a recent benchmark in North American carnivore recovery: the reintroduction of the grizzly bear to historic range through the development of an environmental impact statement (EIS). Although intense emotions have been generated during its six-year development and the planning process that commenced years earlier (MacCracken et al. 1994; Fischer and Roy 1998), we believe this is a model that can be followed for large mammal restoration around the world.

Background

The Grizzly Bear Recovery Plan was developed in 1982 and revised in 1993 (USFWS 1982, 1993). Recovery will be pursued in six "grizzly bear ecosystems" because they contain existing populations or recently were occupied by the species and still contain high-quality habitat. Five of the areas support populations ranging from 10 to 600 individuals: the Yellowstone Ecosystem; the Northern Continental Divide Ecosystem in northwestern Montana; the Cabinet/Yaak Ecosystem in northwestern Montana and northeastern Idaho; the Selkirk Ecosystem in northern Idaho and northwestern Washington; and the Northern Cascades Ecosystem in north-central Washington (Figure 10.1).

The Bitterroot Ecosystem is unique among the six grizzly bear ecosystems because there has been no verified evidence of grizzly bears inhabiting the area for more than 50 years (Melquist 1985; Groves 1987; Serveheen et al. 1990; Kunkel et al. 1991; USFWS 2000a). Nonetheless, three habitat studies (Scaggs 1979; Butterfield and Almack 1985; Davis and Butterfield 1991) suggested that this area could support 200 to 400 grizzly bears (Craighead et al. 1982; Serveheen et al. 1991). Subsequently, the Interagency Grizzly Bear Committee (IGBC) endorsed the Bitterroot Ecosystem as suitable habitat and urged the USFWS to pursue grizzly bear recovery. The IGBC is composed of agency representatives from the USFWS, USDA Forest Service, National Park Service, Montana Bureau of Land Management, U.S. Geological Survey, Wyoming Game and Fish Department, Montana Department of Fish, Wildlife, and Parks, Idaho Department of Fish and Game, Washington Department of Fish and Wildlife, and the British Columbia Wildlife Branch.

In 1992, the USFWS organized an interagency Technical Working Group to develop a Bitterroot Ecosystem chapter for the Grizzly Bear Recovery Plan. Input was sought from a citizen's involvement group and from public meetings conducted in six local communities in central Idaho and western Montana. The Bitterroot Ecosystem Subcommittee of the IGBC, formed in 1993, consisted of agency officials responsible for land and game management decisions in the region. In addition, the Idaho state legislature established a grizzly bear oversight committee comprised of legislators, representatives from Idaho Fish and Game and Wildlife Services, and representatives from the timber, mining, livestock, wildlife, and recreation industries. These groups reviewed 800 public comments received during a 60-day comment period on the draft chapter and made recommendations on how to recover grizzly bears in the Bitterroot Ecosystem.

The USFWS integrated all these views and comments into the Bitterroot

Ecosystem chapter of the Grizzly Bear Recovery Plan (USFWS 1996). The new chapter called for preparation of an environmental impact statement to evaluate a full range of grizzly bear recovery alternatives, including reintroduction of a small number of grizzly bears into the Bitterroot Ecosystem as a nonessential experimental population under section 10(j) of the ESA. The National Environmental Policy Act (NEPA) required preparation of an EIS to analyze and document the proposal's environmental impacts because it was a major federal action with alternatives. Section 102(2) of NEPA requires that "all agencies of the Federal Government shall include in . . . major Federal actions significantly affecting the quality of the human environment" a statement of the environmental impacts.

The USFWS and partner agencies began the NEPA planning process in 1995 and finished in 2000. This chapter summarizes major events and accomplishments and outlines the alternative plan selected to restore a population of grizzly bears to their historic habitat in the Bitterroot Ecosystem.

The Case for Restoration

Lewis and Clark were among the first western explorers to document grizzly bears in the Bitterroot Mountains (Wright 1909; Burroughs 1961). When they traveled through the Bitterroot country in 1806, grizzly bears were abundant and they killed at least seven (Thwaites 1959). As settlement of the West progressed, persecution of the grizzly bear increased. They were trapped for their valuable pelts and pursued by trophy hunters. Even so, grizzly bears remained common through the early 1900s and sustained an annual harvest of 25 to 40 (Wright 1909; Merriam 1922; Burroughs 1961; Moore 1984, 1996).

Following the regionwide forest fires of 1910, Bitterroot homesteaders began grazing thousands of sheep and killing grizzlies to safeguard flocks against bear depredations. Grizzlies had no legal protection in Montana or Idaho during this period and were killed indiscriminately. Beginning in 1927, dams effectively eliminated salmon from the Bitterroots, which caused hungry bears to travel widely in search of food and increased their vulnerability (Moore 1984). By the late 1930s, grizzly bears were rare. The last evidence of grizzly bears in the Bitterroot Mountains was a 1932 death and a set of tracks in 1946 (Moore 1984, 1996). Judging from the lack of written material from 1940 to 1970, few people mourned the apparent passing of the Bitterroot grizzly. With passage of the Endangered Species Act in 1973, however, interest in grizzly bear preservation resulted in its listing as threatened in 1975. Coincidentally, reports of grizzly bear sightings increased (Layser et al. 1979;

Jonkel 1981)—despite the lack of a resident population (Melquist 1985; Groves 1987). No verified tracks or sightings have been documented in more than 50 years.

Large carnivores—grizzly bears in particular—require special consideration because, like humans, they occupy the top of the biotic pyramid. Providing for the largest carnivore is one way to help ensure sufficient habitat for other, less space-dependent species. In this sense, the grizzly bear is an umbrella species with expansive spatial needs, generalized food habits, and broad habitat relations. The present range of the grizzly also encompasses the majority of the remaining range of the endangered Rocky Mountain wolf (*Canis lupus*), wolverine (*Gulo gulo*), lynx (*Lynx lynx*), fisher (*Martes pennanti*), and more common species such as elk (*Cervus elaphus*), mule deer (*Odocoileus hemionus*), and white-tailed deer (*Odocoileus virginianus*).

The Bitterroot Ecosystem of central Idaho and western Montana (Figure 10.1) is one of the largest contiguous blocks of federal land in the United States. The core of the ecosystem contains three wilderness areas that total 15,790 km^2 and form the largest block of wilderness in the Rocky Mountains south of Canada. Wilderness areas include the Selway-Bitterroot (5424 km^2), Frank Church–River of No Return (9553 km^2), and Gospel Hump (810 km^2). This core is surrounded by thousands of square kilometers of public land including all or significant portions of nine national forests (USFWS 2000a). This vast complex of mountains, foothills, canyons, and river valleys

Figure 10.1. The six grizzly bear ecosystems including the centrally located Bitterroot reintroduction area.

is centered on the Idaho Batholith, a large expanse of granite in central Idaho (Alt and Hyndman 1995). The area contains three major mountain ranges: the Salmon River Mountains; the Clearwater Mountains, which extend from the Salmon River north to the upper Clearwater River drainage; and the Bitterroot Mountains, which form the eastern border along the Montana–Idaho state line. The Selway and Salmon rivers originate in the Bitterroot ecosystem. Tributaries of the Snake, Clearwater, Bitterroot, and Clark Fork rivers drain the southern, western, eastern, and northern fringes of the area, respectively.

NEPA Planning and the EIS Process

Planning for the restoration of grizzly bears into the Bitterroot Ecosystem was initiated in July 1994 when the Interagency Grizzly Bear Committee authorized development of an EIS to identify issues and alternatives for recovery, recovery area boundaries, and environmental consequences of implementing the Bitterroot Ecosystem Recovery Plan chapter.

Scoping and Public Participation

The NEPA process officially began with publication of a notice of intent to prepare an EIS in the *Federal Register* on 9 January 1995. The notice formally began the process, as required under NEPA, to obtain input from other agencies and the public on the scope of issues to be addressed in the EIS. The 45-day public comment period resulted in 73 comments, all of which were reviewed by an impartial panel (USFWS 1995a). Most respondents favored grizzly bear reintroduction and discussed issues and alternatives similar to those cited in public comments received on the Bitterroot Ecosystem chapter of the Grizzly Bear Recovery Plan.

The next step was a public participation and interagency coordination program to list the issues related to grizzly bear recovery in the Bitterroot Ecosystem and alternatives to be considered in the EIS. A list of preliminary issues was identified in March 1995 from 800 public comments received on the Bitterroot Ecosystem chapter of the Grizzly Bear Recovery Plan and the 73 public comments on the notice of intent:

- Recovery options and legal classification of the grizzly bear
- Possible restrictions on human uses of public land
- Location and cost of a reintroduction
- Participatory role of citizens in grizzly bear recovery

- Geographic boundaries for recovery
- Illegal killing of grizzly bears
- Concern for human safety
- Control of nuisance grizzly bears

The USFWS used these preliminary issues to formulate three preliminary scoping alternatives:

- Alternative 1: no action (natural recolonization)
- Alternative 2: reintroduction of an experimental population (proposed action)
- Alternative 3: accelerated reintroduction of a standard (fully protected) population

These issues and alternatives were presented in the "Scoping of Issues and Alternatives" brochure (USFWS 1995b), which also provided background information and requested additional ideas and comments. It was mailed to 1100 people and distributed at meetings and open houses during the public scoping process.

Formal public scoping for issues and alternatives began on 5 June 1995 with a 45-day comment period (extended another 30 days due to public requests) and closed on 21 August 1995. Seven public scoping sessions in the form of open houses were held in Grangeville, Orofino, and Boise in Idaho; in Missoula, Helena, and Hamilton in Montana; and in Salt Lake City, Utah. More than 300 people attended these open houses. Written comments on preliminary issues and alternatives were received from more than 3300 individuals, organizations, and government agencies. About 80 percent of written responses were from residents of counties adjacent to the proposed recovery area.

As well, a survey was conducted for the Interagency Grizzly Bear Committee to assess public attitudes on grizzly bear recovery. The survey was performed as part of the initial public involvement process to prepare the EIS. Between 9 June and 24 June, 919 telephone surveys were completed: 311 locally, 306 regionally, and 302 nationally. The public survey indicated that 62 percent of local, 74 percent of regional, and 77 percent of national respondents supported reintroduction of grizzly bears into the Bitterroot Ecosystem (Duda and Young 1995).

A follow-up survey was administered in 1997 to residents of Idaho and three counties in Montana that border the proposed reintroduction site (Selway-Bitterroot Wilderness Area). Although the two surveys were not directly comparable because of slightly different sampling methods, awareness among residents in the region remained about the same and overall support for grizzly bear reintroduction declined from the 1995 level of 62 percent to 46 percent (Duda et al. 1998). Respondents were further asked if they would support reintroduction if four conditions were met: a Citizen Management Committee would have management authority for the reintroduced population; bears moving into populated areas would be relocated; current wildlife protections regarding timber harvest, grazing and recreation on national forest land would be declared adequate for grizzly bear recovery unless the Citizen Management Committee determined otherwise; and costs would be limited to $250,000 a year. Given this scenario, support for reintroduction increased to the same level measured in the 1995 survey (62 percent) and opposition dropped from an initial 45 percent to 30 percent, which is slightly more than in 1995 (Duda et al. 1998). The two most common reasons for support were to save the grizzly bear from extinction and to restore it as a component of the ecosystem. The most popular reason for opposition was concern that grizzly bears are dangerous to humans.

Developing the Draft EIS

The interagency team began the draft EIS in February 1996. A content analysis of the 3300 public scoping comments grouped significant issues into 46 headings (USFWS 1995c). The team reviewed the issues and used them to frame the alternatives and effects analysis for the draft EIS (Table 10.1).

Two new alternatives were suggested during the public scoping period. "The Citizen Management Committee Alternative" was submitted by a coalition of conservation and timber interest groups including the National Wildlife Federation, Defenders of Wildlife, the Resource Organization on Timber Supply, and the Intermountain Forest Industry Association (USFWS 1995c). "The Conservation Biology Alternative" was submitted by an environmental group, the Alliance for the Wild Rockies (USFWS 1995c). Both alternatives were included in the subsequent draft EIS. Because the Citizen Management Committee Alternative was similar to the proposed action, it became the preferred alternative. The draft EIS evaluated four alternatives:

- Alternative 1: restoration of grizzly bears as a nonessential experimental population with citizen management (the preferred alternative). Goal: To accomplish grizzly bear recovery by reintroducing grizzly bears designated as a nonessential experimental population to central Idaho—by implementing provisions in Section 10(j) of the ESA—and conduct grizzly bear management to address local concerns. A Citizen Management Commit-

Table 10.1. Issues Identified Through Public Scoping

Twenty-six issues and impacts were addressed as part of one or more alternatives:
- Management strategies
- Strategies to control nuisance bears
- Illegal killing of grizzly bears
- Recovery time
- Monitoring and evaluation
- Habitat security
- Education
- Political influence
- Private property rights
- Cost of program to taxpayer
- Endangered Species Act (ESA)
- Ecosystem management
- Definition of population viability for grizzly bears
- Effects on grizzly bears from human incursions outside wilderness
- Effects on grizzly bears (genetics, disease, colonization, etc.)
- Grizzly bears as a missing component of ecosystem
- Laws, restrictions, rights, authority
- Federal, state, local, and tribal authority
- Compliance with forest plans
- Are grizzly bears native to the Bitterroots?
- Habitat protection requirements
- Travel corridors and linkages (range of grizzly)
- Nonessential experimental population and area
- Recovery area (boundaries, size, & range)
- Enjoyment of grizzly bears (viewing, etc.)
- Population corridor linkages

Eleven issues (consolidated into seven areas) were analyzed in detail in the EIS because they might be affected by grizzly bear recovery:
- Effects on human health and safety
- Effects on source populations of grizzly bears
- Effects on land use activities (timber harvest, mining, livestock grazing)
- Effects on wildlife
- Effects on public access and recreational use
- Social effects
- Economic effects

Nine issues were not evaluated further in the EIS because they were not significant to the decision being made:[a]
- Consultation with Fish & Wildlife Service
- Interagency Grizzly Bear Committee guidelines
- State or private bear management specialist
- Effects of grizzly bear on other endangered species
- Effects of grizzly bear on other animals, fish, birds
- Miscellaneous
- Visitor use
- Spiritual/cultural issues
- Wilderness Act

[a] Although these issues were not used to formulate alternatives or analyze effects, most are addressed in the EIS.

tee (CMC) composed of local citizens and agency representatives, created under a special rule, would coordinate management of the population. Management would be implemented by federal and state land and wildlife management agencies in Idaho and Montana (Figure 10.2).

- Alternative 2: no action (natural recovery). Goal: To allow grizzly bears to colonize the Bitterroot Ecosystem from their current range in north Idaho and northwestern Montana. Natural recovery of grizzly bears in the Bitterroot Ecosystem would be the ultimate goal.

- Alternative 3: no grizzly bears. Goal: To prevent grizzly bear recovery in the Bitterroot Ecosystem through unregulated killing of any bears that naturally reestablish there.

- Alternative 4: restoration of grizzly bears as a threatened population with habitat restoration and full protection of the ESA. Goal: To achieve recovery through reintroduction of grizzly bears and extensive habitat protection and enhancement to promote natural recovery. A ten-member Scientific Committee would be established to define needs for additional research, develop strategies for reintroduction of bears, and monitor results of the program. The grizzly bear would have full status as a threatened species under the provisions of the ESA.

Public Review of the Draft EIS

The draft EIS was released for public review and comment on 1 July 1997 (USFWS 1997). A notice of availability was published in the *Federal Register,* and the document was widely distributed and published on the internet. Public comments were received through 1 December after a 60-day extension. Comments were also gathered at public hearings and open houses in the towns of Challis, Lewiston, Boise, and Salmon, Idaho, and in Hamilton, Missoula, and Helena, Montana. Approximately 1400 people attended and 293 offered comments. In addition, the USFWS held meetings with local community leaders, state officials, and interest groups located in and near the Bitterroot Ecosystem. Comments on the draft EIS were received from more than 24,000 individuals, organizations, and government agencies.

Public comments were used by the interagency EIS team to identify substantive issues and shape responsive alternatives. They were not used as votes for any particular alternative. Summary statistics were kept during the content analysis process, however, because they can be useful as a barometer of public opinion (USFWS 1998). The majority of comments (75 percent) supported restoration of the grizzly to the Bitterroot Ecosystem. Of the 293 public hearing statements from local communities in Idaho and Montana, 60 percent supported some form of grizzly bear recovery. Statements from hearings held in Montana were 75 percent supportive, whereas 51 percent of the Idaho statements were in opposition.

Developing the Final EIS

Issues raised in public comments on the draft EIS echoed those raised during the initial scoping. Fifty-three major issues were identified, and the USFWS responded to each with an explanation of how issues were addressed in the NEPA process. The final EIS incorporated public comments that resulted in numerous changes to the proposed action (Alternative 1) and the addition of two new alternatives. All six alternatives were developed in response to approximately 28,500 public comments gathered during four comment periods throughout the NEPA process. The two new alternatives in the final EIS were:

- Alternative 1A: restoration of grizzly bears as a nonessential experimental population. Goal: To accomplish grizzly bear recovery by reintroducing grizzly bears designated as a nonessential experimental population—by implementing provisions in Section 10(j) of the ESA—and conduct grizzly bear management to address local concerns. The USFWS would manage the grizzly bear population in the Bitterroot Grizzly Bear Experimental Population Area, which encompasses approximately 65,080 km² (Figure 10.2) including most of central Idaho and part of western Montana.

- Alternative 4A: restoration of grizzly bears as a threatened population with full protection of the ESA. Goal: To achieve recovery through reintroduction of grizzly bears with the USFWS leading and managing recovery. Other federal, state, and tribal agencies would assist with management. The grizzly bear would have full status as a threatened species under the ESA.

The final EIS (USFWS 2000a) was released for a 30-day public review period from 24 March to 24 April 2000. During this time the USFWS received another 14,800 comments. More than 98 percent of the comments supported restoration of the grizzly bear to the Bitterroot Ecosystem.

Development of the Record of Decision

The record of decision to select Alternative 1 was signed on 13 November 2000 by the USFWS regional director of the Mountain-Prairie Region (USFWS 2000b). This decision represented a balance between the biological needs of recovering grizzly bears and public concerns over safety and property. Although local communities may generate the most fear and opposition,

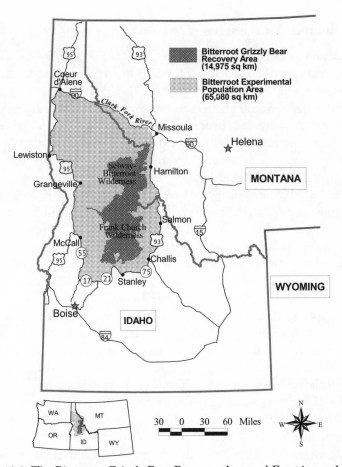

Figure 10.2. The Bitterroot Grizzly Bear Recovery Area and Experimental Population Area.

there is no doubt that local support (or at least tolerance) is crucial to the success of a recovery program (MacCracken et al. 1994; Reading et al. 1991). "Nonessential experimental population" designation under section 10(j) of the ESA allows flexible and responsive management to minimize harm caused by individual grizzly bears to private property, big game populations, other listed or sensitive species, and other natural resource programs on private and public land. Fischer and Roy (1998) say that gaining support depends on giving local citizens an active role in grizzly bear management. Citizen management of the Bitterroot grizzly bear is the centerpiece of the selected alternative.

The Selected Alternative

"Experimental population" designation gives the USFWS flexibility to develop a management approach tailored to specific areas and local conditions. Because reintroduced bears will be classified as experimental, the USFWS can ease concerns about excessive government regulation on private lands, uncontrolled livestock depredation, excessive big game predation, and lack of state government and local citizen involvement in the program. Experimental populations have been successful with other large carnivores, including the gray wolf (*Canis lupus*) in Yellowstone National Park and central Idaho (see chapter 6 in this volume). While gray wolf recovery in the Rocky Mountains remains controversial—especially locally—it has increased tolerance and acceptance levels in all but the most vehement opponents. The USFWS considers the Bitterroot grizzly bear population a "nonessential" experimental population because several additional populations exist within the 48 conterminous United States. Hence, its loss would not be likely to reduce the short-term survival of the species in the wild. The Bitterroot Grizzly Bear Recovery Area includes the Selway-Bitterroot Wilderness and the Frank Church–River of No Return Wilderness—an area of nearly 15,000 km² (23 percent) within the experimental population area (Figure 10.2). Within this recovery area management decisions will facilitate bear recovery and its establishment as core habitat for survival, reproduction, and dispersal.

The CMC's management authority will derive from the interior secretary in consultation with the governors of Idaho and Montana. Fifteen members will be appointed by the secretary based on recommendations of the governors of Montana (five members) and Idaho (seven members) and the Nez Perce Tribe (one member). Montana and Idaho representatives will include one member each from their wildlife agency. Apart from these 13, there will be one member each from the USFWS and the USDA Forest Service. Except for the federal agency representatives, CMC members will be selected from communities within and adjacent to the experimental population area and will consist of a diverse cross section of interests. Two scientific advisors will also be appointed by the interior secretary as nonvoting members. The CMC will base its decisions on the best scientific and commercial data available. These decisions must lead toward recovery of the grizzly bear, minimize social and economic impacts to the extent practicable within the context of the recovery goals for the species, and must not preclude resource extraction.

The CMC will be responsible for recommending changes in land-use standards that enhance grizzly bear management. Recommendations made by the CMC to land and wildlife management agencies will be subject to review, and final decisions on implementation will be made by the responsible agency. The CMC's task is to implement the Bitterroot chapter of the Grizzly Bear Recovery Plan as authorized by the interior secretary in the Final Rule on Establishment of a Nonessential Experimental Population of Grizzly Bears in the Bitterroot Area of Idaho and Montana (USFWS 2000c) and with input from the public.

Setting a Precedent

The CMC concept represents the first time a group of local citizens and agency representatives has been given management responsibility and authority to implement management decisions for recovery of a threatened or endangered species in cooperation with agency representatives. Other attempts to involve stakeholders in endangered species management have always established advisory committees without decision-making authority. Such committees have attained limited success in accomplishing their management recommendations—especially in contentious threatened and endangered species recovery programs. For example, management recommendations for panther recovery provided by the governor-appointed Florida Panther Technical Advisory Council were all but ignored by the Florida Fish and Wildlife Conservation Commission for more than 15 years (Alvarez 1993).

The Idaho Department of Fish and Game and/or the Nez Perce Tribe, the Montana Department of Fish, Wildlife, and Parks, and the USDA Forest Service, in coordination with the USFWS, will exercise day-to-day management responsibility within the experimental population area while implementing the Bitterroot Ecosystem Grizzly Bear Recovery Plan chapter, the special rule, and the policies and plans of the CMC. Day-to-day management involves handling of nuisance bears, answering questions from the public, managing artificial food sources, and similar activities. Bears that move outside the recovery area will not be disturbed unless they demonstrate a serious threat to human safety or livestock.

Subject to funding, grizzly bears will be reintroduced into the Selway-Bitterroot Wilderness during the second year of implementation at the earliest. This release will be no sooner than one year after formation of the CMC and initiation of programs to inform the public and eliminate such attractants as campground garbage. Specific reintroduction sites will be recommended by management agencies to the CMC. The USFWS, in coordination with agency partners and the CMC, will release a minimum of 25 grizzly bears into the recovery area over a period of five years. These bears will come from areas more than 16 km beyond the existing recovery zones in the Yellowstone and the Northern Continental Divide Ecosystems, British Columbia, and Alaska (non-salmon-eating bears). Bears will be translocated only if their absence will have no significant impact on the local population. The tentative recovery goal is approximately 280 grizzly bears (USFWS 1996)—a target that could take 50 to 110 years to attain (USFWS 2000a).

People can continue to kill grizzly bears in self-defense or in defense of others—provided that such taking is reported within 24 hours. Grizzly bears will be managed according to current guidelines (IGBC 1986) except in the case of grizzly bears killing livestock on private land that cannot be captured by management authorities. In such cases, landowners will be issued a permit by the USFWS to harass a grizzly through noninjurious means in order to protect livestock or apiaries. A livestock owner may be issued a permit to kill a grizzly bear if agency efforts fail to capture such a bear or deter its depredations. If significant conflicts occur between experimental grizzly bears and livestock outside the recovery area, these can be resolved by agencies capturing or eliminating the bear. Although there will be no federal compensation program for livestock losses, compensation from private funding sources will be encouraged.

The potential effects associated with a recovered grizzly bear population include the annual losses of 6 cattle (4–8), 25 sheep (5–44), and predation of approximately 504 ungulates per year (USFWS 2000a; Gunther et al. 1995–1998; Madel 1996; Mattson 1997). Nuisance bear incidents could average 37 (0–74) per year based on experience in the Yellowstone (Gunther et al. 1995–1998) and the Northern Continental Divide Ecosystems (Dood and Pac 1993). There are no anticipated impacts on timber harvest, mining, access, and recreational use on either private or public land as a result of the reintroduction. Hunting seasons could be changed, however, if excessive grizzly bear mortality occurs during black bear seasons—based on experience from northern Idaho (Wielgus et al. 1994; Wakkinen 1993; Knick and Kasworm 1989; MacCracken et al. 1994) and as planned for northwestern Montana (Dood et al. 1986). Risk to human health and safety from a recovered grizzly bear population would be less than one injury per year and less than one human fatality every few decades. This risk would be significantly less during the first few decades of reintroduction when grizzly numbers are low (USFWS 2000a).

The issue of human safety was prominent in public comments. Opposition was largely based on fear of injury and death—not unusual for grizzly bear recovery programs. Public comments indicated this "gut-level" fear was exaggerated by a lack of knowledge of bear behavior and how to deal with a grizzly encounter. Polarized comments ranged from those who opposed restoration due to the possibility of injury or loss of life (and their belief that the cost of losing one human life is too high) to those who supported restoration on aesthetic and ecological grounds and were willing to change their behavior and accept some additional risk. The EIS team responded to the array of comments on this issue by adding numerous measures to mitigate the risk of encounters and injury (USFWS 2000a, 2000b).

Implementation Schedule

Following completion of the NEPA process (Figure 10.3), the selected alternative will be implemented as an overlapping staged process beginning with formation of the CMC and installation of bear-proof garbage containers in campgrounds and other facilities. The program to reduce attractants such as garbage will include efforts by the USDA Forest Service, outfitters, and pri-

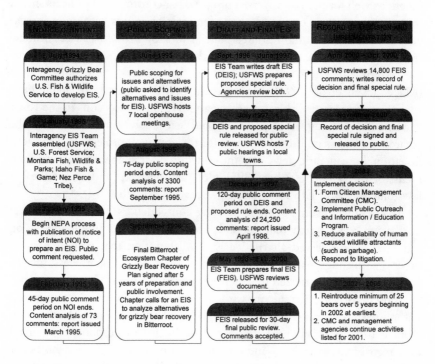

Figure 10.3. Time line and major events in the NEPA planning process for grizzly bear restoration in the Bitterroot ecosystem.

vate landowners in and around the recovery area. The second stage will include several information campaigns: a program targeting recreational users of the recovery area; presentations at public schools in and around the recovery area to teach children about grizzly bears and how to behave in grizzly bear country; presentations to all civic clubs and other interested organizations about grizzly bears and how to behave in grizzly bear country; and informative signs at all trailheads in and around the recovery area. The third stage will be the capture and release of bears in the recovery area. The USFWS published a special rule to establish a nonessential experimental population of grizzly bears on 17 November 2000 (USFWS 2000c).

Results of the NEPA Process

One of the main tenets of the NEPA process is to encourage public involvement in shaping natural resource management decisions. The NEPA process functions well for noncontroversial projects. For controversial projects such as those involving large carnivores, however, interest groups with opposing values vie to change natural resource policies to support their cause. Although federal agencies are required to listen to these polarized and competing views, the resulting decision may be biased by the agency's professional culture (MacCracken et al. 1994). Management of the grizzly bear, an umbrella species, is connected with most other natural resource decisions in the Rocky Mountains—making it a powerful and extremely controversial symbol regardless of one's politics. The result for the gray wolf project was protracted litigation, bureaucratic delay of reintroduction efforts, and alienation of segments of the public (Fischer 1995). Congress delayed funding for the Yellowstone gray wolf EIS for nearly eight years (USFWS 1994).

To circumvent some of these negative consequences, the USFWS developed a recovery strategy that responded to both public comment and biological need. The selected alternative is a collaboration that will support the process with congressional funding through the CMC. Nonetheless, the plan has been soundly criticized by ardent opponents of reintroduction—typified by conservative local residents with utilitarian and dominionistic views of natural resources and wildlife (Kellert 1994) who believe the federal government is forcing its will on them—and also by ardent proponents of grizzly bear recovery such as local and national environmental groups who believe the plan fails to afford adequate protection to the reintroduced population. This case study is a test of the NEPA process in building an implementable decision for a controversial large-mammal restoration project in an atmosphere of polarized public opinion. An EIS process that was expected to take two or three years (Fischer and Roy 1998) was plagued by politically driven delays that ultimately stretched it to almost six years (Figure 10.3).

The future challenge for Bitterroot grizzly restoration lies in the ability of the CMC and the USFWS to build bridges of trust and communication with partner agencies, nongovernment groups, and the public. Although the eight-year planning and NEPA process has established a foundation for success, grizzly recovery in the Bitterroot Ecosystem will depend on the development of common goals and a sense of ownership among all involved. Surely the reestablishment of a viable, self-sustaining grizzly bear population in the Bitterroot Ecosystem would be a tremendous American wildlife success story. Recovery of the grizzly would be a validation of the Endangered Species Act—a legislative expression of the national will to restore threatened and endangered species. And while it is premature to predict an outcome, successful grizzly restoration in the Bitterroot Ecosystem will be as much a product of the species' resilience as it will be the result of a painstaking process that was sensitive both to biological needs and to public opinion.

Literature Cited

Alt, D., and D. Hyndman. 1995. *Northwest Exposures: A Geologic Story of the Northwest.* Missoula: Mountain Press.

Alvarez, K. 1993. *Twilight of the Panther.* Sarasota: Myakka River.

Burroughs, R. D. 1961. *The Natural History of the Lewis and Clark Expedition.* East Lansing: Michigan State University Press.

Butterfield, B. R., and J. Almack. 1985. Evaluation of grizzly bear habitat in the Selway-Bitterroot Wilderness Area. Final report. Idaho Department of Fish and Game Project 04-78-719. Moscow: Cooperative Fish and Wildlife Research Unit, University of Idaho.

Craighead, J. J., J. S. Sumner, and G. B. Scaggs. 1982. A definite system for analysis of grizzly bear habitat and other wilderness resources. Wildlife-Wildlands Institute Monograph 1. Missoula: University of Montana Foundation.

Davis, D., and B. Butterfield. 1991. The Bitterroot grizzly bear evaluation area: A report to the Bitterroot Technical Review Team. Unpublished report. Denver: Interagency Grizzly Bear Committee.

Dood, A., R. D. Brannon, and R. D. Mace. 1986. Final programmatic environmental impact statement: The grizzly bear in northwestern Montana. Helena: Montana Department of Fish, Wildlife, and Parks.

Dood, A. R., and H. I. Pac. 1993. The grizzly bear in northwestern Montana, 1986–1990. Helena: Montana Department of Fish, Wildlife, and Parks.

Duda, M. D., and K. C. Young. 1995. *The Public and Grizzly Bear Reintroduction in the Bitterroot Mountains of Central Idaho.* Harrisonburg, Va.: Responsive Management.

Duda, M. D., S. J. Bissell, and K. C. Young. 1998. *Wildlife and the American Mind: Public Opinion on and Attitudes Toward Fish and Wildlife Management.* Harrisonburg, Va.: Responsive Management.

Fischer, H. 1995. *Wolf Wars.* Helena: Falcon Press.

Fischer, H., and M. Roy. 1998. New approaches to citizen participation in endangered species management: Recovery in the Bitterroot Ecosystem. *Ursus* 10:603–606.

Groves, C. 1987. A compilation of grizzly bear reports for central and northern Idaho. Endangered Species Projects E-III, E-IV. Boise: Idaho Department of Fish and Game.

Gunther, K. A., M. Bruscino, S. Cain, T. Chu, K. Frey, and R. R. Knight. 1995–1998. Grizzly bear–human conflicts, confrontations, and management actions in the Yellowstone Ecosystem, 1994–1997. Interagency Grizzly Bear Committee, Yellowstone Ecosystem Subcommittee report. Yellowstone National Park.

Interagency Grizzly Bear Committee (IGBC). 1986. *Interagency Grizzly Bear Guidelines.* Washington, D.C.: U.S. Forest Service.

Jonkel, C. 1981. The Selway-Bitterroot Wilderness and the Bitterroot Range grizzly bears. Border Grizzly Project Special Report 52. Missoula: University of Montana.

Kellert, S. R. 1994. Public attitudes towards bears and their conservation. *International Conference on Bear Research and Management* 9:43–50.

Knick, S. T., and W. Kasworm. 1989. Shooting mortality in small populations of grizzly bears. *Wildlife Society Bulletin* 17:11–15.

Kunkel, K., W. Clark, and G. Servheen. 1991. A remote camera survey for grizzly bears in low human use areas of the Bitterroot grizzly bear evaluation area. Unpublished report. Boise: Idaho Department of Fish and Game.

Layser, E. F., J. L. Weaver, and D. Carrier. 1979. What's to be done with the grizzly? *Idaho Wildlife* 1(11):3–9.

MacCracken, J. G., D. Goble, and J. O'Laughlin. 1994. Grizzly bear recovery in Idaho. Report 12. Moscow: Idaho Forest, Wildlife, and Range Policy Analysis Group, University of Idaho.

Madel, M. 1996. Rocky Mountain front grizzly bear management program, 4-year progress report. Helena: Montana Department of Fish, Wildlife and Parks.

Mattson, D. J. 1997. Use of ungulates by Yellowstone grizzly bears *Ursus arctos. Biological Conservation* 81:161–177.

Melquist, W. 1985. A preliminary survey to determine the status of grizzly bears in the Clearwater National Forest of Idaho. Moscow: Idaho Cooperative Wildlife Research Unit, University of Idaho.

Merriam, C. H. 1922. Distribution of grizzly bear. *U.S. Outdoor Life* (Dec.):405–406.

Moore, W. R. 1984. Last of the Bitterroot grizzly. *Montana Magazine* (Nov.–Dec.): 8–12.

———. 1996. *The Lochsa Story.* Missoula: Mountain Publishing.

Reading, R. P., T. W. Clark, and S. R. Kellert. 1991. Toward an endangered species reintroduction paradigm, *Endangered Species Update* 8(11):1–4.

Scaggs, G. B. 1979. Vegetation description of potential grizzly bear habitat in the Selway-Bitterroot Wilderness Area, Montana and Idaho. M.S. thesis, University of Montana, Missoula.

Servheen, G., M. S. Nadeau, and C. Queen. 1990. A survey for grizzly bears in the Bitterroot Grizzly Bear Evaluation Area. Unpublished report. Boise: Idaho Department of Fish and Game

Servheen, C., A. Hamilton, R. Knight, and B. McLellan. 1991. Report of the technical review team: Evaluation of the Bitterroot and North Cascades to sustain viable grizzly bear populations. Report to the Interagency Grizzly Bear Committee. Boise: U.S. Fish and Wildlife Service.

Thwaites, G. R. 1959. *Original Journals of the Lewis and Clark Expedition, 1804–1806.* New York: Antiquarian Press.

U.S. Department of Interior, Fish and Wildlife Service (USFWS). 1982. Grizzly bear recovery plan. Denver: U.S. Fish and Wildlife Service.

———. 1993. Grizzly bear recovery plan (revised). Missoula: U.S. Fish and Wildlife Service.

———. 1994. Final environmental impact statement on the reintroduction of gray wolves to Yellowstone National Park and central Idaho. Helena: U.S. Fish and Wildlife Service.

———. 1995a. Summary of public comments on the notice of intent to prepare an environmental impact statement for the reintroduction of grizzly bears recovery to the Bitterroot Ecosystem in east central Idaho and western Montana. Missoula: U.S. Fish and Wildlife Service.

———. 1995b. Scoping of issues and alternatives brochure for grizzly bear recovery in the Bitterroot Ecosystem. Missoula: U.S. Fish and Wildlife Service.

———. 1995c. Summary of public comments on the scoping of issues and alternatives for grizzly bear recovery in the Bitterroot Ecosystem. Missoula: U.S. Fish and Wildlife Service.

———. 1996. Bitterroot ecosystem recovery plan chapter—supplement to the Grizzly Bear Recovery Plan. Missoula: U.S. Fish and Wildlife Service.

———. 1997. Draft environmental impact statement for grizzly bear recovery in the Bitterroot Ecosystem. Missoula: U.S. Fish and Wildlife Service.

———. 1998. Summary of public comments on the draft environmental impact statement for grizzly bear recovery in the Bitterroot Ecosystem. Missoula: U.S. Fish and Wildlife Service.

———. 2000a. Final environmental impact statement for grizzly bear recovery in the Bitterroot Ecosystem. Missoula: U.S. Fish and Wildlife Service.

———. 2000b. Record of decision and statement of findings on the final environmental impact statement for grizzly bear recovery in the Bitterroot Ecosystem. Missoula: U.S. Fish and Wildlife Service.

———. 2000c. Final rule on establishment of a nonessential experimental population of grizzly bears in the Bitterroot area of Idaho and Montana. Missoula: U.S. Fish and Wildlife Service.

Wakkinen, W. L. 1993. Selkirk Mountains grizzly bear ecology project. Threatened and Endangered Species Project E-3-8. Boise: Idaho Department of Fish and Game.

Wielgus, R. B., F. L. Bunnell, W. C. Wakkinen, and P. E. Zager. 1994. Population dynamics of Selkirk Mountain grizzly bears. *Journal of Wildlife Management* 58:266–272.

Wright, W. H. 1909. *The Grizzly Bear.* New York: Scribner.

Case 3. The Paradigm of Grizzly Bear Restoration in North America

CHARLES C. SCHWARTZ

> Alive, the grizzly is a symbol of freedom and understanding—a sign that man can learn to conserve what is left of the earth. Extinct, it will be another fading testimony to things man should have learned more about but was too preoccupied with himself to notice. In its beleaguered condition, it is above all a symbol of what man is doing to the entire planet. If we can learn from these experiences, and learn rationally, both grizzly and man may have a chance to survive. [Craighead 1979:230]

Grizzly bear restoration and recovery is a controversial, highly politicized process. By 1959, when the Craigheads began their pioneering work on Yellowstone grizzly bears, the species had been reduced to a remnant of its historic range. Prior to the colonization of North America by Europeans, the grizzly lived in relatively pristine habitats with aboriginal Native Americans. As civilization expanded, humans changed the face of the landscape, converting grizzly bear habitat to farms and ranches. People killed grizzlies to protect livestock and eliminate a perceived threat to human safety. In concert, habitat loss and direct human-caused mortality had effectively eliminated the grizzly from 95 percent of its historic range in the conterminous United States by the 1920s (Servheen 1989). Grizzly bear numbers had been reduced nearly 98 percent by 1975 when the species was listed as threatened under the Endangered Species Act (ESA) (USFWS 1993).

Today, grizzly bears exist in only five areas in remote and rugged mountainous terrain. These five areas, plus the Bitterroot reintroduction area, are the geographic foci of its recovery. Today, the grizzly bear recovery process is a paradigm bounded by two extremes: pristine habitat (that will never return so long as humans predominate) and extinction—and the slope to extinction increases as one nears a precipice. Because bears are long-lived, intelligent animals, there can be a 10- to 20-year lag between habitat degradation and population decline (Doak 1995). The area beyond the precipice represents a sudden decline in bear numbers associated with such a lag. There are two major forces influencing the bear's position on this slope. On the one hand, there are "user groups" whose demands for resources and other activities tend to push the bear closer to the precipice. Although this group does not want the bear to go extinct, it continues to push the bear to the point of having exactly the smallest number remaining in order to ensure survival. The cumulative impacts from decades of logging and road construction, oil and gas development, livestock grazing, hunting, off-road vehicle activity, railroads, outdoor recreation, exotic species, and land development have created the precipice. These impacts manifest themselves both as degradation and loss of bear habitat and as increased human-caused mortality. Although bears can learn to habituate to humans (Jope 1985; Aumiller and Matt 1994), humans will not tolerate bears in their backyards. Consequently, bears persist worldwide only where humans are sparse or absent (Mattson 1990). On the other hand are those—special interest groups, individuals, and professionals—who place a higher value on the bear and its long-term conservation and survival. It is their efforts and activities that have prevented the bear from slipping over the precipice into extinction.

Those responsible for conservation and recovery of the bear recognize

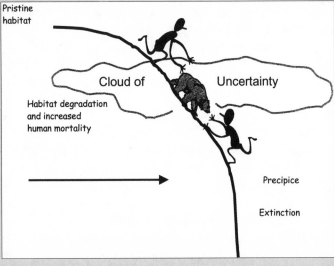

Until grizzly bear populations expand sufficiently in the lower 48 states, they will teeter on the brink of extinction under a cloud of uncertainty.

that there is a "cloud of uncertainty" over this precipice. Because of the many contributing factors, we don't know exactly when the bear will plummet toward extinction. Although scientific information is the foundation of successful recovery, grizzly bears are hard to study. Their low density, secretive nature, and heavily forested environments make research and monitoring difficult at best. Uncertainty is part of science. Adding to the dilemma are equivocal data interpretation and unclear guidelines within the Endangered Species Act (Mattson and Craighead 1994). Crafters of this document could not foresee what the restoration process of threatened and endangered species might entail. Although the ESA requires development of a recovery plan for each listed species, it does not provide clear guidance or define what constitutes a recovered population (Foin et al. 1998). Further, it does not address issues of time, uncertainty, and burden of proof (Mattson and Craighead 1994).

Modern grizzly bear conservation includes population management, habitat preservation, and habitat restoration. Population management focuses on increasing grizzly numbers and reducing human-caused mortality—the former is largely driven by the latter: from 77 to 85 percent of documented bear mortality is caused by people (McLellan et al. 1999). Habitat preservation and restoration have focused on defining recovery zones with a proposed no-net-loss policy in habitat quality, protection of secure habitats, and road removal on public lands (USFWS 1993, 1999).

An underlying management question is: How many bears are necessary for a viable population? Numerous estimates of minimum viable population size have been generated for the Yellowstone Ecosystem using models and population viability analysis (PVA) (Shaffer 1978, 1983; Shaffer and Samson 1985; Suchy et al. 1985; Boyce 1995) under the premise that population size is directly related to likelihood of extinction (Caughley 1994). But PVA is uncertain (Soulé 1987); population size alone is inadequate for evaluating long-term viability (Boyce 1995), and there is no link between current viability analyses and habitat quality, availability, and above all change (Boyce 1992, 1993). And, too, how many bears should live in the United States? Should we settle for the minimum as determined by PVA, or should there be more? Perhaps less?

How much habitat is enough for the grizzly? From a biologist's point of view, more is better. A larger area tends to reduce the impact of random demographic and environmental events. Defining habitat outside of delineated recovery zones, including linkages among them, is an important step to ensure the bear's long-term survival. This issue is particularly important when considering the human impacts on grizzly habitat. In the Yellowstone Ecosystem, for example, introduced exotics threaten whitebark pine, cutthroat trout, and ungulates—important foods of the bear. Attempting to model impacts of blister rust on whitebark pine, global climate change, or other changes to grizzly habitat, however, like PVA, is largely an exercise in educated guessing. Certainly the species withstood tremendous change at the end of the last ice age with colonization of North America by humans and the extinction of some 40 other species of large mammal. Unfortunately, humans cannot predict the future and we lack the details of past extinctions. We can only attempt to predict risk to the species. But by doing so, recovery goes back under the cloud of uncertainty.

For the past 25 years, management of the grizzly bear has been a focus of national attention, controversy, and conflict. The grizzly bear is deified as a symbol of the wilderness and victim of a changing world. Success or failure of grizzly bear restoration is viewed by many as the final testament to the Endangered Species Act. Those trying to push the grizzly bear up the slope to ensure survival represent a diverse group with often conflicting values and little coordination. Too often, those designated with responsibility for restoration have been unresponsive to outside views. This has resulted in management by litigation—a strategy that amplifies polarization and wastes resources. To do what is best for grizzly bear conservation, we must find common ground, cooperate, and communicate. If we don't, the grizzly will slide over the precipice.

Grizzly bear recovery ultimately centers on the social conflicts between groups pushing the bear up or down the slope. People tend to use uncertainty to their advantage when supporting their personal, professional, or political values. Which values predominate, will determine the final fate of the bear.

Literature Cited

Aumiller, L. D., and C. A. Matt. 1994. Management of McNeil River State Game Sanctuary for viewing of brown bears. *International Conference on Bear Research and Management* 9:51–61.

Boyce, M. S. 1992. Population viability analysis. *Annual Reviews of Ecology and Systematics* 23:481–506.

———. 1993. Population viability analysis: Adaptive management for threatened and endangered species. *Transactions of the North American Wildlife and Natural Resources Conference* 58:520–527.

———. 1995. Population viability analysis for grizzly bears (*Ursus arctos horribilis*): A critical review. Report to Interagency Grizzly Bear Committee. Missoula.

Caughley, G. 1994. Directions in conservation biology. *Journal of Animal Ecology* 63:215–244.

Craighead, F. C. 1979. *Track of the Grizzly*. San Francisco: Sierra Club Books.

Doak, D. F. 1995. Source-sink models and the problem of habitat degradation: General models and applications to the Yellowstone grizzly. *Conservation Biology* 9:1370–1379.

Foin, T. C., S. P. D. Riley, A. L. Pawley, D. R. Ayres, T. M. Carlsen, P. J. Hodum, and P. V. Switzer. 1998. Improving recovery planning for threatened and endangered species. *BioScience* 48:177–184.

Jope, K. L. M. 1985. Implications of grizzly bear habitat to hikers. *Wildlife Society Bulletin* 13:32–37.

Mattson, D. J. 1990. Human impacts on bear habitat use. *International Conference on Bear Research and Management* 8:33–56.

Mattson, D. J., and J. J. Craighead. 1994. The Yellowstone grizzly bear recovery program: Uncertain information, uncertain policy. In T.W. Clark, R. P. Reading and A. L. Clarke, eds., *Endangered Species Recovery: Finding the Lessons, Improving the Process.* Washington, D.C.: Island Press.

McLellan, B. N., F. W. Hovey, R. D. Mace, J. G. Woods, D. W. Carney, M. L. Gibeau, W. L. Wakkeinen, and W. F. Kasworm. 1999. Rates and causes of grizzly bear mortality in the interior mountains of British Columbia, Alberta, Montana, Washington, and Idaho. *Journal of Wildlife Management* 63:911–920.

Servheen, C. 1989. *The Status and Conservation of the Bears of the World.* Monograph 2. Victoria: International Conference on Bear Research and Management.

Shaffer, M. L. 1978. Determining minimum viable population size: A case study of the grizzly bear (*Ursus arctos* L.). Ph.D. dissertation, Duke University.

———. 1983. Determining minimum viable population size for the grizzly bear. *International Conference on Bear Research and Management* 5:133–139.

Shaffer, M. L. and F. B. Samson. 1985. Population size and extinction: a note on determining critical population sizes. American Naturalist 125:144–152.

Soulé, M. E., ed. 1987. *Viable Populations for Conservation.* Cambridge: Cambridge University Press.

Suchy, W., L. L. McDonald, M. D. Strickland, and S. H. Anderson. 1985. New estimates of minimum viable population size for grizzly bears of the Yellowstone ecosystem. *Wildlife Society Bulletin* 13:223–228.

U.S. Fish and Wildlife Service (USFWS). 1993. *Grizzly Bear Recovery Plan.* Missoula: U.S. Fish and Wildlife Service.

———. 1999. The draft habitat-based recovery criteria for grizzly bear in the Yellowstone Ecosystem. Draft habitat criteria. Missoula: U.S. Fish and Wildlife Service.

Chapter 11

Mountain Sheep Restoration Through Private/Public Partnership

PAUL R. KRAUSMAN, PETER BANGS, KYRAN KUNKEL,
MICHAEL K. PHILLIPS, ZACK PARSONS, AND ERIC ROMINGER

The mountain sheep (*Ovis canadensis*) is one of the most prized big game species in North America. It is difficult to examine these animals in their habitat without being amazed that they are able to survive in the barren areas that many of them use. But mountain sheep in North America have declined from more than 500,000 in pristine times (Seton 1929; Valdez 1988) to 185,000 in the 1990s (Valdez and Krausman 1999). Moreover, there are fewer than 20,000 desert bighorn sheep in the contiguous United States and several populations are state and federally listed as endangered (such as the peninsular mountain sheep, *O. c. cremnobates*).

Mountain sheep are charismatic. They have a high value to society as a wilderness species and they are a challenging animal to study. One way to determine the social value of a resource is to examine how much someone is willing to pay for an opportunity to use it. Since the 1980s, for example, the Foundation for North American Wild Sheep and state and provincial agencies have auctioned permits for the opportunity to hunt sheep in some western states, Canadian provinces, and in Mexico. The millions of dollars generated from the auction are used for sheep management and research (Table 11.1). Most of this money constitutes the basic budget for sheep management and is used to capture, translocate, and restore mountain sheep to historic range.

Table 11.1. The Maximum Dollar Amount
Generated from Auction of a Single Mountain
Sheep Permit

Area	Amount paid for 1 permit
Alberta, Canada	$405,000
Tiburon Island, Mexico	200,000
North Dakota	47,500
Wyoming	55,000
Colorado	56,000
Utah	72,000
Texas	77,000
Nevada	79,000
Washington	100,000
Idaho	101,000
California	110,000
Oregon	110,000
New Mexico	123,000
Arizona	303,000
Montana	310,000

Mountain sheep populations have declined for an array of human-related reasons (hunting, habitat alteration, competition with livestock) and have made the transition from a locally common species to one of the rarest ungulates in North America (Valdez and Krausman 1999). An entire subspecies, the Audubon or Badlands bighorn (*O. c . auduboni*), which inhabited areas along the Yellowstone and Missouri rivers in eastern Montana, eastern Wyoming, western North and South Dakota, and northwestern Nebraska, has been extirpated. Rocky Mountain (*O. c. canadensis*), California (*O. c. californiana*), and desert races of mountain sheep were also eliminated in parts of their range in the United States and Mexico. The major decline of mountain sheep populations occurred during the latter half of the nineteenth century due to disease transmission from livestock. Heavy grazing in northwestern Mexico and the southwestern United States also occurred in the early 1800s (Holechek et al. 1995).

One successful approach to the restoration of large mammals has been to translocate them into former habitats. In the early 1900s, large mammals were at an all-time low in North America. This situation led to concern by conservationists and politicians that led in turn to North America's early restoration efforts. The history of big game conservation has come in three stages. The first stage occurred when Europeans first arrived in North America: there was little concern for the abundant wildlife; exploitation was the norm. The second stage was realization that this wanton exploitation would cause the demise of large mammals; efforts were initiated to protect the remaining stock. In the third stage, widespread conservation efforts have led to the rise of wildlife restoration and scientific management of this public resource (Mackie 2000). This conservation philosophy includes translocation as a restoration tool.

In 1878, sportsmen translocated 18 white-tailed deer (*Odocoileus virginianus*) from New York to Vermont and ushered in a century of U.S. big game restoration (Mackie 2000). Several years later, in 1892, elk (*Cervus elaphus*) from Yellowstone National Park became the focus of trapping and translocating that helped reestablish elk populations over North America. The translocation of other large mammals followed, but the scientific basis for translocations was not developed until the advent of wildlife management in the 1920s and 1930s (Leopold 1933). Regardless of the rationale behind reestablishing large mammals, such translocations are time consuming, expensive, and logistically and politically challenging (Wolf et al. 1996; Dunham 1997; Fritts et al. 1997). Although guidelines for translocating animals are available (Rowland and Schmidt 1981; IUCN 1995; Wolfe et al. 1996), the successes and failures of many translocations are poorly documented (Short et al. 1992), translocation techniques are rarely tested (Morgart and Krausman 1981; Thompson et al. 2001), and many projects are based partly or even entirely on untested concepts (Hein 1997). At best, North American big game restoration has followed the rather nebulous model of "adaptive management."

Early translocations of mountain sheep suffered from methodological problems such as using padded steel leghold traps. With the advent of the net gun, drop net, safe anesthetizing drugs, and the use of helicopters, however, mountain sheep capture and transport have become more practical. From 1954 to 1978, some 153 mountain sheep in deserts were successfully trapped and translocated. Over the two decades since then, this number has increased by an order of magnitude to more than 2,000 translocations. Although mountain sheep translocations have become commonplace (Bailey 1990; Jessup et al. 1995), most restoration programs have not been successful (Risenhoover et al. 1988). Only 53 percent of 87 translocated populations in nine western states succeeded in the 1980s (Leslie 1980), for example, and only 41 of 100 translocations succeeded in six states between 1923 to 1997 (Singer et al. 2000). Some efforts have succeeded in returning mountain sheep to vacant ranges (Buechner 1960; Trefethen 1975), however, where at least 200 translocations have been made (Bailey and Klein 1997).

Clearly, great effort has gone into restoring mountain sheep populations in the western United States. Numerous state, federal, and private organizations (the Arizona Desert Bighorn Sheep Society, Bighorn Institute, Bighorns Unlimited, Boone and Crockett Club, national and state chapters of the Foundation for North American Wild Sheep, Fraternity of the Desert Bighorn, Grand Slam Club, Rocky Mountain Bighorn Sheep Society, Society for the Conservation of Bighorn Sheep, Texas Bighorn Society, Wild Sheep Society of British Columbia) have supplied financial contributions and labor. And much of this support has been matched with federal funds, especially where restoration has occurred on state or federal land with public access. Despite the deep involvement of the private sector in mountain sheep restoration, most translocations have occurred on public land. This is most likely a reflection of the U.S. policy that recognizes free-ranging wildlife as a public resource. Although private landowners in Sonora, Mexico, are actively involved in wild sheep conservation and translocation (Valdez 1997), placing public resources on private land is not the norm in the United States. In 1985, for example, there were 51 translocations of large indigenous mammals in the United States: 48 were on state and federal property whereas only 3 were on private lands (Nielsen and Brown 1988). We believe that mountain sheep restoration can be furthered through partnerships between the private and public sectors—a model approach that could expand restoration opportunities for mountain sheep. This chapter examines the largest private-land translocation project in New Mexico and evaluates the potential benefits of private involvement to mountain sheep restoration.

Study Area

The transplant occurred in the Fra Cristobal Mountains of Sierra and Socorro counties, New Mexico, 32 km northeast of Truth or Consequences. The mountains are part of the Armendaris Ranch owned by Ted Turner's New Mexico Ranch Properties, Inc. (NMRPI). The mountain range varies from 5 to 8 km wide, is 24 km long, and is 1493 to 2282 m in elevation. The ranch and surrounding area are part of the Upper Chihuahuan Life Zone. Precipitation averages 20.7 cm annually; 68 percent of the rain occurs from May through September (Brown 1982). Water sources include five reservoirs capable of storing 19,000 liters and several rock impressions capable of holding 200 liters. Vegetation consists of Chihuahuan Desert scrub, desert grassland, montane scrub, coniferous woodland, sand scrub, and riparian species (Miller 1999).

Like mountain sheep throughout the West, those in New Mexico were

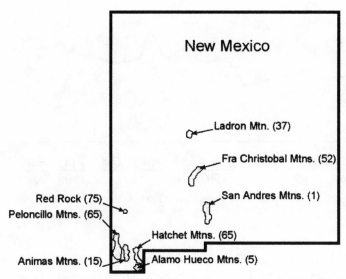

Figure 11.1. Location of desert races of mountain sheep in New Mexico and approximate population size in 2000.

once widespread and inhabited at least 14 mountain ranges. By 1955, however, only two populations existed in the San Andres and Big Hatchet mountains. The population in the San Andres Mountains was reduced to a single female by the year 2000 and the Big Hatchet Mountains now contain fewer than 60 animals. Six other translocated populations number fewer than 80 each (Figure 11.1). The desert population of mountain sheep was listed as endangered in New Mexico in 1980.

The San Andres Mountains supported the largest herd of wild sheep in the state (Hoban 1990) until they were reduced by disease and predation (Rominger and Weisenberger 1999). Other mountain ranges are relatively small, and mountain sheep population growth is static. A translocation was scheduled for the Caballo Mountains (adjacent to the Fra Cristobal Mountains in central New Mexico), but local opposition ended the plans for this mountain sheep restoration (Pederson 1996). The People for the Preservation of the Caballo Mountains (a citizens group in Truth or Consequences) opposed the transplant because they thought it would eliminate human access to the mountain (Pederson 1996). Because the Fra Cristobal Mountains provided potential habitat, were privately owned, and access was already limited, the public supported the translocation there. The decision was in favor of the sheep, not politics, and was supported by an array of concerned citizens—among them the Southwest Consolidated Sportsmen, Safari Club

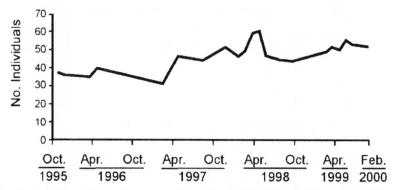

Figure 11.2. Population of mountain sheep in the Fra Cristobal Mountains, New Mexico (1995–2000).

International, Foundation for North American Wild Sheep, Animas Foundation, Southwest Center for Biodiversity, Southwest Environmental Center, and Preservation of Caballo Mountains, Inc. (Pederson 1996). Because the success or failure of restoration efforts is determined by the public (particularly those in decision-making and public-contact positions charged with recovery), the importance of people can become limiting factors in the survival of species (Maehr 1998). It was in the hope that such constraints would be removed that an effort was made to restore mountain sheep to the Fra Cristobal Mountains.

Thirty-seven sheep (24 females, 13 males) were translocated from Red Rock in southwestern New Mexico to the Fra Cristobal Mountains on 25 and 26 October 1995. Seven additional males were translocated from Red Rock in 1997. Red Rock is an enclosure established as a breeding facility where mountain sheep are raised and then translocated throughout vacant habitat in New Mexico. In February 2000 the Fra Cristobal Mountains contained 52 adult and yearling mountain sheep (22 adult females, 18 adult males, 6 yearling females, 6 yearling males) (Figure 11.2).

In New Mexico, income from hunting does not offset the costs of mountain sheep management. Thus private/public partnerships may be instrumental to future mountain sheep restoration work. Ted Turner created the Turner Endangered Species Fund (TESF) in June 1997 to conserve biodiversity by ensuring the persistence of imperiled species and their habitat. Because Turner's NMRPI has a strong commitment to reestablishing native, extirpated, and endangered species, a memorandum of understanding between NMRPI and the New Mexico Department of Game and Fish (NMDGF) was developed to further this work. The memorandum of understanding was designed to allow NMRPI to reestablish a viable self-sustaining mountain sheep population in the privately owned Fra Cristobal Mountains. Once the population exceeds 100 animals, the surplus can be used in other translocation efforts throughout the state and some males could be hunted. The future of mountain sheep in New Mexico depends on the preservation and reoccupation of the isolated ranges that contain suitable habitat.

The Memorandum of Understanding

In 1995, NMDGF and NMRPI approved a plan to translocate mountain sheep into the Fra Cristobal Mountains with the intention of establishing a viable population. Both parties agreed to specific conditions. The New Mexico Department of Game and Fish agreed to:

1. Test all sheep before release (if requested by NMRPI) for diseases and vaccinate each mountain sheep against diseases to the extent that vaccines are available.

2. Make an initial release of disease-free mountain sheep into the Fra Cristobal Mountains in 1995.

3. Coordinate with NMRPI in the management of translocated mountain sheep and their offspring and conduct a postrelease study including population surveys.

4. Coordinate with NMRPI to write a management plan for a minimum viable population that does not exceed carrying capacity.

5. Assume all financial costs associated with the capture, release, management, and monitoring during the restoration effort.

6. Set mountain sheep habitat utilization levels in conjunction with NMRPI.

7. Leave jurisdictional control of private land in the Fra Cristobal Mountains with NMRPI.

Although NMDGF agreed to assume financial responsibility for capture, release, management, and monitoring, many of these expenses have in fact been paid by TESF. In fact, TESF supported research, recaptured 16 female sheep, and hired a full-time biologist and several summer interns to monitor sheep in 1999. As well, TESF developed cooperative agreements with the University of Arizona and the University of California to study population dynamics and diseases of mountain sheep (Boyce et al. 1999). In 1999, a veterinarian was added to the TESF staff. A cooperative agreement was also developed with the Hornocker Wildlife Institute to assess mountain lion impacts on the translocated mountain sheep. Furthermore, TESF and NMDGF paid the Conservation Breeding Specialist Group of the IUCN to conduct a population and habitat viability assessment workshop to assist New Mexico in defining priorities for mountain sheep restoration (Fisher et al. 1999). In all, TESF spent $150,000 on sheep restoration efforts in New Mexico in 1999. Because of TESF and its research arrangements with universities and private organizations, NMDGF was able to direct precious funds and personnel to other pressing wildlife issues. The memorandum of understanding further stipulated that NMRPI agreed to:

1. Provide a site in the Fra Cristobal Mountains for mountain sheep translocation.

2. Permit construction and maintenance of water catchments as needed and determined by NMRPI.

3. Prohibit grazing of domestic sheep and goats in the Fra Cristobal Mountains.

4. Set mountain sheep habitat utilization levels in conjunction with NMDGF.

5. Acknowledge that the State Game Commission has jurisdictional control of the mountain sheep.

The memorandum of understanding also allowed mountain sheep to be translocated elsewhere if the population in the Fra Cristobal Mountains could sustain removals as determined by NMDGF. Moreover, NMRPI would afford NMDGF limited access to the Fra Cristobal Mountains. Access was provided for translocating sheep to or from the Fra Cristobal Mountains, handling dead or diseased mountain sheep, retrieving or replacing radio collars, and monitoring and surveying mountain sheep. NMRPI also provided reasonable access to wildlife interest groups for educational viewing of mountain sheep.

Results and Discussion

Although it is too early to determine the success of the transplant of mountain sheep into the Fra Cristobal Mountains, the cooperative venture between the state and NMRPI has certainly been viewed as a success by both parties. Since 1995 the population of mountain sheep in the Fra Cristobal Mountains is the only population in New Mexico that has grown. Further, it has moved New Mexico closer to removing mountain sheep from the state's endangered status list and could serve as a source for future translocations. If population growth continues, the population may also provide future hunting opportunities—a situation that may create additional revenue for wildlife restoration efforts. Biologists with NMDGF believe that the Fra Cristobal Mountains can support up to 100 sheep. Furthermore, the public supported placing their wildlife resources on private land because that land had uncontested habitat for mountain sheep.

For mountain sheep in New Mexico to be downlisted from endangered to threatened, there must be at least 500 free-roaming sheep in the state with more than 100 in two different areas. To be removed from either list, another metapopulation of 100 must be established (NMDGF 1995). The cooperation between NMDGF and NMRPI has the potential to make their model of mountain sheep restoration a new paradigm for wildlife conservation through private/public partnerships—exceeding the expectations of national strategy sessions conducted decades ago (Brenneman and Bates 1984; Montana Land Reliance and Land Trust Exchange 1982).

The advantages of the private/public partnerships as exemplified by New Mexico mountain sheep restoration differ from those offered by traditional state or federal wildlife management programs. Most notably, the public does not have the opportunity to freely observe animals or utilize them until a huntable population is established. Even then, the number of hunters and other viewers would be limited to only a few. But the benefits to the public are significant. The private landowner has the opportunity to strengthen conservation programs by offering habitat for wildlife that otherwise would not be available and can serve as the nucleus for population expansion by natural colonization and future translocation. Furthermore, the private sector can act as a catalyst to translocation programs when an agency's will is weak and its resources are limited. In the case history presented here, mountain sheep would not have been translocated without private assistance. Although organizations such as TESF can bring considerable financial and real estate resources to restoration programs, the state and federal agencies must be the horse that pulls the buggy because they maintain the long-term view and have the resources to keep programs going. Moreover, changes in landownership patterns may annul a cooperative agreement. The commitments of individuals may change, but state and federal wildlife agencies have a mandate from the public for long-term management.

When TESF completes its planned activities for mountain sheep in the Fra Cristobal Mountains in 2002, it will have accomplished a translocation that has been monitored constantly and is documented with a solid database from which informed decisions can be made. It will then be up to the state to continue the effort. Although private-sector cooperation has been called upon elsewhere to revive small populations such as the Florida panther (*Puma concolor coryi*) (Maehr 1998), success has seldom been achieved. The cooperative mix of public and private in mountain sheep restoration suggests that such efforts should be expanded both geographically and taxonomically.

Literature Cited

Bailey, J. A. 1990. Management of Rocky Mountain bighorn sheep herds in Colorado. Special Report 66. Fort Collins: Colorado Division of Wildlife.

Bailey, J. A., and D. R. Klein. 1997. United States of America. In D. M. Shackleton, ed., *Wild Sheep and Their Relatives, Status Survey and Conservation Action Plan for Caprinae.* Gland, Switzerland: IUCN.

Boyce, W., A. Fisher, H. Provencio, E. Rominger, J. Thilsted, and M. Ahlm. 1999. Elaeophoris in bighorn sheep in New Mexico. *Journal of Wildlife Diseases* 35:786–789.

Brenneman, R. L., and S. M. Bates, eds. 1984. *Land-Saving Action.* Washington, D.C.: Island Press.

Brown, D. C. 1982. Chihuahuan desert scrub. *Desert Plants* 4:169–179.

Buechner, H. K. 1960. *The Bighorn Sheep in the United States: Its Past, Present, and Future.* Wildlife Monograph 4. Bethesda: Wildlife Society.

Dunham, K. M. 1997. Population growth of mountain gazelles, *Gazella gazella*, reintroduced to central Arabia. *Biological Conservation* 81:205–214.

Fisher, A., E. Rominger, P. Miller, and O. Byers. 1999. Population and habitat viability assessment workshop for the desert bighorn sheep of New Mexico (*Ovis canadensis*): Final report. Apple Valley, N.M.: IUCN/SSC Conservation Breeding Specialist Group.

Fritts, S. H., E. E. Bangs, J. A. Fontaine, M. R. Johnson, M. K. Phillips, E. D. Koch, and J. R. Gunson. 1997. Planning and implementing a reintroduction of wolves to Yellowstone National Park and central Idaho. *Restoration Ecology* 5:7–27.

Hein, E. W. 1997. Improving translocation programs. *Conservation Biology* 11:1270–1274.

Hoban, P. A. 1990. A review of desert bighorn sheep in the San Andres Mountains, New Mexico. *Desert Bighorn Council Transactions* 34:14–22.

Holechek, J. L., R. D. Pieper, and C. H. Herbel. 1995. *Range Management.* 2nd ed. Englewood Cliffs, N.J.: Prentice-Hall.

International Union for the Conservation of Nature and Natural Resources (IUCN), Reintroduction Specialists Group. 1995. *Guidelines for Reintroduction.* Gland, Switzerland: IUCN.

Jessup, D. A., E. T. Thorne, M. W. Miller, and D. L. Hunter. 1995. Health implications in the translocation of wildlife. In J. A. Bissonette and P. R. Krausman, eds., *Integrating People and Wildlife for a Sustainable Future.* Bethesda: Wildlife Society.

Leopold, A. 1933. *Game Management.* New York: Scribner.

Leslie, D. R. 1980. Remnant populations of desert bighorn sheep as a source for transplantation. *Desert Bighorn Council Transactions* 24:36–44.

Mackie, R. J. 2000. History of management of large mammals in North America. In S. Demarais and P. R. Krausman, eds., *Ecology and Management of Large Mammals in North America.* Upper Saddle River, N.J.: Prentice-Hall.

Maehr, D. S. 1998. The Florida panther and the Endangered Species Act of 1973. *FAU/FIU Joint Center for Environmental and Urban Problems* (Fall):1–6.

Miller, M. E. 1999. Vegetation of the Fra Cristobal Range, southern New Mexico. Truth or Consequences, N.M.: Turner Endangered Species Fund, Armendaris Ranch.

Montana Land Reliance and Land Trust Exchange. 1982. *Private Options: Tools and Concepts for Land Conservation.* Washington, D.C.: Island Press.

Morgart, J. R., and P. R. Krausman. 1981. The status of transplanted bighorn population in Arizona using an enclosure. *Desert Bighorn Council Transactions* 25:46–49.

New Mexico Department of Game and Fish (NMDGF). 1995. New Mexico's long-range plan for desert bighorn sheep management 1995–2002. Federal Aid in Wildlife Restoration final report. Project W-127-R10, job 1. Santa Fe: New Mexico Department of Game and Fish.

Nielsen, L., and R. D. Brown, eds. 1988. Translocation of wild animals. Milwaukee: Wisconsin Humane Society; Kingsville, Texas: Caesar Kleberg Wildlife Research Institute.

Pederson, J. 1996. The new bighorn sheep of the Fra Cristobal Range. *New Mexico Wildlife* (Jan./Feb.): 14–18.

Risenhoover, K. L., J. A. Bailey, and L. A. Wakelyn. 1988. Assessing the Rocky Mountain bighorn sheep management problem. *Wildlife Society Bulletin* 16:346–352.

Rominger, E. M., and M. E. Weisenberger. 1999. Biological extinction and a test of the "conspicuous individual hypothesis" in the San Andres Mountains, New Mexico. Paper presented at the North American Wild Sheep Conference, Reno.

Rowland, M. M., and J. L. Schmidt. 1981. Transplanting desert bighorn sheep: A review. *Desert Bighorn Council Transactions* 25:25–28.

Seton, E. T. 1929. The bighorn. In E. T. Seton, ed., *Lives of the Game Animals.* Vol. 3. Pt. 2. Garden City: Doubleday.

Short, J., S. D. Bradshaw, J. Giles, R. I. T. Prince, and G. R. Wilson. 1992. Reintroduction of macropods (Marsupiallia:Macrophodoidea) in Australia—a review. *Biological Conservation* 62:189–204.

Singer, F. J., C. M. Papouchis, and K. K. Symonds. 2000. Translocations as a tool for restoring populations of bighorn sheep. *Restoration Ecology* 45:6–13.

Thompson, J. R. , V. C. Bleich, S. G. Torres, and G. P. Mulcahy. 2001. Translocation techniques for mountain sheep: Does the method matter? *Southwestern Naturalist* 46:87–93.

Trefethen, J. B., ed. 1975. *The Wild Sheep of Modern North America.* New York: Boone and Crockett Club and Winchester Press.

Valdez, R. 1988. *Wild Sheep and Wild Sheep Hunters of the New World.* Mesilla, N.M.: Wild Sheep and Goat International.

Valdez, R. 1997. Mexico. In D. M. Shackleton, ed., *Wild Sheep and Their relatives: Status, Survey, and Conservation Action Plan for Caprinae.* Gland, Switzerland: IUCN.

Valdez, R., and P. R. Krausman, 1999. Description, distribution, and abundance of mountain Sheep in North America. In R. Valdez and P. R. Krausman, eds., *Mountain Sheep of North America.* Tucson: University of Arizona Press.

Wolf, G. M., B. Griffith, C. Reed, and S. A. Temple. 1996. Avian and mammalian translocations: Update and reanalysis of 1987 survey data. *Conservation Biology* 10:1142–1154.

PART IV

Abetting Natural Colonization

The resilience of large mammals is difficult to question when the restoration process unfolds with no apparent human assistance. In North America. few native species have exhibited range recolonization as well as the black bear. An expanding population in Texas is not only revealing the rescue process but hints at the importance of landscape structure to restoration potential (Onorato and Hellgren in chapter 12). Such situations are rare, however. More than wishful thinking and serendipity is needed to restore large mammal populations.

In some cases, all that is needed to promote population expansion is a slight adjustment to the landscape—perhaps the removal of human structures (Duke et al. in chapter 13) or the conversion of an artificial landscape barrier to a movement filter (Maehr et al. in chapter 15). Whereas the Florida panther remains effectively trapped in a landscape transformed by water management, highways, and habitat alteration, wolves in the Cascades are taking advantage of a unique experiment to restore the function of a landscape linkage.

Such achievements and potential success stories, however, are the exception. Burgeoning human populations and boat traffic in Florida threaten a manatee population that has recently exhibited the potential for growth (Ackerman and Powell in Case 4). This situation pales in the face of tiger conservation in Indonesia—a region where an impoverished and multiplying human population threatens to grow the tiger out of existence (Tilson et al. in chapter 14). Compared to this Old World dilemma, the obstacles facing large mammal restoration in North America seem easily surmountable. Here we have the luxury of considering such restoration at the scale of the biotic province. That Florida has set aside more than 20 percent of its land in some form of natural area belies the fact that it is one of the fastest-growing regions on the continent. In chapter 16, Harris et al. dare us to imagine large mammal conservation that extends beyond preserve boundaries, beyond regions, and challenges one of the most expensive landscape restoration efforts as insufficient. Ultimately we must view the restoration of large mammals in the evolutionary context from which they arose and under the environmental influences that created the faunas of which they are part. The alternative is to accept a contrived biota that is becoming increasingly homogenized and domesticated.

Chapter 12

Black Bear at the Border: Natural Recolonization of the Trans-Pecos

DAVID P. ONORATO AND ERIC C. HELLGREN

Natural recolonization of historic range (defined as natural reestablishment by a species in an area of past extirpation) by large carnivores is rare because of habitat fragmentation, disturbance, and destruction. Although recolonization by gray wolves (*Canis lupus*) has been documented in eight areas in North America and Europe, wolves crossed extensive areas of farmland to recolonize wildlands in only two of these cases (Wydeven et al. 1998). Genetic data have verified suspected source populations for a recolonizing population of wolves in Montana (Forbes and Boyd 1996). Smith and Clark (1994) have described the reintroduction of black bear to Arkansas as one of the most successful translocations of a large carnivore—an effort that led to recolonization of large tracts of forest far from reintroduction sites. Populations of black bear are increasing in most parts of their geographic range (Brown 1993; Pelton and Manen 1994), and this increase is associated with an expanding distribution (Pelton and Manen 1994). Like wolves, however, black bears can expand their range more effectively through contiguous forest and riparian corridors than across extensive areas of unsuitable matrix such as agricultural lands or desert (Brody and Pelton 1989; Mladenoff et al. 1995; Mollohan and Le Count 1989).

Restoration of black bear populations to the borderlands of western Texas has occurred in the 1990s. This recolonization has resulted from the coalescing of biogeographic, ecological, and sociological factors. In this chapter we discuss the history of the black bear in the Trans-Pecos region, summarize

245

the data on its natural recolonization, and explore factors that have abetted recolonization. The black bear in the western Texas-Mexico border zone provides a useful case study for modeling the spatial and temporal patterns of large carnivore recolonization in a naturally patchy landscape.

The Landscape and Background

The Trans-Pecos region of Texas is dominated by Chihuahuan desert vegetation and scattered mountain ranges that occasionally reach elevations high enough (1500–2000 m) to support woodlands and black bear. Four ranges have elevations exceeding 2000 m that support coniferous, oak, and mixed forests: the Chisos, Davis, Chinati, and Guadalupe (Figure 12.1). Several

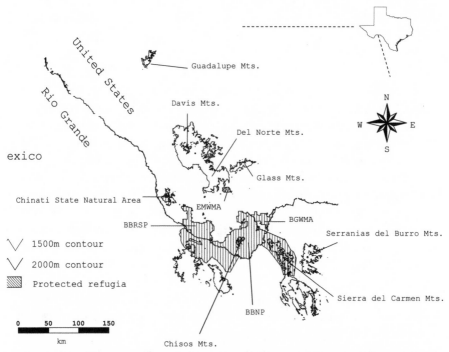

Figure 12.1. Depiction of mainland patches (Serranias del Burro and Sierra del Carmen ranges) and island patches (Chisos Mountains, Glass Mountains, Davis Mountains, Black Gap Wildlife Management Area) available to black bears in the Trans-Pecos region of Texas and Mexico. Acronym definitions are as follows: Big Bend National Park (BBNP), Big Bend Ranch State Park (BBRSP), Black Gap Wildlife Management Area (BGWMA), and Elephant Mountain Wildlife Management Area (EMWMA). The 1500-meter isocline approximates the distribution of woodland habitats in the Chihuahuan Desert matrix.

lower-elevation ranges also occur across the region, including the Glass, Del Norte, Dead Horse (northern extension of the Sierra del Carmen of Coahuila, Mexico), Sierra Diablo, and Rosillos mountains. The result is a landscape of mountain islands of bear habitat set in a sea of Chihuahuan Desert scrub and grassland.

Much of the information on the history of black bear in the Trans-Pecos is available only via conversations with local ranchers and other residents in the area. Two men who have pursued a personal interest in such information are Dr. James F. Scudday and Billy Pat McKinney. Dr. Scudday is a distinguished professor emeritus in biology at Sul Ross State University in Alpine, Texas. His family settled in this small west Texas town in 1926. Over the last 70 years he has developed a good rapport with the local ranchers and hunters of the Trans-Pecos region and has frequently served as a link between state agencies and ranchers when bear incidents arose. His compilation of vast amounts of descriptive and anecdotal information on the black bear makes him an invaluable historian on the regional population.

The McKinney family initially settled the Big Bend region of Texas in 1878. Five generations of McKinneys have lived in the area and have witnessed 120 years of regional change. Billy Pat McKinney is a wildlife specialist for the Texas Parks and Wildlife Department at the Black Gap Wildlife Management Area (BGWMA) in Brewster County. His grandfather was a federal predator control agent in the early 1900s and was involved with the Mexican wolf (*Canis lupus baileyi*). His father too was involved with predator control, dealing mostly with the golden eagle (*Aquila chrysaetos*). The vast oral library compiled over generations, combined with Mr. McKinney's recent personal experiences, provide unique insight into the fate of large carnivores, including black bear, in the Trans-Pecos.

The black bear was prevalent throughout most of western Texas (Davis and Schmidly 1994) in pre-Columbian times but was rarely hunted by indigenous peoples. In the late 1800s, however, bears were hunted in western Texas with firearms (J. Scudday, pers. comm.; Skiles 1995). At the turn of the century, naturalist Vernon Bailey (1905) described bears as "common" in the Chisos, Davis, and Guadalupe mountains. Black bears were common in the adjacent mountain ranges of Mexico as well, such as the Sierra del Carmen and the Serranias del Burro (Figure 12.1). Predator control and heavy recreational hunting in the first half of the twentieth century, however, led to the near extirpation of black bear from the region (Doan-Crider and Hellgren 1996; J. Scudday, pers. comm.).

By 1918, federal predator control agents were using poison bait to control the Mexican wolf in the Trans-Pecos—usually a cow or horse carcass laced with strychnine (J. Scudday, pers. comm.). This technique, although effective, was not species-specific and contributed to the decline of other southwestern carnivores including the grizzly bear (*Ursus arctos;* Brown 1985) and the black bear in the Trans-Pecos. Use of poisoned bait continued until the 1950s, long after most bears were gone (J. Scudday, pers. comm.).

Bear populations in Mexico and the United States were subject to intensive hunting during the late nineteenth and early twentieth centuries. Organized bear hunts that coincided with family outings were common in the Trans-Pecos during the early twentieth century, particularly in the Davis Mountains (J. Scudday, pers. comm.). During this time bears were particularly prized for their meat and lard. Sport hunters continued to take a toll on the black bear in Texas during the 1930s and 1940s. Groups of hunters made their way to western Texas from as far away as Austin and Houston in search of black bear (J. Scudday, pers. comm.).

Populations of black bear in the United States and Mexico exhibited differing population dynamics in the 1930s. With the nationalization of Mexico, American ranchers were evicted. Many new Mexican landowners and ranchers did not have the same hunting technology as their American predecessors and were unable to locate and kill bears as effectively. Moreover, Mexican ranchers have developed a tolerance for predators and are willing to accept greater stock losses (Doan-Crider 1995). Some ranchers in Mexico even view resident bears as a status symbol (B. P. McKinney, Texas Parks and Wildlife Department, pers. comm.). The combination of generally positive rancher attitudes and low hunting pressure has allowed viable bear populations to persist in the Sierra del Carmen and Serranias del Burro mountains (Figure 12.1). Conversely, American ranchers were fervent in their negative attitudes toward predators. Sheep ranchers were particularly ardent when it came to predator control. The sheep industry in Texas was very powerful and produced more wool than any other state in the country during the 1930s and 1940s (Carlson 1982). This disposition, combined with pressure to produce more wool during World War II, sped the decline of large carnivores not only in western Texas but the entire southwestern United States (Brown 1985; B. P. McKinney, pers. comm.).

The persistence of wilderness in the Big Bend region lasted into the twentieth century according to B. P. McKinney and a family tale of two prospectors. Bud Kimble—an experienced hunter from the Big Thicket region of eastern Texas—and John Moss killed a grizzly bear on Pulliam Ridge in what is now Big Bend National Park (BBNP) in 1902 or 1903. Kimble noted that this exceptionally large bear (estimated at 275 kg) had silver-tipped hair, a characteristic noted in other southern grizzlies, and was likely a local resident. The only confirmed record of a grizzly in Texas was a large male killed in the Davis Mountains in 1890 (Brown 1985).

The combination of sport hunting, federal predator control programs, and persecution by ranchers had noticeably reduced the black bear population in western Texas by the middle of the twentieth century. Livestock overgrazing also contributed to habitat degradation (Schmidly 1977). Many of the bear hunts that took place in the 1940s and 1950s were unsuccessful (B. P. McKinney, pers. comm.). By the 1940s, the black bear had been extirpated from the Del Norte and Glass mountains and remnant populations survived only in the Chisos (Borell and Bryant 1942) and Davis mountains (B. P. McKinney, pers. comm.). These populations disappeared by the 1950s.

During the 1950s, the sheep industry in the Trans-Pecos region began to falter (Carlson 1982). The end of World War II, combined with a severe drought, caused many ranchers to abandon wool production (Carlson 1982; B. P. McKinney, pers. comm.)—an economic change that provided the opportunity for the black bear to recolonize the Big Bend region of Texas. When BBNP was established in 1944, bears were rarely seen (Figure 12.2; BBNP black bear sightings database), and those that were observed were likely transient males. This pattern continued into the 1960s when several yearling-sized bears were killed on the Adams Ranch northeast of the park. These were likely young males migrating from the Sierra del Carmens (B. P. McKinney, pers. comm.).

Schmidly (1977) reported no evidence of resident black bears anywhere in the Trans-Pecos during the late 1970s. Although populations in the adjacent

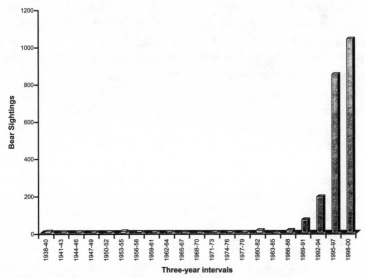

Figure 12.2. Frequency of black bear observations made over three-year intervals in Big Bend National Park (1938–1992).

mountains of Mexico had been reduced (Baker and Greer 1962; Hall 1981; Leopold 1959), a viable population remained (Doan-Crider and Hellgren 1996). After Mexico declared a moratorium on black bear hunts in 1986, the stage was set for local population recovery and recolonization of vacant habitat in the Chisos Mountains.

The Recolonization Process

In discussing the recolonization of black bear in the Big Bend region, it is interesting to compare this event with the factors that inhibited grizzly bear recolonization in the Southwest. Records denote that southwestern grizzly bears once ranged from the San Juan Mountains in southern Colorado to mountain ranges in southern Chihuahua, Mexico (Brown 1985). Unlike the situation with black bear in the Trans-Pecos, grizzly bear populations in the southwest were separated by longer distances and were smaller in size. Thus when human pressures reduced grizzly numbers during the late 1800s and early 1900s, populations were so disjunct, separated by such great distances, and so susceptible to random events that recolonization of vacant historic habitat could not proceed. The result was isolation of the remaining populations and eventual extirpation of the species within the southwestern United States (Brown 1985). In contrast, the proximity of a viable black bear population in Mexico allowed it to return to the Trans-Pecos.

Recolonization by the Trans-Pecos black bear was a slow process that took many decades. Infrequent black bear sightings in the park occurred from 1950 through the 1970s (Figure 12.2). Although two females with cubs were seen in 1969 and 1978, park wildlife specialists discounted the notion of a resident breeding population (Skiles 1995). Over the next ten years, young bears made occasional appearances in the park (Doan-Crider 1995; Skiles 1995). Continued sightings would later confirm that the Chisos Mountains of BBNP contained suitable habitat for the black bear. This habitat was similar in some respects to that occupied by black bears in central Arizona (LeCount et al. 1984). The initial obstacle impeding the establishment of a resident population was the remote possibility of a dispersal or migration by a reproductive female into the Chisos Mountains.

Increasing observations in BBNP during the late 1980s gave credence to the plausibility of a resident population in the park. In 1988, park employees recorded 26 bear sightings. Over the next ten years, employees recorded 2127 observations (Figure 12.2; BBNP black bear sightings database 1999), including numerous sightings of females with cubs and yearlings. The park presently maintains a resident breeding black bear population in the Chisos Mountains (Davis and Schmidly 1994; Skiles 1995).

The metapopulation approach (Hanski and Simberloff 1997) provides a useful conceptual structure for understanding recolonization in the Trans-Pecos. Metapopulations are spatially structured into local breeding populations that can be separated by geography and linked by dispersal. Several possible metapopulation systems may be applicable to the Big Bend Ecosystem—including the source/sink, rescue-effect, and mainland/island metapopulation systems. The black bear population in the Chisos Mountains may once have been part of a source/sink metapopulation with the larger population in Mexico serving as a source. Sinks are populations that typically have a reproductive rate (R_o) that is less than 1 (Stacey et al. 1997) and would go extinct without immigration. Due to the human pressures in the early twentieth century, the entire Trans-Pecos region became a sink for Mexican bears. The key characteristic of this type of metapopulation is that migration is directionally skewed (Stacey et al. 1997). The presence of a growing population in BBNP (Davis and Schmidly 1994; D. P. Onorato, unpublished data) and the migration of bears from the Trans-Pecos back into Mexico (Bonnie R. McKinney, Texas Parks and Wildlife, unpublished data; D. P. Onorato, unpublished data) demonstrate that the source/sink model is no longer at work in this region.

The rescue-effect metapopulation system entails migration that averts local extinction and can vary in direction and rate of dispersal (Stacey et al. 1997). Consequently, a population may alternate as a source or sink through time (Harrison 1991; Stacey and Taper 1992). At present there is little doubt that Mexico continues to be the source of immigrating black bears for the Trans-Pecos region of Texas. Thus we believe that the mainland/island metapopulation model is the most appropriate for this region, where small, suitable habitat patches (BBNP and BGWMA) are located within dispersal distance from a very large occupied habitat patch (Mexico) (Hanski and Simberloff 1997). A large bear population resides in the Sierra del Carmen and Serranias del Burro ranges of Mexico (Doan-Crider and Hellgren 1996). As these populations have grown, bears have dispersed across the Rio Grande to BBNP and BGWMA (Figure 12.1). These two habitat patches may serve as stepping stones for future colonizations.

Mountain ranges of the Trans-Pecos that once supported black bear are the next likely recolonization sites. These include the Davis, Del Norte, and Glass mountains located north of the Chisos (Figure 12.1). Other regions of potential recolonization include the Chinati and Housetop mountains (B. P.

McKinney, pers. comm.), and Val Verde County near Amistad National Recreational Area (J. Scudday, pers. comm.). Thus black bears are returning to western Texas. But whether they will be tolerated or persecuted as they venture from public land onto private land remains to be seen.

The Sociology of Recolonization

At least four factors have contributed to recolonization of the Trans-Pecos region: increased human tolerance to bears, law enforcement, the existence of suitable habitat in Texas, and the presence of a mainland population (refugia). Historically, the black and grizzly bear were extirpated from this region because of the threats, real and imagined, they posed to human interests (Brown 1985; Doan-Crider and Hellgren 1996). Although negative attitudes toward bears still exist, their prevalence and intensity have diminished. Doan-Crider (1995) indicates that landowners in northern Mexico were willing to resolve bear/livestock conflicts without wide-scale lethal control. She remarks that landowners were crucial in the success of black bear management in the region and recommends education on bear biology and behavior to aid in this process.

Certainly the presence of designated and de facto refugia (Pelton and Manen 1994) in the Trans-Pecos region also has aided the recolonization process (Figure 12.1). Big Bend National Park comprises more than 3200 km² of protected land and is surrounded on two sides by large (more than 400 km²) state-owned natural areas (Big Bend Ranch State Park [BBRSP] and BGWMA). Private landowners in the Serranias del Burro also provide a de facto refuge of nearly 3600 km² of land that is protected from poaching (Doan-Crider and Hellgren 1996). Further, the northern end of the Sierra del Carmen range in Mexico has been proposed as part of an international biosphere reserve (Cummings and Mallan 1996). Although it is not all suitable black bear habitat, these areas also provide dispersal corridors that facilitate colonization of the Chisos and adjacent mountains.

Ecological Constraints

Large carnivores and other species that are characterized by limited female dispersal and have been extirpated by human or natural causes are unlikely to recolonize former range quickly (Avise 1995). The black bear exhibits male dispersal and female philopatry (Rogers 1987b) as territorial mothers shift ranges to accommodate female offspring. Without the advantages of maternal tolerance and the utilization of portions of her territory, female offspring may have difficulty setting up land tenure during dispersal (Rogers 1987b). Natural selection favors philopatry if the reproductive success of a mother's daughter is at least double the amount that the mother's reproductive success is reduced by sharing portions of her territory (Hamilton 1964; Wilson 1975).

An examination of dispersal and philopatry in the black bear helps to explain why it took more than 40 years for a female to return as a resident breeder in the Chisos Mountains. Three studies of dispersal in black bear indicate that most subadult (one–three-year-old) males disperse but subadult females rarely do. Results are remarkably consistent: rates and distances of male dispersal are 100 percent and 13–219 km (n = 20; Rogers 1987a), 100 percent and 30–200 km (n = 8; Elowe and Dodge 1989), and 83 percent (n = 21; Schwartz and Franzmann 1992). Conversely, rates for females are 10 percent and 3–11 km (n = 31; Rogers 1987b), 8 percent and 15 km (n = 13; Elowe and Dodge 1989), and 3 percent (n = 30; Schwartz and Franzmann 1992). Only 7 percent of females disperse, and only one record of female dispersal exceeds 15 km (54 km by a subadult female; Maehr 1997). Thus the likelihood of a breeding female traversing 35 km of open Chihuahuan desert is remote. Mathematically, the probability of a female dispersal event is

$$P_{chisos} = (P_d)(0.25)(P_{35}) = 0.002$$

where

P_{chisos} = probability that a female bear from Mexico will disperse to the Chisos

P_d = probability of female dispersal (0.07)

0.25 = probability of a female dispersing north (assuming a random dispersal direction), and

P_{35} = probability of a female dispersing 35 km (0.14; one of seven documented female dispersal events exceeded 35 km)

Doan-Crider and Hellgren (1996) have estimated the density of female bears to be 0.17 animal/km^2 in the nearby Serranias del Burro mountain chain. At any given time, only a segment of female bears in the population are the appropriate age for dispersal (one to three years). Therefore, even large populations covering hundreds of square kilometers have an extremely low probability of supplying a single female disperser to cover that distance in any single year. Conversely, a dispersing male would be quite capable of dispersing to the Chisos.

An alternative view of recolonization is that the initial females that settled in the park may have arrived during a large-scale autumnal migration from Mexico to find unexploited resources. Returns to natal ranges may have been discouraged by abundant food, adequate habitat, and the onset of hibernation. Autumnal shifts in home ranges have been reported for both genders in numerous black bear studies (Garshelis and Pelton 1981; Hellgren and Vaughan 1990; Maehr 1997; Rogers 1987a). LeCount et al. (1984) reported that four radio-collared Arizona black bears (two females) made temporary fall excursions of 25 to 40 km from summer ranges. The historical records of females with cubs in BBNP may have resulted from such autumnal peregrinations.

Conservation and Research Implications

Do patterns of southwestern U.S. black bear recolonization have regional and global implications? Regionally the temporal scale at which the black bear recolonized disjunct mountains in Trans-Pecos Texas took several decades. This process could be shortened, however, by translocating females to suitable habitat that is within male dispersal distance. Similar management could facilitate Florida panther (*Puma concolor coryi*) colonization that has been attempted by males only (chapter 15 in this volume). In contiguous habitat, black bear and probably many other solitary carnivores spread by incremental range expansion by related females (Rogers 1987a). The Chinati, Davis, and Glass mountains would make good translocation sites because they have suitable habitat and indeed have supported bears in post-Columbian times. Bears have been sighted in the latter two of these ranges in the past few years (B. McKinney, pers. comm.). Restoration through translocation in these areas would enlarge and stabilize the functional mainland/island black bear metapopulation presently found in the Trans-Pecos.

Natural recolonization in the black bear provides an opportunity to investigate conservation genetics, metapopulation theory, and social organization in the species. Rogers (1987a), for example, demonstrated female philopatry and a female land-tenure system in a regional black bear population in northern Minnesota. His work predicts that the distribution of female bears in a recolonized area will expand as a series of partially overlapping ranges composed of female offspring of the original residents. Conversely, male offspring should either disperse or become residents during early periods of low-density recolonization. Although females may exhibit

very high relatedness if a single founder is responsible for an extant breeding population, dispersal by related males and immigration of unrelated
males would eventually ameliorate potential inbreeding effects. The spatial
relationships among females of known genetic relationships would also
prove interesting. These data could test Rogers's (1987a) observations that
female philopatry should be expressed in genetic relatedness among females
found in adjacent home ranges. Initial work on this question in contiguous
occupied bear habitat shows no relationship between female relatedness and
range overlap (Schenk et al. 1998). A newly reestablished population may
exhibit a different pattern, especially if female colonization is infrequent, as
it appears to be in the Trans-Pecos.

Natural recolonization by the black bear will also shed light on what constitutes a barrier to dispersal in this and other ursids. Analysis of maternally
and biparentally inherited genetic markers, for example, can affirm the identity of a source population. Among brown bears, nuclear microsatellite markers have shown that water barriers of 2–4 km were sufficient to reduce or
eliminate female dispersal, whereas 7 km was adequate to hinder male dispersal (Paetkau et al. 1998). These barriers led to genetic differences among
insular and continental populations. Similarly, recolonization of the Trans-
Pecos by black bear has occurred despite a 35-km terrestrial barrier (Chihuahuan desert). Analysis of the mtDNA control region (maternal) and
microsatellites (biparental) also have been used for studies of regional genetic
variation (Paetkau and Strobeck 1994) and phylogeography (Paetkau and
Strobeck 1996) in black bears. A more field-oriented approach to understanding black bear colonization could involve the use of GPS collars on dispersal-age Mexican bears.

Black bear recolonization in the Trans-Pecos represents a useful case
study for modeling the spatial and temporal rate of unassisted large carnivore
recolonization in a naturally fragmented landscape. The historical evidence in
the Trans-Pecos supports Avise's (1995) contention that large carnivores with
limited female dispersal will be slow to recolonize disjunct portions of their
range from which they have been extirpated. Species with common female
dispersal—such as gray wolves (Boyd-Heger and Pletscher 1999) or mountain lions (Sweanor et al. 2000)—are more likely to recolonize vacant range
even if that range has been fragmented by natural or human forces. Female
dispersal in geographically expanding bear populations may be greater than
previously believed, however. Swenson et al. (1998) suggest that dispersal
behavior in continuous populations such as Fennoscandian brown bear may
be different than along the periphery of populations expanding into vacant
habitat. Increased female dispersal in the latter situation would benefit the

conservation and restoration of these species and help maintain metapopulations (Swenson et al. 1998). Such events may be viewed as more socially and politically acceptable because the process tends to be slow and allows local residents and managers to acclimate gradually to a changing large-mammal community.

Acknowledgments

We thank R. K. Skiles for his persistence in supporting the fieldwork that will form the supplement to this chapter. We thank D. L. D. Crider and B. R. McKinney, who contributed samples to the genetic analyses, and B. P. McKinney and J. Scudday for relating their historical experience on black bears in the region. This study was supported by funding from the Natural Resources Preservation Program, through the U.S. Geological Survey and the National Park Service, and logistical support from the Oklahoma Cooperative Fish and Wildlife Research Unit (U.S. Geological Survey–Biological Resources Division, Oklahoma State University, Oklahoma Department of Wildlife Conservation, and Wildlife Management Institute, cooperating).

Literature Cited

Avise, J. C. 1995. Mitochondrial DNA polymorphism and a connection between genetics and demography of relevance to conservation. *Conservation Biology* 9:686–690.

Bailey, V. 1905. *Biological Survey of Texas*. Vol. 25. Washington, D.C.: Department of Agriculture, Bureau of Biological Survey.

Baker, R. H., and J. K. Greer. 1962. Mammals of the Mexican state of Durango. *Publications of the Museum, Michigan State University, Biological Series* 2:25–154.

Borell, A. E., and M. D. Bryant. 1942. *Mammals of the Big Bend Area of Texas*. University of California Publication of Zoology 48. Berkeley: University of California.

Boyd-Heger, D. K., and D. H. Pletscher. 1999. Characteristics of dispersal in a colonizing wolf population in the central Rocky Mountains. *Journal of Wildlife Management* 63:1094–1108.

Brody, A. J., and M. R. Pelton. 1989. Effects of roads on black bear movements in western North Carolina. *Wildlife Society Bulletin* 17:5–10.

Brown, D. E. 1985. *The Grizzly in the Southwest*. Norman: University of Oklahoma Press.

Brown, G. 1993. *The Great Bear Almanac*. New York: Lyons & Burford.

Carlson, P. H. 1982. *Texas Woolybacks*. College Station: Texas A&M University Press.

Cummings, J., and C. Mallan. 1996. *Mexico Handbook*. Chico, Calif.: Moon Publications.

Davis, W. B., and D. J. Schmidly. 1994. *The Mammals of Texas*. Austin: Texas Parks and Wildlife Press.

Doan-Crider, D. L. 1995. Population characteristics and home range dynamics of the black bear in northern Coahuila, Mexico. M.S. thesis, Texas A&M University–Kingsville.

Doan-Crider, D. L., and E. C. Hellgren. 1996. Population characteristics and winter ecology of black bears in Coahuila, Mexico. *Journal of Wildlife Management* 60:398–407.

Elowe, K. D., and W. E. Dodge. 1989. Factors affecting black bear reproductive success and cub survival. *Journal of Wildlife Management* 53:962–968.

Forbes, S. H., and D. K. Boyd. 1996. Genetic variation of naturally colonizing wolves in the central Rocky Mountains. *Conservation Biology* 10:1082–1090.

Garshelis, D. L., and M. R. Pelton. 1981. Movements of black bears in the Great Smoky Mountains National Park. *Journal of Wildlife Management* 45:912–925.

Hall, E. R. 1981. *The Mammals of North America*. New York: Wiley.

Hamilton, W. D. 1964. The genetical theory of social behaviour, pts. 1 and 2. *Journal of Theoretical Biology* 7:1–52.

Hanski, I. A., and D. Simberloff. 1997. The metapopulation approach, its history, conceptual domain, and application to conservation. In I. A. Hanski and M. E. Gilpin, eds., *Metapopulation Biology: Ecology, Genetics, and Evolution*. San Diego: Academic Press.

Harrison, S. 1991. Local extinction in a metapopulation context: An empirical evaluation. *Biological Journal of the Linnaen Society* 42:73–88.

Hellgren, E. C., and M. R. Vaughan. 1990. Range dynamics of black bears in Great Dismal Swamp, Virginia–North Carolina. *Proceedings of the Annual Conference of the Southeast Association of Fish and Wildlife Agencies* 44:268–278.

LeCount, A. L., R. H. Smith, and J. R. Wegge. 1984. Black bear habitat requirements in Central Arizona. Special Report 14. Tucson: Arizona Game and Fish Department.

LeCount, A. L., and C. Mollohan. 1995. Big Bend National Park bear reconnaissance. Big Bend National Park, Texas: National Park Service.

Leopold, A. S. 1959. *Wildlife of Mexico: The Game Birds and Mammals*. Berkeley: University of California Press.

Maehr, D. S. 1997. The comparative ecology of bobcat, black bear and Florida panther in south Florida. *Bulletin of the Florida Museum of Natural History* 40:1–176.

Mladenoff, D. J., T. A. Sickley, R. G. Haight, and A. P. Wydeven. 1995. A regional landscape analysis and prediction of favorable gray wolf habitat in the northern Great Lakes region. *Conservation Biology* 9:279–294.

Mollohan, C. M., and A. L. LeCount. 1989. Problems of maintaining a viable black bear population in a fragmented forest. In *Conference on Multiresource Management of Ponderosa Pine Forests*. Flagstaff: Northern Arizona University.

Paetkau, D., and C. Strobeck. 1994. Microsatellite analysis of genetic variation in black bear populations. *Molecular Ecology* 3:489–495.

———. 1996. Mitochondrial DNA and the phylogeography of Newfoundland black bears. *Canadian Journal of Zoology* 74:192–196.

Paetkau, D., G. F. Shields, and C. Strobeck. 1998. Gene flow between insular, coastal and interior populations of brown bears in Alaska. *Molecular Ecology* 7:1283–1292.

Pelton, M. R., and F. T. v. Manen. 1994. Distribution of black bears in North America. *Eastern Workshop on Black Bear Research and Management* 12:133–138.

Rogers, L. L. 1987a. Effects of food supply and kinship on social behavior, movements, and population growth of black bears in northeastern Minnesota. *Wildlife Monographs* 97:1–72.

———. 1987b. Factors influencing dispersal in the black bear. In B. D. Chepko-Sade and Z. T. Halpin, eds., *Mammalian Dispersal Patterns*. Chicago: University of Chicago Press.

Schenk, A., M. E. Obbard, and K. M. Kovacs. 1998. Genetic relatedness and home-range overlap among female black bears (*Ursus americanus*) in northern Ontario, Canada. *Canadian Journal of Zoology* 76:1511–1519.

Schmidly, D. J. 1977. *The Mammals of Trans-Pecos Texas: Including Big Bend National Park and Guadalupe Mountains National Park*. College Station: Texas A&M University Press.

Schwartz, C. C., and A. W. Franzmann. 1992. Dispersal and survival of subadult black bears from the Kenai Peninsula, Alaska. *Journal of Wildlife Management* 56:426–431.

Skiles, J. R. 1995. Black bears in Big Bend National Park—the Tex-Mex Connection. *Proceedings of the Western Black Bear Workshop* 5:67–73.

Smith, K. G., and J. D. Clark. 1994. Black bears in Arkansas: Characteristics of a successful translocation. *Journal of Mammalogy* 75:309–320.

Stacey, P. B., and M. Taper. 1992. Environmental variation and the persistence of small populations. *Ecological Applications* 2:18–29.

Stacey, P. B., M. L. Taper, and V. A. Johnson. 1997. Migration within metapopulations: The impact upon local population dynamics. In I. A. Hanksi and M. E. Gilpin, eds., *Metapopulation Biology: Ecology, Genetics and Evolution*. San Diego: Academic Press.

Sweanor, L. L., K. A. Logan, and M. G. Hornocker. 2000. Cougar dispersal patterns, metapopulation dynamics, and conservation. *Conservation Biology* 14:798–808.

Swenson, J. E., F. Sandegren, and A. Soderberg. 1998. Geographic expansion of an increasing brown bear population: Evidence for presaturation dispersal. *Journal of Animal Ecology* 67:819–826.

Wilson, E. O. 1975. *Sociobiology*. Cambridge: Harvard University Press.

Wydeven, A. P., T. K. Fuller, W. Weber, and K. MacDonald. 1998. The potential for wolf recovery in the northeastern United States via dispersal from southeastern Canada. *Wildlife Society Bulletin* 26:776–784.

Chapter 13

Restoring a Large-Carnivore Corridor in Banff National Park

DANAH L. DUKE, MARK HEBBLEWHITE, PAUL C. PAQUET,
CAROLYN CALLAGHAN, AND MELANIE PERCY

Concern over wildlife population persistence and restoration in human-dominated landscapes has become a priority in conservation. Disturbances that fragment natural habitats often result in small and isolated wildlife populations that are likely to go extinct. (See Wilcox and Murphy 1985; Glenn and Nudds 1989; Ims et al. 1993; Newmark 1993, 1995; McNally and Bennett 1996; Oehler and Litvaitis 1996; Swart and Lawes 1996; Burkey 1997; Gurd and Nudds 1999). The adverse effects of fragmentation are particularly evident in areas dominated by humans and where habitat occurs only in disjunct patches. For populations to persist in these conditions, individuals must move freely between habitat islands that collectively supply their life requisites. Land uses, however, often conflict with species' requirements and limit the use of potential travel routes. The presence of a settlement or highway, for example, may alter traditional wildlife movements or lead to permanent abandonment of habitat. Further, some individuals may be forced to navigate through areas of severe human disturbance to find food, disperse, migrate, mate, and simply use their home ranges.

Wildlife corridors can provide connectivity in human-dominated landscapes (Maehr 1990; Saunders and deRebeira 1991; Beier 1995; Odette and Thomas 1996; Dunning et al. 1995) and reduce the adverse effects of fragmentation (Newmark 1993; Walker and Craighead 1997). A corridor is a linear two-dimensional landscape element that links two or more patches of habitat that once were connected (Soulé and Gilpin 1991). Corridors are especially important for wide-ranging species such as wolves (*Canis lupus*) and grizzly bears (*Ursus arctos*). Key characteristics include width and length, vegetative cover, habitat quality, location, human influences, noise, light, edge effects, degree of connectivity, and the presence of barriers (Harrison 1992; Newmark 1993; Soulé and Gilpin 1991; Lindenmayer et al. 1993; Fleury and Brown 1997). Individual animals are more likely to use pathways that include key components of their preferred habitat (Rosenberg et al. 1997). Ecological factors that determine the availability and quality of wildlife corridors are dynamic and change seasonally and from year to year.

Corridors function at scales ranging from large regional corridors linking many watersheds to small local corridors linking patches of residual habitat. A large-scale regional corridor may comprise a network of smaller corridors. Small-scale corridors are often remnant strips of land near residential areas, agricultural zones, recreational areas, highways, and railways. If wildlife do not use these small-scale corridors, the regional corridor and habitat network may be jeopardized. As habitat fragmentation increases with human activity, a finer network of small-scale corridors is required to ensure the persistence of certain species.

In the central Canadian Rocky Mountains, steep, rugged terrain is not conducive to the movements of certain species. The natural fragmentation of the landscape confines the movements of large mammals to low-elevation valley bottoms. Major rivers, creeks, and interconnecting passes function as local and regional travel networks (Paquet 1993). In Banff National Park, Alberta, the Bow River Valley comprises the highest-quality habitat for gray wolves in the central Canadian Rockies (Paquet et al. 1996; Green et al. 1996) and permits interchange between the United States and Canada (Boyd et al. 1998; Paquet et al. 1996; Boyd and Pletscher 1999).

Although gray wolves naturally recolonized the Bow River Valley of Banff National Park in the mid-1980s, multiple-lane highways, secondary roads, recreation facilities, and heavily populated urban areas divide the valley into isolated habitats and small unconnected movement corridors that are dysfunctional for wolves (Paquet 1993; Heuer 1995; Paquet et al. 1996). Because human activity influences these corridors, wolves rarely use them (Heuer 1995; Stevens et al. 1996; Stevens and Owchar 1997; Heuer et al. 1998; Duke 1999). This loss of connectivity has compelled wolves to adopt alternative travel routes and abandon high-quality habitats. The alternative routes are circuitous and energetically expensive (Paquet et al. 1996).

The Cascade Corridor has the greatest potential for use by large carnivores. It is one of three travel routes linking wildlife habitat to the east and west past the Banff townsite. The other routes are within areas of high human use and contain subdivisions, commercial developments, and a golf course. In 1993, the government of Canada commissioned us to assess the ecological status of the Bow River Valley and evaluate the potential for restoring carnivore movement around the Banff townsite. Our empirically derived habitat/movement model, a result of this study, showed that reducing travel impediments and human disturbance would permit wolves to return to shorter traditional pathways through more hospitable terrain (Paquet et al. 1996). Consequently, we recommended that Parks Canada should remove all human structures and reduce human activities within the Cascade Corridor (Parks Canada 1997). By reducing human disturbance, we argued, a dysfunctional movement corridor could be rehabilitated for wolves. In the fall of 1997, human presence within the "experimental" corridor was reduced to a level shown by the model to be compatible with usage by wolves. This chapter describes the unprecedented restoration of a gray wolf corridor.

Study Area

Our study area encompassed Banff National Park (BNP) and portions of the Bow River Valley east of the national park. Banff National Park is in the Continental Range of the southern Canadian Rocky Mountains, about 110 km west of the city of Calgary, Alberta. The Bow River Valley Regional Corridor (BVRC) extends from the confluence of the Bow and Kananaskis rivers to Bow Summit.

The BVRC is a conduit for wildlife movements originating in many other major valleys. The portion of the BVRC used by the Cascade wolf pack is 2 to 6 km in width, is oriented on a northwest to southeast axis, and covers approximately 90 km². Habitat within the BVRC is montane, subalpine, or alpine depending on elevation. Low-elevation montane is the most productive and biologically diverse habitat found in Banff National Park, yet it only accounts for 3 percent of the area (Holroyd and Van Tighem 1983). Approximately 82 percent of the park's montane habitat is in the Bow River Valley (Holroyd and Van Tighem 1983). Large mammal associations include moose (*Alces alces*), elk (*Cervus elaphus*), white-tailed deer (*Odocoileus virginianus*), mule deer (*Odocoileus hemionus*), bighorn sheep (*Ovis canadensis*), mountain goat (*Oreamnos americanus*), mountain caribou (*Rangifer tarandus*) wolf (*Canis lupus*), coyote (*Canis latrans*), cougar (*Puma concolor*), grizzly bear (*Ursus arctos*), black bear (*Ursus americanus*), and lynx (*Lynx lynx*).

The Cascade Corridor is 1 km north of the Banff townsite and links the Forty-Mile Valley, the Cascade Valley, and the Bow River Valley. The Trans-Canada Highway—a fenced, four-lane high-speed freeway—defines the southern boundary and Cascade Mountain the northern boundary. The Cascade Corridor is approximately 6 km long and ranges in width from 350 m to 1.5 km. Vegetation within the corridor is composed of approximately 20 percent open meadow, 50 percent open forest, and 30 percent closed forest (Heuer 1995). The territory of the Cascade wolf pack includes portions of the lower Bow River Valley east of the Banff townsite. The Cascade Corridor is in the extreme southwestern corner of the territory and forms the border between the Cascade pack and Bow Valley pack. In winter, when prey are concentrated at lower elevations, the Cascade pack uses the lower Bow River Valley. During the remainder of the year, the pack is mostly in remote regions of the park (Paquet et al. 1996).

Before 1997, wolves rarely used the Cascade Corridor despite its location in high-quality wolf habitat (Paquet et al. 1996). Levels of human activity in the Cascade Corridor were moderate to high (Banff Bow Valley Study 1996). Infrastructure within the corridor included a hotel, a ski access road, a reservoir with access road, a fenced wildlife facility (buffalo paddock), barns, horse corrals, an active airfield, and a fenced military Cadet Camp training facility (Figure 13.1). Human use in the corridor included hiking, vehicular traffic, horse traffic around the corrals and along the base of Cascade Mountain, ice and rock climbing on Cascade Mountain, and airfield traffic.

In the fall of 1997, Parks Canada removed the buffalo paddock, barns, and horse corrals and closed the airstrip to all but emergency traffic (Figure 13.1). This step substantially reduced human activity within the corridor, though hiking, vehicle traffic (Norquay Road), and climbing of Cascade Mountain was still permitted.

Methods

We captured wolves in modified leghold traps (McBride No. 4 Wolf Trap) or by helicopter darting via Cap-Chur rifle (Palmer). Wolves were immobilized with Telazol or with combinations of Telazol/Xylazine, or Ketamine/Xylazine. The Canadian Council for Animal Care and the Bow River Valley Cumulative Effects Research Panel approved the techniques used to capture and handle wolves. Once immobilized, wolves were fitted with LMRT-3 VHF radio-collars (Lotek Engineering, Aurora, Ontario). Paquet (1993) presents a complete description of the wolf capture and handling procedures used in this study.

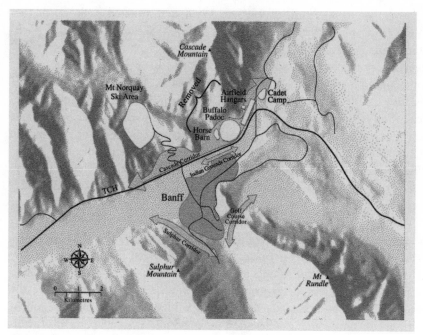

Figure 13.1. Composition of the Cascade Corridor, Banff National Park, Alberta, before restoration in 1997. Restoration included the removal of barns, corrals, a fenced buffalo paddock, an airfield, and hangars.

When wolves were in the Bow River Valley, we located them daily using ground or aerial radiotelemetry (White and Garrott 1990). Otherwise, weekly telemetry locations were obtained using rotary or fixed-wing aircraft. Linear track transects were established next to areas of high human activity throughout the Bow River Valley in 1993 for monitoring large-mammal use and movement patterns within narrow zones and potential travel corridors (Heuer 1995). Transects in the Cascade Corridor extended from valley bottom up the sides of the valley until the first significant natural obstacle was encountered. Transects were divided into 100-m intervals, and wildlife presence or absence was documented within each interval. Transects were sampled within a three-day period after each snowfall. When snow conditions permitted, we backtracked all wolf tracks. We recorded wolf tracking sequences on 1:20,000 corrected air photos or used GPS locations and 1:50,000 topographic maps. Tracking sequences were then digitized as line segments using geographic information system (GIS) software.

Wolves are social animals. Thus we determined use of the Cascade Corridor and the BVRC by packs rather than individuals. During the winter

(1 November to 30 April) of 1993 to 1999, we estimated pack use of the BVRC and the Cascade Corridor by using the number of wolf pack radiotelemetry days (PTD) as our sample unit for analysis. This represents the number of days that a radio-collared wolf and all or parts of the pack were in the area. Three wolf relocations within the Cascade Corridor on a given day, for example, represented one PTD. Their frequent withdrawal to remote areas often impeded our ability to find wolves. Due to difficulties in locating wolves in backcountry areas, our results underrepresent the number of backcountry PTDs. This bias, however, was consistent throughout the study. When the Cascade pack was divided between the backcountry and Bow River Valley with radio-collared wolves in each, we counted this as one PTD in each area.

Wolf pack home ranges were determined with Calhome (Kie 1996) using the 100 percent minimum convex polygon (MCP) to describe the wolf territories before, during, and after restoration. We used one radiolocation per day selected at random as the sample to determine home range (White and Garrot 1990). In mountainous environments, however, standard home range estimators fail to describe spatial habitat use by wildlife accurately because much of the area is unusable high-elevation rock and ice. Since 95 percent of wolf relocations at Banff National Park occur below 1850 m (Paquet 1993; Callaghan, unpublished data), we restricted home range descriptors to areas below this altitude.

We determined the number of PTDs that occurred within the entire home range of the Cascade pack for each year. We then determined the number of PTDs that occurred in the backcountry, the BVRC, and the Cascade Corridor and expressed them as a proportion of total use of their home range. To assess whether the Cascade Corridor served as a movement passage or as functional habitat, we recorded the number and type of kills made by wolves in the corridor before and after restoration. In addition, we tested for differences in snow depth and prey density between corridors and between winters. Snow depth was measured to the nearest centimeter at each 100-m interval along transects in the Cascade Corridor and the BVRC. Mean snow depth per transect was calculated for the Cascade Corridor and a control transect (Penstock Corridor) within the BVRC. (The Penstock Corridor is approximately 5 km east of the Cascade Corridor in the BVRC.) We determined differences in ungulate abundance in the Cascade and Bow River Valley regional corridors before and after restoration by comparing the number of ungulate track crossings of the Cascade and Penstock survey transects over the four-year period of study. Crossing frequencies of elk, bighorn sheep, and white-tailed deer and mule deer (combined) were recorded for each transect at 100-m intervals. But as elk tracks were much more abundant than other ungulate species and because elk are the most important prey of wolves in BNP (Huggard 1993; Paquet 1993; Paquet et al. 1996), we only analyzed elk crossings.

We derived expected use by wolves of the Cascade and the Bow River Valley regional corridors based on corridor area. To assess whether wolf use of the Cascade Corridor changed after restoration, we used a chi-square goodness-of-fit test to compare observed use with expected use (Sokal and Rohlf 1995). Our analysis assumed that each corridor had the same habitat value and that habitat quality was constant throughout the study. Habitat quality is better in the BRVC corridor (Holroyd and Van Tighem 1983; Paquet et al. 1996), however, so our analysis is conservative. We compared elk crossing indices before restoration (1995–1996 and 1996–1997) and after restoration (1997–1998 and 1998–1999), using the mean number of ungulate tracks per year and per transect in a two-factor 2 × 4 ANOVA. We compared differences in mean snow depth for each transect location (Penstock and Cascade corridors) for each year before (1995–1996 and 1996–1997) and after restoration (1997–1998 and 1998–1999) in a two-factor 2 × 4 ANOVA. Where data variances were unequal, an SQRT transformation was applied (Sokal and Rohlf 1995). To examine where significant differences occurred, we used post hoc Bonferroni multiple-comparison tests adjusted for experimentwise error rates. Significant trends were examined by using simple correlation analyses.

Results

We combined results from home range use, kill sites, snow depth, and relative elk abundance pre- and postrestoration to determine the impact of the Cascade Corridor restoration on wolf movements.

Cascade Pack Home Range

From 1993 through 1997, the home range of the Cascade pack comprised the Cascade watershed and the Bow River Valley watershed east of the town of Banff and north of the Trans-Canada Highway (607 km²). Between 1997 and 1999, the home range increased to include the Panther, Dormer, and Red Deer valleys and portions of the Clearwater Valley (1847 km²). The portion of the BVRC used between 1993 and 1999 was about 96 km² including the Cascade Corridor.

Before corridor restoration, we found no statistical difference between use of the Cascade Corridor and use of the BVRC (Table 13.1; χ^2_3 = 3.13; $p >$ 0.20). Following restoration, the Cascade Corridor was used more than expected (χ^2 = 40.22; $p < 0.001$) (Figure 13.2). Chi-square residuals suggest

Table 13.1. Chi-Square Statistics for Expected Wolf Use (Based on Area in km^2) of Cascade and Bow Valley Corridors vs. Observed Use Measured in Wolf Pack Radiotelemetry Days

| | WOLF PACK TELEMETRY DAYS | | | | | |
| | Cascade Corridor | | Bow Valley Corridor | | | |
	Observed	Expected	Observed	Expected	Total	χ^2
Prerestoration (1993–1997)	5	9	145	141	150	3.96
Postrestoration (1997–1999)	30	4	37	62	67	40.22*

Note: Critical value for prerestoration was $\chi^2_{a = 0.05, (3)} = 7.815$; for the postrestoration the critical value was $\chi^2_{a = 0.05, (1)} = 3.841$.
*Significant at $\alpha = 0.05$.

that wolf use of the Cascade Corridor was greater than expected during the postrestoration period, whereas use in the BVRC was less.

Before 1997–1998, wolf use of the Bow River Valley was higher (mean proportion PTD was 63 percent) than use of the backcountry (mean proportion PTD was 34.8 percent). After 1997, use of the Bow River Valley declined to 29.6 percent of total PTDs with a corresponding increase in time spent in the backcountry (mean = 52.5 percent)—thus the percentage of time wolves spent in the Bow River Valley regional corridor decreased substantially after 1997. Nevertheless, proportional use of the Cascade Corridor increased from 3.5 percent of PTDs before restoration to 21.6 percent of PTDs after restoration.

Kill Sites

During the four winters before corridor restoration (1993–1994 to 1996–1997), the Cascade pack killed one adult mule deer in the Cascade Corridor. The wolves also scavenged one female elk killed by a cougar. In the first winter following restoration of the corridor (1997–1998), the Cascade wolf pack killed one adult elk (sex unknown), one yearling male elk, three adult female elk, one adult bighorn sheep (sex unknown), and one adult mule deer (sex unknown). In the second year following restoration (1998–1999), the pack killed one adult mule deer (sex unknown).

Snow Depth

Variation in mean snow depth between years and location was unequal (Levene's $F_{7,121} = 5.016$; $p < 0.0005$). Using the SQRT $(x + 0.5)$ to transform

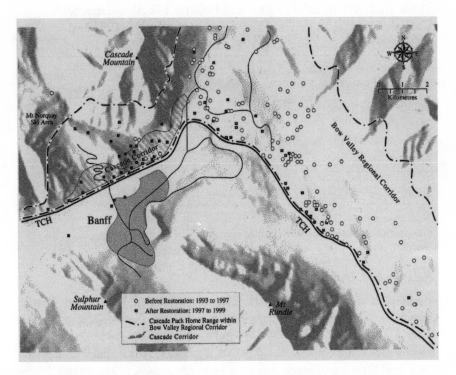

Figure 13.2. Radiotelemetry locations of the Cascade wolf pack within the Bow Valley Regional Corridor and Cascade Corridor before and after restoration.

snow depth data (Sokal and Rohlf 1995), we improved variance equality (Levene's $F_{7,121}$ = 2.711; p = 0.12), and found no significant differences in mean snow depth between the BVRC and Cascade Corridor ($F_{1,121}$ = 0.639; p = 0.426), and no interaction between corridor and year related to mean snow depth ($F_{3,121}$ = 0.221; p = 0.881). Snow depth differed significantly between years ($F_{3,121}$ = 10.457; p = 0.0005). Post hoc multiple comparisons revealed significant differences between 1995–1996 and 1996–1997 (p = 0.01), and between 1996–1997 and 1997–1998 (p < 0.005), and 1998–1999 (p < 0.005). Because no significant differences existed between corridor locations, we report only the snow depths in the Cascade Corridor (Table 13.2). Finally, we found no significant correlation between snow depth and time (Pearsons R = −0.789; p = 0.211).

Table 13.2. Mean Snow Depths and Relative Elk Abundance for Cascade Corridor Over Four-Year Study Period: 1995–1996 and 1996–1997 Prerestoration and 1997–1998 and 1998–1999 Postrestoration

Year	Elk crossings per 100 km	SD	N	Mean snow depth (cm)	SD	N
PRERESTORATION						
95/96	3.007	2.811	17	25.117	10.86	18
96/97	0.474	0.554	19	31.667	13.74	19
POSTRESTORATION						
97/98	1.345	2.040	18	19.932	8.40	18
98/99	1.315	1.963	12	12.556	4.64	12

Note: Years with same letters indicate significant differences revealed by posthoc Bonferroni multiple-comparison tests (experimentwise $\alpha = 0.05$).

Relative Elk Abundance

Analysis of variance of mean elk track counts between years and location showed no significant differences in counts between the BVRC and Cascade Corridor ($F_{1,120} = 2.034$; $p = 0.156$), and no interaction between corridor and year on elk track counts ($F_{3,120} = 1.266$; $p = 0.289$). Elk track counts did, however, differ significantly between years ($F_{3,120} = 4.110$; $p = 0.008$). Post hoc multiple comparisons indicated that this significant difference was between elk track counts during 1995–1996 and 1996–1997 ($p = 0.04$). Because no significant differences existed between corridor locations, we report only the elk track counts in the Cascade Corridor (Table 13.2). Finally, no significant correlation was found between elk track counts and time (Pearsons $R = -0.512$; $p = 0.488$).

Discussion

Wolf packs select travel routes that offer an optimal combination of security, habitat quality, and energetic efficiency. Wolves usually avoid human facilities and activities, terrain that is difficult to negotiate, and habitat of low quality. In the central Canadian Rockies, natural landforms and the condensed arrangement of potential habitats make wolves susceptible to the adverse effects of corridor disturbances. In less topographically complex environments such as the Great Lakes region, multiple travel routes link blocks of wolf habitat (D. Mech, pers. comm.; D. Mladenoff, pers. comm.). In such areas, destruction or degradation of one or two routes is not usually critical because safe alternatives are available. Wolves in mountainous environments, in contrast, are unable to avoid valley bottoms or use alternative travel routes without affecting their fitness (Paquet et al. 1996). Under these circumstances, tolerance of disruption may be lower than in other human-dominated environments (Minnesota, Wisconsin) where wolves can avoid disturbed sites without jeopardizing their survival.

We evaluated the response of wolves by comparing movements of packs and individuals before and after restoration. As predicted, use of the corridor increased significantly after human presence declined. The Cascade pack was in the restored Cascade Corridor every time we located them in the Bow River Valley. During the prerestoration period wolves inhabited the Bow River Valley for weeks at a time without using the Cascade Corridor. In 10 years of monitoring, packs of wolves were recorded moving through the Cascade Corridor in only the two years following restoration.

Snow depth did not explain the increase in use. In winter, deep snow in the backcountry restricted wolf movements to lower-elevation montane habitat such as the Bow River Valley (Paquet et al. 1996). Therefore, deep snow enhances wolf use of the Cascade Corridor whereas shallow snow reduces it. Contrary to our expectation, however, wolves increased their use of the Cascade Corridor despite shallow snow. Nor did ungulate density explain the increase in use. Wolves select areas of high elk density for hunting (Hebblewhite, unpublished data; Huggard 1993; Weaver 1994; Paquet et al. 1996). If ungulate abundance influenced wolf use of the corridor, we would expect the proportion of ungulates in the Cascade Corridor (relative to the BVRC) to have increased after restoration. Correspondingly, we would expect to see an increase in track abundance along the Cascade survey transects following corridor restoration. The only significant change in ungulate abundance was a decline in 1996–1997. Nevertheless, wolf use of the corridor increased.

Restoration of wolf movements through the Cascade Corridor may promote dispersal between the Cascade pack and the Bow River Valley pack and assist in maintaining local wolf populations. On the regional scale, restoration increases the function of the entire Bow River Valley Regional Corridor by reducing the resistance of an arterial connection—much as blood flow is improved with bypass surgery. This improvement will likely encourage regional dispersal of wolves in and around Banff National Park.

This landscape-scale experiment provides compelling evidence that a reduction in human structures and activity can promote wolf use of a wildlife corridor. Although a clear cause-and-effect relation is difficult to demonstrate under our rugged field conditions, we accounted for potential confounding factors such as shifting wolf packs, snow depths, and ungulate abundance. We have demonstrated that an experimental approach on a sub-regional scale not only furthers our understanding of the value of wildlife corridors but represents a valuable conservation tool. We agree with Beier and Noss (1998) that a truly experimental approach to corridor research is neither practical nor necessary to improve our knowledge of corridor structure and function. Experimentation on a scale relevant to wolf pack home ranges would require millions of dollars and replication across multiple landscapes. Although additional monitoring is desirable, our success seems to confirm that corridor restoration is achievable given ample understanding of the target species and a commitment to act on such knowledge. The results also suggest that habitat/movement models can be used successfully to indicate travel linkages and determine levels of human activity that impede wildlife movements. In sum, then, this process is a very promising approach for identification and restoration of wildlife movement corridors and should be applied elsewhere.

Acknowledgments

We appreciate generous financial support from Parks Canada, Alberta Human Resources and Employment, Human Resources Development Canada, Canadian-Pacific Foundation, Paquet Wildlife Fund, Peter W. Busch Family Foundation, Yellowstone to Yukon Initiative, and World Wildlife Fund Canada. We thank T. Hurd, C. White, and K. Heuer for support in all aspects of the research. M. Johnson and T. Shury provided veterinary advice and assistance for immobilization. M. Dupuis and L. Cooper provided safe fixed-wing and rotary wing aircraft support. We received invaluable field assistance from A. Adamson, S. Anderson, S. Antonation, C. Atkinson, C. Bates, M. Dragon, J. Elliott, S. Goulet, A. L. Horton, O, Huito, T. Kaminski, A. Kortello, M. Mauro, C. Mueller, C. Nietvelt, D. Robertson, S. Stevens, J. Vos, N. Waltho, and J. Wasylyk. We dedicate this chapter to warden Keith Everts, whose example inspired our efforts to protect national parks.

Literature Cited

Banff Bow Valley Study. 1996. *Banff Bow Valley, At the Crossroads*. Technical Report of the Banff Bow Valley Task Force (R. Page, S. Bayley, J. D. Cook, J. E. Green, J. R. Brent Ritchie) prepared for the Honorable Sheila Copps. Ottawa: Minister of Canadian Heritage.

Beier, P. 1995. Dispersal of juvenile cougars in fragmented habitat. *Journal of Wildlife Management* 59:228–237.

Beier, P., and R. Noss. 1998. Do habitat corridors provide connectivity? *Conservation Biology* 12:1241–1252.

Boyd, D. K., P. C. Paquet, S. Donelon, R. R. Ream, D. H. Petscher, and C. C. White. 1995. Transboundary movements of a recolonizing wolf population in the Rock Mountains. In L. N. Carbyn, S. H. Fritts, and D. R. Seip, eds., *Ecology and Conservation of Wolves in a Changing World.* Edmonton: Canadian Circumpolar Institute.

Boyd, D. K., and D. H. Pletscher. 1999. Characteristics of dispersal in a colonizing wolf population in the central rocky mountains. *Journal of Wildlife Management* 63:1094–1108.

Burkey, T. 1997. Metapopulation extinction in fragmented landscapes: Using bacteria and protozoa communities as model ecosystems. *American Naturalist* 15:568–591.

Duke, D. 1999. Wildlife corridors around developed areas in Banff National Park, Winter 1997/1998. Progress report. Banff: Parks Canada Banff Warden Service.

Dunning, J., R. Borgella, K. Clements, and G. Meffe. 1995. Patch isolation, corridor effects and colonization by a resident sparrow in a managed pine woodland. *Conservation Biology* 9:542–550.

Fleury, A., and R. Brown. 1997. A framework for the design of wildlife conservation corridors with specific application to southwestern Ontario. *Landscape and Urban Planning* 37:163–186.

Glenn, S. M., and T. D. Nudds. 1989. Insular biogeography of mammals in Canadian parks. *Journal of Biogeography* 16:261–268.

Green, J., C. Pacas, L. Cornwell, and S. Bayley, eds. 1996. *Ecological Outlooks Project: A Cumulative Effects Assessment and Futures Outlook of the Banff Bow Valley.* Report prepared for the Banff Bow Valley Study. Ottawa: Department of Canadian Heritage.

Gurd, D. B., and T. D. Nudds. 1999. Insular biogeography of mammals in Canadian parks: A re-analysis. *Journal of Biogeography* 26:973–982.

Harrison, R. 1992. Toward a theory of inter-refuge corridor design. *Conservation Biology* 6:293–295.

Heuer, K. 1995. Wildlife corridors around developed areas in Banff National Park. Banff: Parks Canada Banff Warden Service.

Heuer, K., R. Owchar, D. Duke, and S. Antonation. 1998. Wildlife corridors around developed areas in Banff National Park Winter 1996/97. Banff: Parks Canada Banff Warden Service.

Holroyd, G. L., and K. J. Van Tighem. 1983. Ecological (biophysical) land classification of Banff and Jasper National Parks—Vol. III: The Wildlife Inventory. Edmonton: Environment Canada.

Huggard, D. J. H. 1993. Prey selectivity of wolves in Banff National Park. I: Prey species. *Canadian Journal of Zoology* 71:130–139.

Ims, R. A., J. Rolstad, and P. Wegge. 1993. Predicting space use responses to habitat fragmentation: Can voles *Microtus oeconomus* serve as an experimental model system (EMS) for capercaille grouse *Tetrao urogallus* in boreal forest? *Biological Conservation* 63:261–268.

Kie, J. 1996. CALHOME: A program for estimating animal home ranges. *Wildlife Society Bulletin* 24:342–344.

Lindenmayer, D. B., R. B. Cunningham, C. F. Donnelly, M. T. Tanton, and H. A. Nix. 1993. The conservation of arboreal marsupials in the montane ash forests of the central highlands of Victoria, south-east Australia. IV: The distribution and abundance of arboreal marsupials in retained linear strips (wildlife corridors) in timber production forests. *Biological Conservation* 68:207–221.

Maehr, D. S. 1990. The Florida panther and private lands. *Conservation Biology* 4:167–170.

McNally, R., and A. F. Bennett. 1996. Species-specific predictions of the impact of habitat fragmentation: local extinction of birds in the Box-Ironbark Forests of Central Victoria, Australia. *Biological Conservation* 82:147–155.

Newmark, W. D. 1993. The role and design of wildlife corridors with examples from Tanzania. *Ambio* 22:500–504.

———. 1995. Extinction of mammal populations in western North American national parks. *Conservation Biology* 9(4):512–526.

Odette, L., and C. Thomas. 1996. Open corridors appear to facilitate dispersal by ringlet butterflies (*Aphantopus hyperantus*) between woodland clearings. *Conservation Biology* 10:1359–1365.

Oehler, J., and J. Litvaitis. 1996. The role of spatial scale in understanding responses of medium-sized carnivores to forest fragmentation. *Canadian Journal of Zoology* 74:2070–2079.

Paquet, P. C. 1993. Summary reference document—ecological studies of recolonizing wolves in the Central Canadian Rocky Mountains. Prepared by John/Paul & Associates for Parks Canada. Banff: Banff National Park Warden Service.

Paquet, P. C., J. Wierzchowski, and C. Callaghan. 1996. Effects of human activity on gray wolves in the Bow River Valley, Banff National Park, Alberta. In J. C. Green, C. Pacas, L. Cornwell, and S. Bayley, eds., *Ecological Outlooks Project: A Cumulative Effects Assessment and Futures Outlook of the Banff Bow Valley.* Prepared for the Banff Bow Valley Study. Ottawa: Department of Canadian Heritage.

Parks Canada. 1997. *Banff National Park Management Plan.* Banff: Banff National Park.

Rosenberg, D., B. R. Noon, and E. C. Meslow. 1997. Biological corridors: Form, function, and efficacy. *Bioscience* 47:677–686.

Saunders, D. A., and C. P. deRebeira. 1991. Values of corridors to avian populations in a fragmented landscape. In D. A. Saunders and R. J. Hobbs, eds., *Nature Conservation 2: The role of Corridors*. Chipping Norton: Surrey Beatty and Sons.

Sokal, R., and F. J. Rohlf. 1995. *Biometry: The Principles and Practice of Statistics in Biological Research*. New York: Freeman.

Soulé, M., and M. Gilpin. 1991. The theory of wildlife corridor capability. In D. A. Saunders and R. J. Hobbs, eds., *Nature Conservation 2: The Role of Corridors*. Chipping Norton: Surrey Beatty and Sons.

Stevens, S., C. Callaghan, and R. Owchar. 1996. A survey of wildlife corridors in the Bow Valley of Banff National Park, Winter 1994/95. Banff: Parks Canada Banff Warden Service.

Stevens, S., and R. Owchar. 1997. A survey of wildlife corridors in the Bow Valley of Banff National Park Winter 1995/96. Banff: Parks Canada Banff Warden Service.

Swart, J., and M. J. Lawes. 1996. The effect of habitat patch connectivity on samango monkey (*Cercopithecus mitis*) metapopulation persistence. *Ecological Modeling* 93:57–74.

Walker, R., and L. Craighead. 1997. Analyzing wildlife movement corridors in Montana using GIS. Report 11. The Green Papers. Bozeman: American Wildlands.

Weaver, J. 1994. Ecology of wolf predation amidst high prey diversity in Jasper National Park, AS. Ph.D. dissertation, University of Montana, Missoula, MT.

White, G. C., and Garrott, R. A. 1990. *Analysis of Wildlife Radio-Tracking Data*. Toronto: Academic Press.

Wilcox, B., and D. Murphy. 1985. Conservation strategy: The effects of fragmentation on extinction. *American Naturalist* 125:879–887.

Chapter 14

Tiger Restoration in Asia: Ecological Theory vs. Sociological Reality

RONALD TILSON, PHILIP NYHUS, NEIL FRANKLIN, SRIYANTO,
BASTONI, MOHAMMAD YUNUS, AND SUMIANTO

It is generally agreed that large mammal conservation requires large inter-connected tracts of habitat (Simberloff et al. 1999). Isolated reserves may be unable to support long-term viable populations of wide-ranging large mammals, whereas a well-connected network of reserves might be a surrogate for unfragmented landscapes (Harris 1984; Noss and Cooperrider 1994; Noss et al. 1996). Large mammal restoration has been mostly attempted in landscapes where human population densities are low, where land available for conservation is extensive, where poaching is rare, and where political and financial support for conservation is strong (Breitenmoser et al. 2001). Many examples are in North America—such as panther (*Puma concolor coryi*) habitat in southern Florida (Maehr 1997) and Greater Yellowstone Ecosystem grizzly bear (*Ursus arctos*) and wolf (*Canis lupus*) habitats (Noss and Cooperrider 1994; Fritts et al. 1995; Breitenmoser et al. 2001). Much of the world's biological diversity, however, occurs in fragmented landscapes where human population densities are high, little land is available for conservation, poaching is common, and financial and political support for conservation is weak (Table 14.1). How these ecological and sociological considerations are balanced in regions like Asia will determine the future of large mammals.

The Sumatran tiger (*Panthera tigris sumatrae*) highlights what we believe

Table 14.1. Comparison of Factors Influencing Large Mammal
Conservation and Restoration in North America and Southeast Asia

Characteristic	North America	Southeast Asia
Human population density	Low	High
Land available for conservation	Extensive	Limited
Poaching	Rare	Frequent
Political and financial support for conservation	High	Low

are important—and often overlooked—components of restoring large mammals. Our purpose here is to stimulate discussion about the capacity of current ecological paradigms to address the challenges of large mammal restoration in Asia that are caused by pressing sociological realities. Populations of tigers and other large mammals continue to decline across much of Asia, and the success of efforts to conserve and restore them depends in no small part on how human-dominated landscapes are incorporated into conservation plans. As Howard Quigley of the Hornocker Wildlife Institute has often stated, "Good conservation is based on good science" (pers. comm.). Unfortunately for the tiger and other large forest mammals in Asia, we are far from good science and conservation: we lack basic data on their distribution and ecology, we poorly understand their habitat requirements, and we have yet to identify all the threats to their existence (Seidensticker et al. 1999).

Background

Indonesia is the only country to have experienced the recent extinction of two tiger subspecies: the Bali tiger (*P. t. balica*) (Hoogerwerf 1970) and the Javan tiger (*P.t. sondaica*) (Seidensticker 1987). Sumatran tigers still remain, but in the last 20 or 30 years their population has dwindled as habitat has been converted and degraded across the island (Tilson et al. 1994). The future of the Sumatran tiger is far from secure. The situation is no better for many other large mammals that share the tigers' habitat, including the rhino, elephant, orangutan, and tapir. We emphasize this because we are concerned that major tiger conservation management decisions that impact the future of wild large-mammal populations will be made without adequate data or sufficient attention to the interplay of ecological theory and human needs.

Sumatran tigers once numbered in the thousands and were found across the island, but today they are increasingly restricted to a handful of isolated protected areas (Tilson et al. 1994). An estimated 500 Sumatran tigers remain, but their distribution is spread over an unknown number of small, disjunct populations in eight provinces. The decline of these populations and their increasingly isolated distribution is directly related to the rapid human population growth on the island and the large-scale loss and degradation of optimal habitat (Tilson et al. 1994). At a population and habitat viability analysis (PHVA) workshop held in Sumatra in 1992 (Tilson et al. 1994), human population growth, transmigration programs, and agriculture were cited as reasons for the decline in nonprotected tiger habitat. There are two primary threats to the Sumatran tiger: a small population size and the unnatural removal of tigers from small populations. Although poaching and other causes may remove only a few tigers, these losses may be problematic (Seal et al. 1994; Tilson et al. 1994). We still do not know exactly how many tigers remain, their distribution, or the extent of tiger/human conflict in Sumatra (Tilson 1999). Since 1995 we have examined these questions at Way Kambas National Park as part of the Sumatran Tiger Project (Franklin et al. 1999; Nyhus et al. 1999; Tilson et al. 1996, 1997; Tilson 1999) and are now extending our approach and findings to other areas of Sumatra.

There is growing agreement among tiger conservationists that a set of basic factors are crucial to the survival of tigers in the wild. These factors include sufficient habitat area, sufficient prey, low human disturbance, and genetic viability (Norchi and Bolze 1995; Nowell and Jackson 1996; Seidensticker 1997; Seidensticker et al. 1999). Where these conditions prevail, tigers can be resilient because their populations can offset losses and their offspring can colonize new areas (Smith 1993; Sunquist et al. 1999). Because of this adaptability and resilience, tigers were among the most widely distributed cat species (Nowell and Jackson 1996). Today, their distribution is increasingly confined to protected areas that are small and fragmented (Dinerstein et al. 1997; Nowell and Jackson 1996). The recent extinctions of the Javan and Caspian (*P. t. virgata*) tigers are reminders that restricted distributions, persecution by humans, and loss of prey populations can be terminal (Seidensticker 1987; Seidensticker et al. 1999; Sunquist et al. 1999).

The rate of land cover change in many areas of Asia is high. In most of Sumatra, little baseline information exists and the status of certain nonprotected forests remains virtually unknown. Government maps frequently identify forests when in fact they have been converted to other uses such as plantations, farms, and settlements. In 1997 alone, an estimated 15,000 km^2 were burned by major fires in Sumatra (Levine et al. 1999). Without field verification or remote monitoring from aerial and satellite imagery, we can only speculate how much land is really available for conservation and restoration. In the meantime, habitat continues to be lost.

Even less is known about the type and extent of other threats facing tigers and many other large mammals. Illegal hunting is a clear and present danger across all of the tiger's range. There is little information about poaching in most of Asia. A growing body of literature suggests that retribution for attacks on humans and livestock may be a significant reason for the tiger's decline (McDougal 1987; Tilson and Nyhus 1998). A review of reports and press accounts suggests that between 1978 and 1997 as many as 146 people were killed by tigers and more than 350 tigers (approximately 17.5 per year) were killed or captured across the entire island of Sumatra (Nyhus 1999).

Study Area and Methods

The island of Sumatra is one of 17,000 islands in the republic of Indonesia—the world's largest archipelago, fourth most populous country, and home to some of the richest biological diversity on the planet (MSPE 1992; Whitten et al. 1987). Sumatra is the fifth-largest island in the world and the second most populous in Indonesia after Java. Covering 474,000 km², an area just larger than the state of California, it is also home to 45 million people (BPS 1999)—and rapid deforestation. Today less than 20 percent of its once abundant lowland forest remains (Collins et al. 1991; Whitten et al. 1987).

Sumatra's protected-area system contains many small reserves and few large ones. The total land area managed for protection covers approximately 17 percent of the island. Of its 230 protected areas, 75 percent are smaller than 300 km² and only ten (4 percent) are greater than 1000 km². Large reserves account for 44 percent of the total protected area in Sumatra. The three largest national parks—Kerinci Seblat (13,680 km²), Gunung Leuser (7927 km²), and Bukit Barisan (3650 km²)—account for 31.3 percent of protected land. More than half of Sumatra's total protected-area system is not managed primarily for ecosystem protection, let alone for tigers. Some 83 percent of the total number and 54 percent of the total area of the island's protected area system is classified as protection forest (*Hutan Lindung*) where management features erosion control, watershed protection, and timber harvest (Table 14.2). The only strictly protected areas of significance are the island's six national parks (*Taman Nasional*), which represent 6.2 percent of the island's total area.

To address the lack of data about tiger ecology, distribution, habitat, and survival threats in Sumatra, we assessed potential habitat and threats by examining 52 mapped tracts of potential tiger habitat (Franklin et al. 1999; Nyhus 1999). First, we consulted forestry officials to eliminate areas that were known to have been converted or otherwise unlikely to contain tigers. Second, we searched the remaining 15 areas for tiger signs (scrapes, foot-

Table 14.2. Number and Size of Protected Areas in Sumatra by Official Indonesian and IUCN Categories

Category	N	% total N	Sum area (km²)	Mean area (km²)	SD	% total area
Grand forest park	1	0.4	222	222.0	0	0.3
Hunting park	4	1.7	1,149	287.3	343	1.4
Recreation park	5	2.2	223	44.6	52	0.3
National park	6	2.6	29,461	4,910.2	4992	36.6
National reserve	9	3.9	567	63.0	76	0.7
Game reserve	13	5.6	5,261	404.7	322	6.5
Protection forest	192	83.1	43,689	227.5	308	54.2
TOTAL	231	100.0	80,572	348.8	1095	100.0

prints, urine sprays, or feces) and set up motion-activated cameras along trails to document tiger presence and prey identity. Third, we interviewed active and former poachers to better understand the extent and rate of illegal tiger losses over the last ten years.

Results and Discussion

Of the 15 sites that supported extensive forest, signs of tigers were found at six, including Sumatra's two national parks and two adjacent areas (Sumatran Tiger Project 1999). At Way Kambas National Park at least 37 tigers were identified (Franklin et al. 1999). Signs of known tiger prey species were found at nine sites and included sambar deer (*Cervus unicolor*), barking deer (*Muntiacus muntjak*), and wild pig (*Sus scrofa*) (Figure 14.1).

The actual tiger density in most of the protected areas in Sumatra is unknown, but it is likely to be between one and four tigers per 100 km² (Franklin et al. 1999; Griffiths 1994). Using the more conservative estimate of one tiger per 100 km², only two protected areas could have at least 50 tigers and none would have 250—an estimated 100-year minimum viable population (Seal et al. 1994). At four tigers per 100 km², all of the national parks could theoretically contain from 50 to 250 tigers (Figure 14.2). These estimates do not consider the effective habitat available, which may be considerably less than the total due to inholdings, disturbance, hunting pressures, and edge effects. Mountainous topography, rivers, roads, and other landscape features further reduce the amount of suitable tiger habitat. Moreover, not all tigers will be of breeding age in a given population (too young or too old). Thus, even the largest protected areas in Sumatra, in isolation, are unlikely to

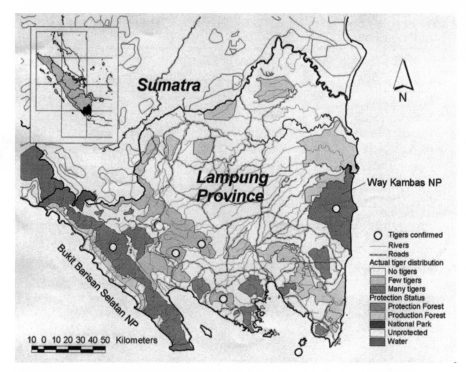

Figure 14.1. Forest habitat in Lampung province, Sumatra, and distribution of tigers based on preliminary field assessments.

contain 100 breeding tigers. Based on the results of a population viability analysis, only two of Sumatra's national parks are capable of supporting a viable population of tigers for more than 100 years. Each of the remaining protected areas is unlikely to support a viable population for more than 100 years, even if poaching, habitat loss, and disturbance are controlled (Seal et al. 1994).

The actual area available for tiger conservation in Sumatra is almost certainly more isolated, fragmented, and degraded than is suggested on paper. In Lampung province, for example, more than 11,456 km^2 is theoretically available for conservation. Many of these reserves appear to be linked on paper to other protected areas to form a connected network of protected habitat (Figure 14.1). But the low occupancy rate in these reserves suggests that this paper tiger metapopulation is only that. All of these areas, large and small, are close to or bisected by road networks, towns and cities, and agriculture. Further, none of them is effectively protected from poaching. In 12 of the 15 areas where tigers were found, we observed signs of illegal hunting. Even in

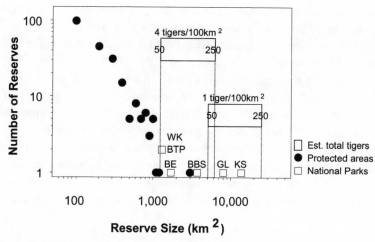

Figure 14.2. Number and size of protected areas in Sumatra and hypothetical tiger populations.

the national parks, game hunting, fishing, bird collecting, and other exploitation is common (Nyhus 1999; O'Brien et al. 2000). Interviews with 35 active and retired poachers revealed that close to 400 tigers have been killed over an eight-year period in southern Sumatra—and that the number of tigers killed each year has increased.

Sociological Considerations

The status of tigers and tiger habitat in Sumatra becomes clearer when examined in the context of recent social and political crises. The human population increased by at least 3.5 percent between 1971 and 1980 and 3 percent between 1981 and 1990 to over 75/km^2 on the island of Sumatra and 200/km^2 in Lampung (Nyhus 1999). This rapid growth comes from government-sponsored immigration of largely poor, rural people (Fearnside 1997). And most new settlement has occurred in fertile lowland forests adjacent to national parks (Whitten 1987). This development in tiger habitat is expected to continue for several decades. Expansion of cultivation, estate crops, and tree plantations has had a dramatic impact on Sumatra's landscape. Oil palm (*Elaeis guineensis*) alone accounts for at least 1.8 million hectares and is increasing due to Indonesia's efforts to become the world's largest producer of palm oil (Potter and Lee 1998).

Economic and political upheaval in Indonesia and the Asian economic collapse have resulted in high unemployment in some rural areas. In early

1998, massive student protests in Jakarta brought down Indonesia's president of more than 30 years and plunged the country into political confusion. This political and economic turmoil coincided with drought caused by the El Niño Southern Oscillation (ENSO) events of 1997–1998 (Levine et al. 1999). Fires set to clear new areas for cultivation often burned out of control. As a result, at least 15,000 km² of forest were consumed (Levine et al. 1999). Even in Way Kambas National Park, thousands of villagers were encouraged by a lucrative seafood market to illegally enter the park near the southwest boundary to dig waterways for shrimp farms. More than 400 people moved into the park and occupied 1225 ha before they were evicted by park staff, military, and police. In Lampung, several of the protection forests were cleared during 1998–1999, and more are likely to follow in the wake of political and economic instability.

Conservation and Restoration Challenges

Sumatra is typical of Asia, where protected areas are small (Dinerstein and Wikramanayake 1993). Compared to large reserves in North America, many of Sumatra's protected areas are unlikely to maintain viable populations of large mammals and are too small to be considered for ecosystem restoration. If size alone is considered, only two reserves are likely to maintain demographically and genetically viable populations of tigers over the next 100 years. Potential tiger habitat could be expanded if the 50 largest (more than 300 km²) of Sumatra's protection forests could be secured immediately. If tiger hunting is sustainable and prey is abundant, demographically viable tiger populations can survive in reserves as small as 300 km² (Karanth and Stith 1999). The magnitude of this task, however, is daunting and the obstacles are tremendous because of the great number of people living adjacent to these forests and extracting resources from them.

Small reserves may be easier to protect than large reserves (Peres and Terborgh 1995). At Way Kambas National Park, a small reserve, 70 percent of the reserve is surrounded by water and boat patrols can monitor most of the boundary. Land patrols can easily traverse the park and its upland border. In the large Kerinci Seblat National Park, law enforcement is more difficult due to inholdings along access roads, poorly demarcated boundaries, and easy entry (MacKinnon 1997). Future management of an expanding tiger reserve system will need to incorporate effective regulation of access (Kramer et al. 1997).

In keeping with modern conservation theory (Harris 1984; Noss and Harris 1986; Noss and Cooperrider 1994; Dobson et al. 1999), landscape linkages are needed to facilitate the movement of animals and their genes within a metapopulation. Ideally, with adequate restoration of forest corridors between and among core protected areas, tigers would colonize new areas. The most likely location of such a network would be along the island's western Barisan Mountain chain connecting three of Sumatra's largest national parks (Gunung Leuser, Kerinci Seblat, and Bukit Barisan Selatan) and stretching east through the center of the island to Bukut Tigapuluh and Berbak national parks.

The reality of Sumatra's rapidly changing landscape, however, is that few protected areas today can be connected effectively by habitat corridors. Way Kambas National Park, for example, is surrounded by 27 villages that support nearly 500,000 people within 10 km (Nyhus 1999). A corridor linking Way Kambas National Park to Bukit Barisan Selatan National Park would need to cross 100 km of densely populated and developed lands with human densities averaging 200/km^2 (Nyhus 1999). Such landscape retrofitting would involve the relocation of thousands of families and the restoration of heavily used farmlands.

Such challenges do not obviate the need to restore connectivity to tiger habitat and tiger populations. But where habitat connectivity is not currently feasible, gene flow can be maintained with "virtual corridors." By this we mean that tigers can be moved among core protected areas or, in extreme cases, among captive facilities and wild populations. Translocation of tigers from one area to another could create gene flow among small but demographically healthy populations (Dobson et al. 1999). The global captive breeding community has already made significant progress in managing small populations of tigers through managed reproduction to maximize genetic diversity (Tilson and Christie 1999). Further convergence of ex-situ and in-situ programs is possible. With proper monitoring and safeguards in place, for example, captive tigers could be returned to the wild. While reintroduction of problem tigers is unlikely, the reintroduction of their offspring may be more acceptable. Such links may forestall the effects of human population growth and continued habitat loss and allow for the development of a tiger habitat restoration strategy.

Buffer zones (Sayer 1991; Shafer 1990) extend the available habitat for plants and animals (extension buffering) and provide resources and services to people (socio-buffering) (MacKinnon et al. 1986). The challenge is to identify land use that fills both roles (Salafsky 1993). In the Tropics—and particularly in Southeast Asia—buffers that incorporate complex (multispecies) agroforestry systems into forest preserves are a promising approach. (See Siebert 1989; Michon and d'Foresta 1990; Salafsky 1993; Van Shaik and Terborgh 1993; Potter and Lee 1998; Vandermeer et al. 1998; Johns 1999.)

But complex agroforestry creates conditions that may increase conflict between wildlife and people (Tilson and Nyhus 1998). Our observations (Nyhus 1999) support those of Woodroffe and Ginsberg (1998) who found that conflicts with people on reserve borders are the major cause of large-carnivore mortality. If large carnivores are to survive, more attention must be given to reducing human/wildlife conflicts at the edges of protected areas. Although the role of conflict in conservation planning has not received as much attention as habitat loss and poaching (Tilson and Nyhus 1998), it may need to play a more central role in efforts to expand or restore corridors and buffer zones adjacent to protected core tiger habitat.

Many specialist, forest-interior species in the tropics are vulnerable to logging, edge effects, and other disturbances (Bierregaard et al. 1992; Lovejoy et al. 1986). Tigers, however, can thrive in secondary growth and edges where primary productivity supports a high ungulate biomass (Nowell and Jackson 1996). Way Kambas National Park is a good example of this. It was commercially logged between 1954 and 1974, resulting in the clearing of at least 75 percent of the reserve (ANZDEC and AMYTHAS 1995; Ministry of Forestry 1995). Today the park's secondary-growth forest and mixed forest–grasslands contain what may be some of Sumatra's highest densities of ungulates and tigers (Franklin et al. 1999). This suggests that degraded tiger habitat in Sumatra can in fact recover from intensive human uses. Given that many primary forests are being intensively used and converted to early successional stages, we suggest that future tiger management integrate secondary forests into an islandwide system of preserves and multiple-use areas (Noss and Harris 1986).

Managers hoping to restore degraded habitats face considerable obstacles: cost, evicting and moving illegal settlers, and lack of political will. Sites that now have permanent settlements are unlikely to recover sufficiently to support large mammals, let alone tigers. Recently cleared areas can revert to forest, but each day without protection increases restoration costs and enhances the likelihood of conversion to farms and settlements. Nevertheless, some areas such as agricultural and tree plantations, if left to revert to forest, could support tigers within decades. Although such a scenario may seem optimistic, declines in the value of palm oil, rubber, timber, and other Sumatran agricultural products could create opportunities for restoring tiger habitat.

Incorporating Social and Political Considerations

Ecological theory and reserve design are important aspects of large mammal conservation and restoration. But successful restoration of wide-ranging species such as the tiger must also address many social and political challenges, especially in regions such as Southeast Asia (Seidensticker et al. 1999). Although field researchers are expected to give policymakers conservation recommendations that are based on good science, ecological decision making is often made in a vacuum or against long odds. The feasibility of connecting large reserves to form a connected network currently pales in the face of an economy that encourages forest settlement and conversion to cultivation.

We believe it is time to find out whether restoration theory in areas like North America is—or is not—appropriate to the challenge of restoring large-mammal habitat in the human-dominated landscapes of Asia. When the most ecologically appropriate options are not available, what is the next best option? This is a central dilemma in the fight to save the tiger in Indonesia. While law enforcement, real and virtual corridors, buffer zones, restoration of degraded habitat, and better integration of social and political considerations are all components of successful tiger management, a paradigm that incorporates them all has yet to be implemented. The foundation for effective tiger conservation must synthesize ecological theory, sociological reality, and accurate field data. How we address this challenge today will preserve the Sumatran tiger—or seal its fate with the extinct tigers of Java and Bali.

Literature Cited

ANZDEC and AMYTHAS. 1995. National conservation plan for Indonesia: A review and update of the 1982 national conservation plan for Indonesia, Vol. 3H: Lampung Province. Jakarta: Ministry of Forestry, Directorate General of Forest Protection and Nature Conservation

Bierregaard, R. O. J., T. E. Lovejoy, V. Kapos, A. A. dos Santos, and R. W. Hutchings. 1992. The biological dynamics of tropical rainforest fragments. *BioScience* 42:859–866.

BPS. 1999. Statistics by regions: Country's statistical profile and provincial profile. http://www.bps.go.id/profile/index.html.

Breitenmoser, U., C. Breitenmoser-Würsten, L. N. Carbyn, and S. M. Funk. 2001. Assessment of carnivore reintroductions. In J. L. Gittleman, S. M. Funk, D. W. Macdonald, and R. K. Wayne, eds., *Carnivore Conservation*. Cambridge: Cambridge University Press.

Collins, N. M., J. A. Sayer, and T. C. Whitmore, eds. 1991. *The Conservation Atlas of Tropical Forests: Asia and the Pacific*. London: Macmillan.

Dinerstein, E., and E. D. Wikramanayake. 1993. Beyond "hotspots": How to prioritize investments in biodiversity in the Indo-Pacific region. *Conservation Biology* 7:53–65.

Dinerstein, E., E. Wikramanayake, J. Robinson, U. Karanth, A. Rabinowitz, D. Olson, T. Mathew, P. Hedao, M. Connor, G. Hemley, and D. Bolze. 1997. *A Framework for Identifying High Priority Areas and Actions for the Conservation of Tigers in the Wild.* Washington, D.C.: World Wildlife Fund–U.S. and Wildlife Conservation Society.

Dobson, A., K. Ralls, M. Foster, M. E. Soulé, D. Simberloff, D. Doak, J. A. Estes, L. S. Mills, D. Mattson, R. Dirzo, H. Arita, S. Ryan, E. A. Norse, R. F. Noss, and D. Johns. 1999. Connectivity: Maintaining flows in fragmented landscapes. In M. E. Soulé and J. Terborgh, eds., *Continental Conservation: Scientific Foundations of Regional Reserve Networks.* Washington, D.C.: Island Press.

Fearnside, P. 1997. Transmigration in Indonesia: Lessons from its environmental and social impacts. *Environmental Management* 21:553–570.

Franklin, N., Bastoni, Sriyanto, S. Dwiatmo, J. Manansang, and R. Tilson. 1999. Last of the Indonesian tigers: A cause for optimism. In J. Seidensticker, S. Christie, and P. Jackson, eds., *Riding the Tiger: Tiger Conservation in Human-Dominated Landscapes.* Cambridge: Cambridge University Press.

Fritts, S. H., E. E. Bangs, J. A. Fontaine, W. G. Brewster, and J. F. Gore. 1995. Restoring wolves to the northern Rocky Mountains of the United States. In L. N. Carbyn, S. H. Fritts, and D. R. Seip, eds., *Ecology and Conservation of Wolves in a Changing World.* Edmonton: Canadian Circumpolar Institute.

Griffiths, M. 1994. Population density of Sumatran tigers in Gunung Leuser National Park. In R. L. Tilson, K. Soemarna, W. Ramono, S. Lusli, K. Traylor-Holzer, and U. S. Seal, eds., *Sumatran Tiger Population and Habitat Viability Analysis Report.* Apple Valley, Minn.: IUCN/SSC Captive Breeding Specialist Group.

Harris, L. D. 1984. *The Fragmented Forest: Island Biogeography Theory and the Preservation of Biotic Diversity.* Chicago: University of Chicago Press.

Hoogerwerf, A. 1970. *Udjung Kulon: The Land of the Last Javan Rhinoceros.* Leiden: E. J. Brill.

Johns, N. 1999. Conservation in Brazil's chocolate forest: The unlikely persistence of the traditional cocal agroecosystem. *Environmental Management* 23:31–47.

Karanth, K. U., and B. M. Stith. 1999. Prey depletion as a critical determinant of tiger populations. In J. Seidensticker, S. Christie, and P. Jackson, eds., *Riding the Tiger: Tiger Conservation in Human-Dominated Landscapes.* Cambridge: Cambridge University Press.

Kramer, R., C. v. Schaik, and J. Johnson, eds. 1997. *Last Stand: Protected Areas and the Defense of Tropical Biodiversity.* New York: Oxford University Press.

Levine, J. S., T. Bobbe, N. Ray, R. G. Witt, and A. Singh. 1999. Wildland fires and the environment: A global synthesis. UNEP/DEIAEW/TR.99-1. Nairobi: United Nations Environment Program.

Lovejoy, T. E., J. R. O. Bierregaard, A. B. Rylands, J. R. Malcolm, C. E. Quintela, L.

H. Harper, J. K. S. Brown, A. H. Powell, G. V. N. Powell, H. O. R. Schubart, and M. B. Hays. 1986. Edge and other effects on isolation on Amazon forest fragments. In M. E. Soulé, ed., *Conservation Biology: The Science of Scarcity and Diversity.* Sunderland, Mass.: Sinauer Associates.

MacKinnon, J., C. MacKinnon, G. Child, and J. Thorsell. 1986. *Managing Protected Areas in the Tropics.* Gland, Switzerland: IUCN.

MacKinnon, K. 1997. The ecological foundations of biodiversity protection. In R. Kramer, C. V. Schaik, and J. Johnson, eds., *Last Stand: Protected Areas and the Defense of Tropical Biodiversity.* New York: Oxford University Press.

Maehr, D. S. 1997. *The Florida Panther: Life and Death of a Vanishing Carnivore.* Washington, D.C.: Island Press.

McDougal, C. 1987. The man-eating tiger in geographic and historical perspective. In R. L. Tilson and U. S. Seal, eds., *Tigers of the World: The Biology, Biopolitics, Management, and Conservation of an Endangered Species.* Park Ridge, N.J.: Noyes.

Michon, G., and H. d'Foresta. 1990. Complex agroforestry systems and the conservation of biological diversity in harmony with nature. In *International Conference on Tropical Biodiversity.* Kuala Lumpur.

Ministry of Forestry. 1995. *Way Kambas National Park Management Plan: 1994–2019.* Bogor: Ministry of Forestry, Directorate General of Forest Protection and Nature Conservation.

MSPE. 1992. Draft report of the Indonesian country study on biological diversity. Jakarta: Ministry of State for Population and Environment.

Norchi, D., and D. Bolze. 1995. *Saving the Tiger: A Conservation Strategy.* New York: Wildlife Conservation Society.

Noss, R. F., and L. D. Harris. 1986. Nodes, networks, and MUMs: Preserving diversity at all scales. *Environmental Management* 10:299–309.

Noss, R. F., and A. Y. Cooperrider. 1994. *Saving Nature's Legacy: Protecting and Restoring Biodiversity.* Washington, D.C.: Island Press.

Noss, R. F., H. B. Quigley, M. G. Hornocker, T. Merrill, and P. C. Paquet. 1996. Conservation biology and carnivore conservation in the Rocky Mountains. *Conservation Biology* 10:949–963.

Nowell, K., and P. Jackson, eds. 1996. *Wild Cats: Status Survey and Conservation Action Plan.* Gland, Switzerland: IUCN.

Nyhus, P. J. 1999. Elephants, tigers, and transmigrants: Conflict and conservation at Way Kambas National Park, Sumatra, Indonesia. Ph.D. dissertation, University of Wisconsin, Madison.

Nyhus, P., Sumianto, and R. Tilson. 1999. The tiger human dimension in southeast Sumatra, Indonesia. In J. Seidensticker, S. Christie, and P. Jackson, eds., *Riding the Tiger: Tiger Conservation in Human-Dominated Landscapes.* Cambridge: Cambridge University Press.

O'Brien, T. G., H. T. Wibisono, and M. F. Kinnaird. 2000. The influence of a declining prey base and human encroachment on a Sumatran tiger population (abstract). In *Program and Abstracts*. Missoula: Society for Conservation Biology.

Peres, C. A., and J. W. Terborgh. 1995. Amazonian nature reserves: An analysis of the defensibility status of existing conservation units and design criteria for the future. *Conservation Biology* 9:34–46.

Potter, L., and J. Lee. 1998. Tree planting in Indonesia: Trends, impacts, and directions. Bogor: Center for International Forestry Research.

Salafsky, N. 1993. Mammalian use of a buffer zone agroforestry system bordering Gunung Palung National Park, West Kalimantan, Indonesia. *Conservation Biology* 7:928–933.

Sayer, J. 1991. *Rainforest Buffer Zones: Guidelines for Protected Area Managers*. Gland, Switzerland: IUCN.

Seal, U., K. Soemarna, and R. Tilson. 1994. Population biology and analyses for Sumatran tigers. In R. L. Tilson, K. Soemarna, W. Ramono, S. Lusli, K. Traylor-Holzer, and U. S. Seal, eds., *Sumatran Tiger Population and Habitat Viability Analysis Report*. Apple Valley, Minn.: IUCN/SSC Captive Breeding Specialist Group.

Seidensticker, J. 1987. Bearing witness: Observations on the extinction of *Panthera tigris balica* and *Panthera tigris sondaica*. In R. L. Tilson and U. S. Seal, eds., *Tigers of the World: The Biology, Biopolitics, Management, and Conservation of an Endangered Species*. Park Ridge, N.J.: Noyes.

———. 1997. Saving the tiger. *Wildlife Society Bulletin* 25:6–17.

Seidensticker, J., S. Christie, and P. Jackson, eds. 1999. *Riding the Tiger: Tiger Conservation in Human-Dominated Landscapes*. Cambridge: Cambridge University Press.

Shafer, C. L. 1990. *Nature Reserves: Island Theory and Conservation Practice*. Washington, D.C.: Smithsonian Institution Press.

Siebert, S. F. 1989. The dilemma of a dwindling resource: Rattan in Kerinci, Sumatra. *Principes* 33:79.

Simberloff, D., D. Doak, M. Groom, S. Trobulak, A. Dobson, S. Gatewood, M. E. Soulé, M. Gilpin, C. Martinez del Rio, and L. Mills. 1999. Regional and continental conservation. In M. E. Soulé and J. Terborgh, eds., *Continental Conservation: Scientific Foundations of Regional Reserve Networks*. Washington, D.C.: Island Press.

Smith, J. L. D. 1993. The role of dispersal in structuring the Chitwan tiger population. *Behavior* 24:195.

Sumatran Tiger Project. 1999. *Sumatran Tiger Project Report*. Minneapolis: Sumatran Tiger Project.

Sunquist, M., K. U. Karanth, and F. Sunquist. 1999. Ecology, behavior and resilience of the tiger and its conservation needs. In J. Seidensticker, S. Christie, and P. Jackson, eds., *Riding the Tiger: Tiger Conservation in Human-Dominated Landscapes*. Cambridge: Cambridge University Press.

Tilson, R. 1999. Sumatran tigers: From PHVA to conservation action. *Cat News* 31:3–6.

Tilson, R. L., K. Soemarna, W. Ramono, S. Lusli, K. Traylor-Holzer, and U. S. Seal. 1994. Sumatran tiger population and habitat viability analysis report. Apple Valley, Minn.: IUCN/SSC Captive Breeding Specialist Group.

Tilson, R., N. Franklin, P. Nyhus, Bastoni, Sriyanto, D. Siswomartono, and J. Manansang. 1996. In situ conservation of the Sumatran Tiger in Indonesia. *International Zoo News* 43:316–324.

Tilson, R., D. Siswomartono, J. Manansang, G. Brady, D. Armstrong, K. Traylor-Holzer, A. Byers, P. Christie, A. Salfifi, L. Tumbelaka, S. Christie, D. Richardson, S. Reddy, N. Franklin, and P. Nyhus. 1997. International co-operative efforts to save the Sumatran tiger *Panthera tigris sumatrae. International Zoo Yearbook* 35:129–138.

Tilson, R., and P. Nyhus. 1998. Keeping problem tigers from becoming a problem species. *Conservation Biology* 12:261–262.

Tilson, R., and S. Christie. 1999. Effective tiger conservation requires cooperation: Zoos as a support for wild tigers. In J. Seidensticker, S. Christie, and P. Jackson, eds., *Riding the Tiger: Tiger Conservation in Human-Dominated Landscapes.* Cambridge: Cambridge University Press.

Vandermeer, J., M. v. Noordwijk, J. Anderson, and C. Ong. 1998. Global change and multi-species agroecosystems: Concepts and issues. *Agricultural Systems and Environment* 67:1–22.

Van Shaik, C. P., and J. Terborgh. 1993. Production forests and protected forests: The potential for mutualism in the tropics. *Tropical Biodiversity* 1:183.

Whitten, A. J. 1987. Indonesia's transmigration program and its role in the loss of tropical rain forests. *Conservation Biology* 1:239–246.

Whitten, A. J., S. J. Damanik, J. Anwar, and N. Hisyam. 1987. *The Ecology of Sumatra.* Yogyakarta: Gajah Mada University Press.

Wikramanayake, E. D., E. Dinerstein, J. G. Robinson, K. U. Karanth, A. Rabinowitz, D. Olson, T. Mathew, P. Hedao, M. Connor, G. Hemley, and D. Bolze. 1999. Where can tigers live in the future? A framework for identifying high-priority areas for the conservation of tigers in the wild. In J. Seidensticker, S. Christie, and P. Jackson, eds., *Riding the Tiger: Tiger Conservation in Human-Dominated Landscapes.* Cambridge: Cambridge University Press.

Woodroffe, R., and J. R. Ginsberg. 1998. Edge effects and the extinction of populations inside protected areas. *Science* 280:2126–2128.

Chapter 15

The Florida Panther: A Flagship for Regional Restoration

DAVID S. MAEHR, THOMAS S. HOCTOR, AND LARRY D. HARRIS

The Florida panther (*Puma concolor coryi*) is listed as endangered by the state of Florida and the U.S. government (USFWS 1987). Although loss of habitat is the single most frequently cited cause of its endangerment (Pearlstine et al. 1995), the principal tools aimed at recovery and management sidestep the issue of disappearing forest cover. Unlike conspecifics in western North America, the panther is an obligate forest creature: the rugged topography and often treeless expanses of western mountain lion range are replaced in Florida by remote sea-level swamps and dense upland forests (Maehr 1997a). Southern Florida's forests provide panthers with virtually all of their life history requirements from natal dens (Maehr et al. 1989) and day beds (Maehr et al. 1990a) to ungulate prey (Maehr et al. 1990b). Individual panthers maintain home ranges that span ecosystems, and dispersing animals can roam over hundreds of kilometers. Yet managers of public preserves are forced to satisfy the panther's needs on increasingly isolated and disjointed patches of forest as recovery protocols become increasingly parochial. What was once a holistic approach that considered the Florida panther's health, genetics, demographics, land management, and reintroduction (USFWS 1987) now targets a single solution, genetic restoration, as its recovery focus.

Since the publication of *The Theory of Island Biogeography* (MacArthur and Wilson 1967), the spatial attributes of autecology have received increasing attention by researchers. Theoretical and applied approaches to mammalian landscape ecology emerged in the early 1970s as the result of work primarily on rodent pests (Hansson 1995). In recent years, numerous papers, an entire book, and a special journal section have been devoted to the theory and practice of mammalian landscape ecology and the importance of this new field to conservation efforts (Lidicker 1995; Bowers 1997). Although the relation of the Florida panther to the landscape has been examined (Maehr and Cox 1995), the simple discovery that the subspecies is inextricably linked to forest cover has been insufficient to expand land management beyond the boundaries of present-day public preserves. Harris (1984) outlined the dangers to wide-ranging mammals from deforestation, fragmentation, and the ecological isolation of parks and preserves more than a decade ago.

This chapter examines the evolution of restoration efforts for a geographically isolated large carnivore and looks beyond current recovery plans in pursuit of a landscape solution to the Florida panther's tenuous existence. Recent recovery planning by the Florida Panther Interagency Committee (composed of two state and two federal natural resource agencies) has reduced the possibility that a landscape approach will be used to manage the Florida panther. Original recovery objectives—such as the establishment of a captive breeding program and the reintroduction of panthers to north Florida—have been scuttled in favor of a more expedient approach that primarily involves the introduction of individuals from another subspecies into the single known population (Maehr and Caddick 1995). These actions spell trouble for the eventual return of an animal that now occupies only 5 percent of its original range (Maehr 1997a).

The Park Paradigm

The global adoption of the park paradigm, based on the Yellowstone model (Wright and Mattson 1996), reflected an effort to save vignettes of outstanding scenic beauty and provide spiritual benefits to park users (Muir 1916). Planning for the long-term viability of complete species assemblages or for the integration of each park with the surrounding landscape was seldom attempted. (See Wright et al. 1933, Wright and Thompson 1934, and Shelford 1936 for early criticism of this approach.) Although the park mission has grown to include the protection of biota, U.S. parks and parklike preserves are primarily set-asides for human recreation (Wright and Mattson 1996). From a landscape perspective, the traditional park paradigm—managing a discrete area in isolation from the surrounding landscape—is insufficient to provide for the long-term security of species and biotic communities (Harris et al. 1996). Today, at least in a design sense, there is growing interest in the integration of parks with larger-scale plans that incorporate landscape connectivity and population considerations spanning entire states and continents (Harris and Hoctor 1992; Foreman et al. 1992; Soulé and Noss 1998; Hoctor et al. 2000). Natural resource agencies have been slow to embrace such strategies (Grumbine 1992:171).

The principal wildlife policymaker in the United States, the U.S. Fish and Wildlife Service (USFWS), is traditionally a steward of biodiversity from a single-species management perspective. Although much of its effort involves habitat protection, this emphasis has stemmed largely from a mandate to focus on individual game species, nuisance species, and especially endangered species (Foin et al. 1998). Recently this tradition has grown to include entire communities, ecosystems, and multi-species recovery plans (USFWS 1998)—changes that reflect a move toward ecosystem-level thinking (Agee 1996). Nonetheless, this approach is still founded on the species and the Florida panther will continue to be dealt with on an individual basis. From the perspective of biodiversity managers, wide-ranging carnivores offer a compromise. Clearly the protective shadow of an umbrella species must be evaluated cautiously, but the Florida panther offers the opportunity to study and conserve a landscape species that can be utilized to protect and restore entire regions such as the southeastern United States.

Historical Perspective

Administrative protection of the Florida panther was provided by the Florida Fish and Wildlife Conservation Commission (FWC) in 1958, setting off two decades of speculation regarding the statewide population. Estimates ranged from 15 to 300 panthers (Layne and McCauley 1976). No temporal or spatial trends are apparent in these *educated* guesses, except that most observations of individuals occurred in southern Florida. Before the first panther was captured and fitted with a radio-collar in 1981, a cloud of speculation obscured the status of the subspecies. Jenkins (1971:87) suggested that the panther existed in "countable numbers in Florida where it is reasonably common in some locations." After intensive searches, however, enough evidence was accumulated to produce an estimate of fewer than 30 (Nowak and McBride 1973). The listing of the subspecies as federally endangered in 1967 preceded the creation of the first Florida Panther Recovery Team in 1976 when there was still doubt about its very existence. Only sporadic sightings and occasional roadkills in extreme southern Florida suggested otherwise.

A paucity of tangible evidence, let alone details of the species' natural history, created an atmosphere that fostered false impressions of the panther among natural resource agencies. The earliest radiotelemetry studies revealed a population that appeared skewed toward older individuals, reproductively stagnant, and susceptible to mortality ranging from malnutrition and highway collision to intraspecific aggression and capture accidents (Belden 1986; Roelke et al. 1986). These impressions, based on fewer than ten animals, resulted in the development of management and recovery guidelines designed for a crisis situation. Captive breeding was believed to be the surest way to avoid complete extinction: it would create opportunities for both population augmentation and reintroduction (USFWS 1987). Although ten panther kittens were removed from the wild and placed in captivity by 1992—removals that had little impact on the reproductive output of the population (Maehr 1997b)—the captive breeding program was abandoned two years later in favor of in situ genetic augmentation. During their stay in captivity, none of the captive panthers were permitted to breed. Two females died in captivity, and two males died after they were released in southern Florida after genetic augmentation had begun. The remainder are still in captivity. In 1994 eight adult female Texas cougars (*Puma concolor stanleyana*) were introduced into southern Florida panther range. To date these females have produced nearly 40 hybrid kittens (Land and Lacy 2000).

Maehr (1998:182) summarized the demographic status of the Florida panther before the introduction of Texas cougars as follows:

> The long-term demographic studies of the Florida panther can be boiled down to a few generalities. Florida panthers inhabit an area of about 881,000 ha that is bounded on four sides by water and human developments. Not all of southern Florida is equally capable of supporting panthers: habitat quality declines from northwest to southeast. The largest public expanses of southern Florida (Everglades National Park and eastern Big Cypress National Preserve) support the fewest panthers per unit area of any known occupied range. Immigration into the southern Florida population is unknown, but search efforts to the north of the Caloosahatchee River have been insufficient to rule out the possibility of ties to another population. Demographics of the Florida panther are similar to those of unhunted western cougar populations: overall survival is high and the bulk of mortality likely occurs in the dispersing, non-resident, non-reproducing segment of the population. These conclusions are supported by the observation that the greatest known cause of mortality among radio-collared study animals is intraspecific aggression (i.e. resident adult males killing dispersing subadult males), not highways. Females are

readily recruited into the population, and they conceive their first litters before they are two years of age. Litter size averages two—the norm for the species. Adult turnover in the population is low, and home ranges are stable unless a vacancy is created by death or abandonment of the occupant. Reproduction occurs at a rate that exceeds deaths of resident adults. These are not characteristics of a disappearing population.

The only demographic dysfunction that emerges from an examination of the panther research record is that the population appears to be effectively landlocked south of latitude 26°60' N—a line that coincides with a virtually impenetrable swath of obstacles: the artificially dredged Caloosahatchee River, the unforested south rim of Lake Okeechobee, intensive agriculture, and urban sprawl. Krebs et al. (1969) found that reduced dispersal in voles resulted in abnormal population dynamics. Gaines et al. (1991) and Lidicker (1995a) have suggested that such "frustrated dispersal" can lead to population crashes due to the saturation of marginal habitat and negative feedback on source habitat. Although the Florida panther would certainly appear to be susceptible to such problems, successful reproduction in the panther's core range (source habitat) has been sufficient to compensate for the loss of juveniles that were sent to the aborted captive breeding program and has continued to force dispersal-age males into inhospitable fringes (sink habitat) of the subspecies' distribution. These young males continuously attempt to escape the confines of the local population—as in *Puma concolor* throughout its range—but have been mostly blocked by human changes to the landscape. Today the path of least resistance is the Caloosahatchee River itself, the artificially dredged channel that now severs a 30-km-wide land bridge which linked southern Florida with central Florida until the late 1800s.

Landscape Context

MacArthur (1972) noted that dispersing individuals will be rewarded with reproductive success if they can colonize nearby superior habitat. Such dispersers are also the most likely to maintain genetic variability and reclaim vacant range (Joule and Cameron 1975; Lidicker 1975). Because much of the landscape to the north of the Caloosahatchee River is considered good panther habitat (Maehr et al. 1992; Cox et al. 1994), dispersing panthers would experience increased "reproductive value" (MacArthur 1972:151) if they could routinely access this area of abundant food and suitable cover. For most of the research record, Florida panther dispersal has been frustrated (Maehr 1997c) in the sense of Lidicker (1988). Three panthers have crossed the

Caloosahatchee River in the last three years, but none has found a mate. Present-day landscape filters (the river, human infrastructure, and scattered forest on either side of the river) apparently allow only the occasional male disperser into an area without a proved population. A valuable enhancement to panther recovery might be to retrofit the landscape to encourage male dispersal and facilitate female colonization. The result would be the creation of a naturally functioning metapopulation that could tolerate the natural ebbs and flows of its population segments.

That Florida panthers have not fluctuated out of existence may be due in part to a long history of local adaptation, the elimination of lethal alleles, and a tolerable arrangement of habitat patches in southern Florida. Within the area immediately adjacent to the habitat core (Maehr 1997b) one finds, as expected, high reproductive rates and dispersal success that is limited by a lack of vacant range and intolerant residents. Moving as little as 10 km away from the core in any direction, dispersing panthers encounter increasingly patchy forests (Maehr 1997b, 1998) and a predominantly hostile landscape matrix. Whether it is due to urbanization, agriculture, or plant communities dominated by sedges and grasses, panthers in these peripheral areas suffer higher individual mortality rates and exist in subpopulations that are more susceptible to extinction than panthers in the core population (Bass and Maehr 1991).

Maehr and Cox (1995) and Kerkhoff (1997), using more than ten years of radiotelemetry data, have demonstrated that panther habitat is not dependent on a particular type of vegetation but is related almost solely to the extent of available forest cover. When southern Florida panther range is examined along the forest gradient recognized by Maehr (1997b) and three representative areas along this gradient are subjected to ROMPA analysis (ratio of optimal to marginal patch area; Lidicker 1988), the relation between forest cover and the ephemeral status of panther occupation away from the core becomes clear. We used Landsat Thematic Mapper satellite imagery (Kautz et al. 1993) and ArcView GIS (Environmental Systems Research 1997) to delineate forest cover in areas of occupied and unoccupied panther habitat in Florida. Representative, 64-km^2 landscape blocks were then selected and a 1-km^2 network was superimposed over these projections. Optimal habitat units (1 km^2 subblocks) were considered those that contained a minimum of 50 percent forest cover. This value has been found to be relevant for small mammals, and is used here in lieu of experimentally obtained data for the panther. Panthers seem reluctant to travel across unforested habitat that is greater than 90 m in width, however, a measure that may be linked to forest habitat fragmentation (Maehr and Cox 1995). Subblocks with less than 50 percent forest cover were considered marginal. The size of blocks was convenient from an analytical perspective, but this size also approximates the minimum size of adult female home ranges in good-quality habitat (Maehr et al. 1991).

In the area where the panther population has remained stable for decades, ROMPA values are well above 0.50. A fluctuating population typifies the area of moderately low forest cover (ROMPA = 0.48), and a population susceptible to extinction (Bass and Maehr 1991) occasionally inhabits an area of very low ROMPA (0.03) (Figure 15.1). These findings agree with the suggestion by Lidicker (1988:229) that for populations in landscapes with high ROMPA "multiannual cycles are less likely because so much optimal habitat is available that harsh season populations are not reduced sufficiently to prevent complete recovery of the population during the following breeding season." Because panthers are longer lived and do not appear to respond spatially to seasonal variation (Maehr et al. 1991), they interact with their landscape at a different temporal scale than the microtine populations for which ROMPA was first applied. In this sense the panther is less likely to experi-

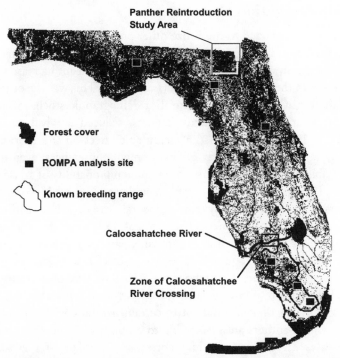

Figure 15.1. Forest cover and the distribution of Florida panther in Florida (adapted from Maehr 1997c).

ence superannual cycles in extensive forests. On the other hand, one would expect that in marginal habitat patches, panthers respond to fluctuations in environmental conditions such as food supply more severely than in optimal patches. Indeed, food (ungulate prey) appears to be limited in lower-ROMPA landscapes such as the eastern Big Cypress Swamp and the Everglades (Maehr et al. 1990b) where white-tailed deer (*Odocoileus virginianus*) densities are only a fraction of those in high-ROMPA landscapes (McCown et al. 1991; Smith and Bass 1994). Clearly, then, high-ROMPA panther habitat represents that part of the subspecies' range that serves as a population source (Pulliam 1988) and the origin of unsuccessful dispersers. This area of high ROMPA was referred to as the population core by Maehr (1997b) and covers approximately 1320 km^2; less than 15 percent of known occupied range (880,970 ha; Maehr 1990). Thus, more than 85 percent of the panther-occupied landscape functions as a population sink. Interestingly, more extensive forests occur in northen Florida where local political opposition has paralyzed reintroduction (Belden and Hagedorn 1993).

Other Small-Population Concerns

This discussion is not intended to minimize other problems associated with small populations. The best demographic and genetic prospects for the recovery of a small population would stem from a rapid and significant increase in numbers—whether in the wild or in captivity (Lacy 1994). This was the original intent of the captive breeding program. But demographic stochasticity (skewed sex ratios at birth, reproductive failures, reduced survival, etc.) (Goodman 1987a, 1987b) has not had a measurable effect on the Florida panther up to now. Although a variety of environmental influences, even in the absence of habitat loss, could increase the extinction probability of an isolated population (Brussard and Gilpin 1989; Caughley and Gunn 1996), the panther appears little influenced by environmental variance. Fluctuating water levels do not affect its movements (Maehr et al. 1991), and if a litter is lost due to a short-term event such as a hurricane, panther reproduction is not locked into a single seasonal pulse as it is with the black bear (*Ursus americanus*). Panthers can reproduce throughout the year. The lack of mortality and detectable behavioral response by panthers to Hurricane Andrew in 1992 supports the notion that they are resilient to certain natural environmental extremes. And while panthers are susceptible to a number of endemic and naturalized diseases in southern Florida (Forrester 1992; Forrester et al. 1985), none has caused epizootic mortality as has been observed in a nearby bobcat (*Lynx rufus*) population (Wassmer et al. 1988). Apparently the panther's inherently low density and infrequent encounters with conspecifics likely reduces the rate of disease transmission in the subspecies.

Perhaps the most difficult elements to evaluate with regard to panther management are those related to genetics, although concerns over highway mortality, disease, and environmental contamination have also surfaced (Jordan 1991). While some arguments have been compelling enough to replace captive breeding plans with genetic augmentation, there is some sense that the rate at which these genetic changes are taking place may be unwarranted (Maehr and Caddick 1995; Maehr 1998). The essence of the debate is simply this: observed patterns of genetic variability in the Florida panther are not paralleled by the demographic performance of the population. The Florida panther exhibits high kitten survival (80 percent) and overall demographics that are remarkably similar to unhunted populations of *Puma concolor* in the western United States (Maehr and Caddick 1995). Further, Maehr and Caddick (1995:1296) observe that "the panther displays characteristics of an expanding population that appears unencumbered (for the present) by low genetic variability." Neither highway mortality, disease, nor contaminants have had a measurable effect on the panther's population structure or reproductive performance (Maehr 1998).

Without knowing the historical genetic makeup of the population—and considering the isolated nature of any population at a distributional extreme (Simpson 1964) (and, in this case, at the end of a peninsula)—it is difficult to conclude just how different contemporary Florida panthers are from pre-Columbian stock. Perhaps many deleterious recessive alleles were purged long before recovery efforts began. Or perhaps the panther is not yet sufficiently inbred to have experienced detectable demographic consequences (Caughley 1994). Although inbreeding has not been shown to affect the demographics of the Florida panther (Maehr and Caddick 1995), any significant increase in its numbers would certainly reduce the potential for genetics (or any other factor) to threaten its survival. Because occupied range south of the Caloosahatchee River is likely saturated (Maehr 1997a), the best prospects for population increase are either in captivity or north of the river.

A Regional Landscape Approach

Despite the shift from a holistic approach (USFWS 1987) to a local, symptom-oriented approach to panther recovery (Captive Breeding Specialist Group 1992) and a tendency for agencies to deal with individual panther management rather than the root of the problem, there has been a growing focus on a landscape-scale perspective (Harris and Atkins 1991), even if such an approach has not been officially adopted. The importance of corridors and private landownership was recognized well before genetic augmentation was implemented (Maehr 1990). Moreover, the relation between landscape productivity and panther occurrence has been understood for more than a decade (Maehr et al. 1989; Smith and Bass 1994). Even so, although there is no doubt that the panther is a wide-ranging species, it has yet to be managed as a creature of the landscape. As official plans evolved to manage a single population within the confines of an artificial landscape island, empirical analyses demonstrated the existence of panther habitat outside of southern Florida (Cox et al. 1994; Maehr et al. 1992; Maehr and Cox 1995) and suggested the presence of other populations in the state (Maehr 1996; Belden et al. 1994). Although habitat preservation has been identified as a key component of panther recovery (Logan et al. 1994), little progress has been made toward the expansion of protected areas or a metapopulation approach to panther management. The single land acquisition proposal that acknowledges the panther as a possible beneficiary of land purchases near the Caloosahatchee River (Eller et al. 1996) was given a low priority by a recent project assessment (Florida Natural Areas Inventory 1997) and the pursuit of this landscape connection has been advocated primarily by only one panther recovery agency.

The Florida Panther Habitat Preservation Plan identified approximately 300,000 ha of private land as potentially significant panther habitat in southern Florida (Logan et al. 1994). Most of this habitat leads north from Big Cypress National Preserve (BCNP) and surrounds the Caloosahatchee River basin. Telemetry studies centered on the northern Big Cypress Swamp have regularly demonstrated panther use of habitat on private lands directly north of the BCNP complex (Maehr and Cox 1995). Despite the evidence of panthers noted earlier, however, the status of the panther in this area is equivocal and a reproducing population may not exist (Maehr et al. 1992) even though forest cover appears adequate to support a population.

Reintroducing Panthers in Northern Florida

One Florida panther recovery objective has been nearly forgotten: reestablishment of two other populations within its historic range in the southeastern United States (USFWS 1987). The first site chosen was the Okefenokee National Wildlife Refuge/Osceola National Forest complex in northeastern Florida and southern Georgia (Belden 1986) where 15 Texas cougars (*Puma concolor stanleyana*) were experimentally released into northern Florida from 1988 through 1994 to test the potential for this large forested landscape to support a panther population (Belden and Hagedorn 1993; Belden 1994). This study indicated that reintroduction was ecologically feasible and there was a much larger area capable of supporting panthers than just the original target zone. Some of the cougars settled to the west of the release site in the Suwannee River basin—an area of abundant prey and cover. Several other individuals engaged in long-distance dispersal, including several that entered southeastern and south-central Georgia (Belden 1994)—suggesting that additional habitat existed beyond the state line.

The experiment led Belden (1994:161) to conclude: "An initial stocking of at least 10 mountain lions can be used as a successful technique for establishing a population in northern Florida." Limited space was not a problem. The most significant obstacle to reintroduction was a local minority of people who fear panther effects on livestock, land use, and children. Otherwise, even though road densities and fragmentation due to major highways are general concerns, north-central Florida and southern Georgia offer sufficient resources to support a panther population.

The Florida Greenways Initiative

As efforts to address panther recovery from a landscape perspective have limped along, the state of Florida has initiated an unprecedented, integrated statewide conservation program that is unrelated to this endangered subspecies. The program was preceded by independent activities to educate Floridians on habitat fragmentation issues and the need for an integrated habitat conservation system. (See Harris 1984, 1985; Harris and Gallagher 1989; Harris and Scheck 1991; Harris and Atkins 1991; Noss 1991.) After years of baseline funding for acquisition of ecologically sensitive lands, the state began a $3-billion, ten-year program in 1990 called Preservation 2000. At the urging of conservation groups, the governor established the Florida Greenways Commission to consider the development of an ecological and recreational greenways system (Florida Greenways Commission 1994; Hoctor et al. 2000). In their concluding report in 1994, the commission recommended the establishment of an ecological greenways system that:

- Conserves critical elements of Florida's green infrastructure of native ecosystems and landscapes

- Restores and maintains essential connectivity among diverse native ecological systems and processes

- Helps these ecosystems and landscapes to function as dynamic systems

• Maintains the evolutionary potential that will allow the components of Florida's ecosystems and landscapes to adapt to future environmental changes

Clearly the needs of the Florida panther can be neatly integrated into this statewide conservation effort.

Based on these recommendations, the Florida legislature agreed to the development of a statewide greenways system in 1995. As part of an effort to develop a plan before the year 2000, Hoctor et al. (2000) developed a statewide ecological network consisting of large areas of ecological significance and landscape linkages. The network incorporates core reserves, corridors, and buffers through an increasingly hostile landscape matrix (Harris 1984; Noss and Harris 1986; Noss 1992), and builds upon an earlier proposal by Noss (1991) to use riparian corridors to link the entire state. Ecological greenways modeling has offered an opportunity to identify a physically connected ecological conservation system composed of both public and private lands that can meet biological diversity conservation objectives despite explosive human population growth. (Florida currently has a population of 14 million people and from 1980 to 1990 grew at a rate of more than 300,000 people per year; Fernald and Purdum 1992.)

Getting Panthers Out of Southern Florida

Analysis of lands north of the core population of panthers in southwestern Florida indicates fair to excellent opportunities for range expansion and the reestablishment of other populations. Florida is fortunate with respect to the relative wealth of information on its ecologically significant lands and biodiversity. These data are primary inputs into the ecological greenways model.

Obviously not all forest cover nor all conservation lands qualify as primary panther habitat. Corridors need not serve as permanent habitat for panthers, for example, but merely act as travel conduits between subpopulations. Given that at least some dispersing panthers will utilize marginal habitat for travel, clearly there is an opportunity to establish a population north of the Caloosahatchee River or enhance an established population if one exists. Reports of panthers have occurred regularly from Myakka River State Park on the west coast to Corbett Wildlife Management Area on the east coast and the upper Kissimmee River basin to the north (Belden 1994). The panther could benefit from the reestablishment of landscape linkages across the river by enhancing the fitness of potential subpopulations or encouraging the formation of new ones in former range.

Colonization Potential

Although panthers have periodically been documented north of the Caloosahatchee River (Maehr 1997a), none of their origins was known, and only one was captured and radio-collared in the area before 2001 (Maehr et al. 1992). The panthers that recently crossed the river and left the area of frustrated dispersal have shed new light on the potential for the population to expand and take advantage of a developing greenways system. ROMPA analyses indicate that the six 64-km^2 habitat blocks bordering the Caloosahatchee River where the three panthers crossed offer little forested cover (2 to 34 percent) relative to occupied range in southern Florida but suggest that scattered forest may support a dispersal event (Figure 15.1). That these subadult males apparently used the same crossing point in the heaviest forest cover available suggests that open habitats are avoided whenever possible during dispersal. After crossing the Caloosahatchee River, one of the panthers traveled approximately 250 km in proximity to the proposed landscape linkage paralleling the Lake Wales Ridge and Kissimmee River basin. Its movements have included trips through patchy forests in the vicinity of one of the world's busiest entertainment attractions—Disney World—without mishap or encounters with humans.

Though it is impossible to confirm the cause of these long-distance movements, they may well be related to high population density in the source population and a lack (or extremely low density) of female panthers north of the Caloosahatchee River. Dispersing males in southern Florida generally leave occupied range, but eventually they return to seek a mate (Maehr 1997b). Females are philopatric and usually disperse no farther than the width of their mother's home range. Regardless of the cause of this dispersal, panther recovery might benefit greatly if both males and females were encouraged to make the Caloosahatchee crossing. This possibility would be enhanced by:

- Protecting and restoring forest along a strategic stretch of the waterway, so that it obtains a ROMPA value greater than 50 percent, and establishing protected forested corridors that link panther habitat on both sides of the river

- Reducing the effective width of the dredged channel by installing vegetation-protected bank-cutout ramps on either side of the river

- Using the natural surplus of offspring that the southern Florida population produces on an annual basis, and reintroducing Florida panther females to suitable habitat north of the river

- Constructing a permanent river-spanning structure that supports vegetation suitable for panther travel
- Adopting a combination of these options

Such actions would be valuable not only for restoring a population in south-central Florida but for creating a stepping-stone in a statewide metapopulation that would encourage dispersal and the creation of new populations to the north. The northern half of the state still has extensive forested uplands and contains several large hubs of public conservation lands. Based on availability of large patches of potentially suitable habitat, existing conservation areas, roadless areas, and generally low human population density, much of northern Florida may be capable of supporting panthers (Cox et al. 1994). A reintroduced population in the Okefenokee National Wildlife Refuge and Osceola National Forest could recolonize suitable areas in the Florida panhandle, central Georgia, and possibly the Ocala National Forest in north-central Florida.

The Florida Panther: Flagship of a Statewide Conservation System

Although much of the work on Florida panther recovery has rightly concentrated on the only documented population in southwestern Florida, there has been a compulsion to ignore the regional landscape considerations that will assure the panther's future. Continuing to focus on micromanagement of the existing population on public land denies several facts: (1) much of the panther's habitat is still on private land; (2) there are suitable lands north of the Caloosahatchee River that could support a supplemental population; (3) panther dispersal is frustrated by artificial landscape features; (4) the population produces an annual surplus of offspring; (5) there is a need (and recovery plan mandate) to establish additional populations within the panther's historic range; and (6) the panther is a classic flagship species for regional habitat restoration (Miller et al. 1999).

Panther conservation provides a serious opportunity for designing and implementing a Florida statewide conservation system. Cox et al. (1994) cite at least 53 animal and plant species of conservation interest within the strategic habitat conservation areas identified for panthers near the Caloosahatchee River. Efforts to address private property issues for panther conservation could also pave the way for dealing with similar problems that will be encountered in a statewide conservation system. But action must be quick. Though there is ample opportunity to restore the panther to significant parts of its former range through a statewide conservation system, today's pace of development demands that land protection efforts happen soon. Private lands in southwestern and south-central Florida are threatened by conversion to agriculture and residential/urban development. Lands in northern Florida face pressure, too, as development shifts from southern and central Florida. If panther conservation is to fulfill its promise as a biodiversity-saving tool, a landscape approach must be adopted by natural resource management agencies. The current recovery paradigm is no less a threat to the panther's existence than its inability to escape the confines of a finite southern Florida landscape.

Literature Cited

Agee, J. K. 1996. Ecosystem management: An appropriate concept for parks? In R. G. Wright, ed., *National Parks and Protected Areas: Their Role in Environmental Protection.* Cambridge, Mass.: Blackwell Science.

Bass, O. L., and D. S. Maehr. 1991. Do recent panther deaths in Everglades National Park suggest an ephemeral population? *National Geographic Research and Exploration* 7:427.

Belden, R. C. 1986. Florida panther recovery plan implementation—a 1983 progress report. In S. D. Miller and D. D. Everett, ed., *Cats of the World: Biology, Conservation and Management.* Washington, D.C.: National Wildlife Federation.

———. 1994. Florida panther reintroduction feasibility study. In D. B. Jordan, ed., *Proceedings of the Florida Panther Conference.* Gainesville, Fla.: U.S. Fish and Wildlife Service.

Belden, R. C., and B. W. Hagedorn. 1993. Feasibility of translocating panthers into northern Florida. *Journal of Wildlife Management* 57:388–397.

Belden, R. C., W. B. Frankenberger, and J. C. Roof. 1994. Florida panther distribution. In D. B. Jordan, ed., *Proceedings of the Florida Panther Conference.* Gainesville, Fla.: U.S. Fish and Wildlife Service.

Bowers, M. A. 1997. Mammalian landscape ecology. *Journal of Mammalogy* 78:997–998.

Brussard, P. F., and M. E. Gilpin. 1989. Demographic and genetic problems of smell populations. In U. S. Seal, E. T. Thorne, M. A. Bogan, and S. H. Anderson, eds., *Conservation Biology and the Black-Footed Ferret.* New Haven: Yale University Press.

Captive Breeding Specialist Group. 1992. Genetic management strategies and population viability of the Florida panther (*Felis concolor coryi*). Report of a workshop. U.S. Fish and Wildlife Service Contract 14-16-0004-92-983. Apple Valley, Minn.: Captive Breeding Specialist Group.

Caughley, G. 1994. Directions in conservation biology. *Journal of Animal Ecology* 63:215–244.

Caughley, G., and A. Gunn. 1996. *Conservation Biology in Theory and Practice.* Cambridge, Mass.: Blackwell Science.

Cox, J., R. Kautz, M. MacLaughlin, and T. Gilbert. 1994. *Closing the Gaps in Florida's Wildlife Habitat Conservation System.* Tallahassee: Florida Game and Fresh Water Fish Commission.

Eller, A. C., Jr., C. E. Hilsenbeck, W. J. Caster, R. A. Hilsenbeck, and R. L. Burns. 1997. Caloosahatchee Ecoscape: Glades and Hendry counties—a cooperative proposal. Tallahassee: Florida Department of Environmental Protection.

Environmental Systems Research. 1997. ArcView GIS Version 3.0a. Redlands, Calif.: Environmental Systems Research, Inc.

Fernald, E. A., and E. D. Purdum. 1992. *Atlas of Florida.* Gainesville: University Press of Florida.

Florida Greenways Commission. 1994. Creating a statewide greenways system: For people, for wildlife, for Florida. Tallahassee: Florida Department of Environmental Protection.

Florida Natural Areas Inventory. 1997. Caloosahatchee Ecoscape project assessment. Tallahassee: Florida Natural Areas Inventory.

Foin, T. C., S. P. D. Riley, A. L. Pawley, D. R. Ayres, T. M. Carlsen, P. J. Hodum, and P. V. Switzer. 1998. Improving recovery planning for threatened and endangered species. *BioScience* 48:177–184.

Foreman, D., J. Davis, D. Johns, R. F. Noss, and M. Soulé. 1992. The Wildlands Project mission statement. *Wild Earth* (special issue):3–4.

Forrester, D. J. 1992. *Parasites and Diseases of Wild Mammals in Florida.* Gainesville: University Press of Florida.

Forrester, D. J., J. A. Conti, and R. C. Belden. 1985. Parasites of the Florida panther (*Felis concolor coryi*). *Proceedings of the Helminthological Society of Washington* 52:95–97.

Gaines, M. S., N. C. Stenseth, M. L. Johnson, R. A. Ims, and S. Bondruop-Nielsen. 1991. A response to solving the enigma of population cycles with a multifactorial perspective. *Journal of Mammalogy* 72:627–631.

Goodman, D. 1987a. The demography of chance extinction. In M. E. Soulé, ed., *Viable Populations for Conservation.* Cambridge: Cambridge University Press.

———. 1987b. Consideration of stochastic demography in the design and management of biological reserves. *Natural Resources Modeling* 1:205–234.

Grumbine, R. E. 1992. *Ghost Bears: Exploring the Biodiversity Crisis.* Washington, D.C.: Island Press.

Hansson, L. 1995. Development and application of landscape approaches in mammalian ecology. In W. Z. Lidicker, Jr., ed., *Landscape Approaches in Mammalian Ecology and Conservation*. Minneapolis: University of Minnesota Press.

Harris, L. D. 1984. *The Fragmented Forest: Island Biogeography Theory and the Preservation of Biotic Diversity*. Chicago: University of Chicago Press.

————. 1985. *Conservation Corridors: A Highway System for Wildlife*. ENFO. Winter Park: Florida Conservation Foundation.

Harris, L. D., and P. Gallagher. 1989. New initiatives for wildlife conservation: The need for movement corridors. In G. Mackintosh, ed., *In Defense of Wildlife: Preserving Communities and Corridors*. Washington, D.C.: Defenders of Wildlife.

Harris, L. D., and K. Atkins. 1991. Faunal movement corridors in Florida. In W. Hudson, ed., *Landscape Linkages and Biodiversity*. Washington, D.C.: Island Press.

Harris, L. D., and J. Scheck. 1991. From implications to applications: The dispersal corridor principle applied to the conservation of biological diversity. In D. A. Saunders and R. J. Hobbs, eds., *Nature Conservation 2: The Role of Corridors*. Sydney: Surrey Beatty & Sons.

Harris, L. D., and T. Hoctor. 1992. Cross Florida greenbelt state recreation and conservation area management plan. Vol. 4: Report on biological issues. Gainesville: University of Florida.

Harris, L. D., T. Hoctor, D. Maehr, and J. Sanderson. 1996. The role of networks and corridors in enhancing the value and protection of parks and equivalent areas. In R. G. Wright, ed., *National Parks and Protected Areas: Their Role in Environmental Protection*. Cambridge, Mass.: Blackwell Science.

Hoctor, T. S., M. H. Carr, and P. D. Zwick. 2000. Identifying a linked reserve system using a regional landscape approach: The Florida ecological network. *Conservation Biology* 14:984–1000.

Jenkins, J. H. 1971. The status and management of bobcat and cougar in the southeastern states. In S. E. Jorgenson and L. D. Mech, eds., *Proceedings of the Symposium on Native Cats of North America*. Washington, D.C.: U.S. Fish and Wildlife Service.

Jordan, D. B. 1991. A proposal to establish a captive breeding population of Florida panthers. Supplemental environmental assessment. Gainesville, Fla.: U.S. Fish and Wildlife Service, University of Florida.

Joule, J., and G. N. Cameron. 1975. Species removal studies. I: Dispersal strategies of sympatric *Sigmodon hispidus* and *Reithrodontomys fulvescens* populations. *Journal of Mammalogy* 56:378–396.

Kautz, R. S., D. T. Gilbert, and G. M. Mauldin. 1993. Vegetative cover in Florida based on 1985–1989 Landsat thematic mapper imagery. *Florida Scientist* 56:135–154.

Kerkhoff, A. J. 1997. Toward a panther-centered view of the forests of South Florida. M.S. thesis, University of New Mexico, Albuquerque.

Krebs, C. J., B. L. Keller, and R. H. Tamarin. 1969. *Microtus* population biology: Demographic changes in fluctuating populations of *M. ochrogaster* and *M. pennsylvanicus* in southern Indiana. *Ecology* 50:587–607.

Lacy, R. C. 1994. Managing genetic diversity in captive populations of animals. In M. L. Bowles, and C. J. Whelan, eds., *Restoration of Endangered Species: Conceptual Issues, Planning, and Implementation*. Cambridge: Cambridge University Press.

Land, E. D., and R. C. Lacy. 2000. Introgression level achieved through Florida panther genetic restoration campaign. *Endangered Species Update* 17:99–104.

Layne, J. N., and M. N. McCauley. 1976. Biological overview of the Florida panther. In P. C. Pritchard, ed., *Proceedings of the Florida Panther Conference*. Casselberry: Florida Audubon Society.

Lidicker, W. Z. 1975. The role of dispersal in the demography of small mammals. In F. B. Golley, K. Petrusewicz, and L. Ryszkowski, ed., *Small Mammals, Their Productivity and Population Dynamics*. Cambridge: Cambridge University Press.

———. 1988. Solving the enigma of microtine "cycles." *Journal of Mammalogy* 69:225–235.

———, ed. 1995a. *Landscape Approaches in Mammalian Ecology and Conservation*. Minneapolis: University of Minnesota Press.

———. 1995b. The landscape concept: something old, something new. In W. Z. Lidicker, Jr., ed., *Landscape Approaches in Mammalian Ecology and Conservation*. Minneapolis: University of Minnesota Press.

Logan, T., A. C. Eller Jr., R. Morrell, D. Ruffner, and J. Sewell. 1994. Florida panther habitat preservation plan. Gainesville: Florida Panther Interagency Committee, U.S. Fish and Wildlife Service.

MacArthur, R. H. 1972. *Geographical Ecology*. Princeton: Princeton University Press.

MacArthur, R. H., and E. O. Wilson. 1967. *The Theory of Island Biogeography*. Princeton: Princeton University Press.

Maehr, D. S. 1990. The Florida panther and private lands. *Conservation Biology* 4:167–170.

———. 1996. The Florida panther: All present and accounted for? In J. W. Tischendorf and S. J. Ropski, ed., *Proceedings of the Eastern Cougar Conference 1994*. Erie, Pa.: American Ecological Research Institute.

———. 1997a. The comparative ecology of bobcat, black bear, and Florida panther in South Florida. *Bulletin of the Florida Museum of Natural History* 40:1–176.

———. 1997b. The Florida panther and the Endangered Species Act of 1973. *Environmental and Urban Issues* 24(4):1–8.

———. 1997c. *The Florida Panther: Life and Death of a Vanishing Carnivore*. Washington, D.C.: Island Press.

———. 1998. The Florida panther in modern mythology. *Natural Areas Journal* 18:179–184.

Maehr, D. S., E. D. Land, J. C. Roof, and J. W. McCown. 1989. Early maternal behavior in the Florida panther (*Felis concolor coryi*). *American Midland Naturalist* 122:34–43.

Maehr, D. S., J. C. Roof, E. D. Land, and J. W. McCown. 1990a. Day beds, natal dens, and activity of Florida panthers. *Proceedings of the Annual Conference of the Southeastern Association of Fish and Wildlife Agencies* 44:310–318.

Maehr, D. S., R. C. Belden, E. D. Land, and L. Wilkins. 1990b. Food habits of panthers in southwest Florida. *Journal of Wildlife Mangement* 54:420–423.

Maehr, D. S., E. D. Land, and J. C. Roof. 1991. Social ecology of the Florida panther. *National Geographic Research and Exploration* 7:414–431.

Maehr, D. S., J. C. Roof, E. D. Land, J. W. McCown, and R. T. McBride. 1992. Home range characteristics of a panther in south central Florida. *Florida Field Naturalist* 20:97–103.

Maehr, D. S., and G. B. Caddick. 1995. Demographics and genetic introgression in the Florida panther. *Conservation Biology* 9:1295–1298.

Maehr, D. S., and J. A. Cox. 1995. Landscape features and panthers in Florida. *Conservation Biology* 9:1008–1019.

McCown, J. W., M. E. Roelke, D. J. Forrester, C. T. Moore, and J. C. Roboski. 1991. Physiological evaluation of 2 white-tailed deer herds in southern Florida. *Proceedings of the Annual Conference of the Southeastern Association of Fish and Wildlife Agencies* 45:81–90.

Miller, B., R. Reading, J. Strittholt, C. Carroll, R. Noss, M. Soulé, O. Sanchez, J. Terborgh, D. Brightsmith, T. Cheeseman, and D. Foreman. 1999. Using focal species in the design of nature reserve networks. *Wild Earth* 8(4):81–92.

Muir, J. 1916. *A Thousand-Mile Walk to the Gulf.* Boston: Houghton Mifflin.

Noss, R. F. 1991. Landscape connectivity: Different functions at different scales. In W. E. Hudson, ed., *Landscape Linkages and Biodiversity.* Washington, D.C.: Island Press.

———. 1992. The Wildlands Project: Land conserving strategy. *Wild Earth* (special issue):10–25.

Noss, R. F., and L. D. Harris. 1986. Nodes, networks, and MUMs: Preserving diversity at all scales. *Environmental Management* 10:299–309.

Nowak, R. M., and R. T. McBride. 1973. Feasibility of a study on the Florida panther. Report to the World Wildlife Fund. Mimeo.

Pearlstine, L. G., L. A. Brandt, W. M. Kitchens, and F. J. Mazzotti. 1995. Impacts of citrus development on habitats of southwest Florida. *Conservation Biology* 9:1020–1032.

Pulliam, H. R. 1988. Sources, sinks, and population regulation. *American Naturalist* 132:652–661.

Roelke, M. E., E. R. Jacobson, G. V. Kollias, and D. J. Forrester. 1986. Medical management and biomedical findings on the Florida panther, *Felis concolor coryi*, July 1,

1985 to June 30, 1986. Annual report. Tallahassee: Florida Game and Fresh Water Fish Commission.

Shelford, V. E. 1936. Conservation of wildlife. In A. E. Parkins and J. R. Whitaker, ed., *Our Natural Resources and Their Conservation.* New York: Wiley.

Simpson, G. G. 1964. Species density of North American Recent mammals. *Systematic Zoology* 13:57–73.

Smith, T. R., and O. L. Bass, Jr. 1994. Landscape, white-tailed deer, and the distribution of Florida panthers in the Everglades. In S. Davis, and J. Ogden, ed., *Everglades: The Ecosystem and Its Restoration.* Delray Beach, Fla.: St. Lucie Press.

Soulé, M., and R. Noss. 1998. Rewilding and biodiversity: Complementary goals for continental conservation. *Wild Earth* 8(3):18–28.

U.S. Fish and Wildlife Service (USFWS). 1987. Florida panther recovery plan. Atlanta.

———. 1998. Multi-species recovery plan for the threatened and endangered species of South Florida. Vol. 1: The species. Technical/Agency draft. Vero Beach, Fla. Compact disc.

Wassmer, D. A., D. D. Guenther, and J. N. Layne. 1988. Ecology of the bobcat in south-central Florida. *Bulletin of the Florida Museum of Natural History* 33:159–228.

Wright, G. M., J. D. Dixon, and B. H. Thompson. 1933. *Fauna of the National Parks of the United States Series 1.* Washington, D.C.: Government Printing Office.

Wright, G. M., and B. H. Thompson. 1934. *Fauna of the National Parks of the United States.* Series 2. Washington, D.C.: Government Printing Office.

Wright, R. G., and D. J. Mattson. 1996. The origin and purpose of national parks and protected areas. In R. G. Wright, ed., *National Parks and Protected Areas: Their Role in Environmental Protection.* Cambridge, Mass.: Blackwell Science.

Case 4. Can Manatee Numbers Continue to Grow in a Fast-Developing State?

BRUCE B. ACKERMAN AND JAMES A. POWELL

The conservation of the endangered Florida manatee (*Trichechus manatus latirostris*), despite its aquatic lifestyle, differs little from that for other large mammals: habitat loss and human-related mortality are the primary focuses of its recovery plan (USFWS 1996). Manatees are long-lived, have few off-spring, and require high adult and juvenile survival to maintain healthy populations. Exploited in the past and reduced to low numbers, they now are managed in areas where human activity is often high. Although the manatee has exhibited population growth over two decades, habitat and mortality issues remain obstacles to its complete recovery.

Natural and Unnatural History

The manatee is a large, slow-moving marine mammal that grazes on aquatic plants in shallow bays, lagoons, rivers, and springs in the southeastern United States, primarily in Florida. It is unusual among marine mammals in that it uses low-energy foods in a high-energy environment. Manatees were hunted by prehistoric Americans and were noted by fifteenth-century European explorers for their good flavor (Reynolds and Odell 1991). They are believed to have been at low numbers until the middle of the twentieth century (O'Shea 1988; Reynolds and Odell 1991; USFWS 1996), and little was known about their biology and distribution until the 1960s (Hartman 1979). Female manatees mature at about 5 years of age, have one calf every 2 or 3 years, and may live 50 to 60 years (Eberhardt and O'Shea 1995; USFWS 1996; Marmontel et al. 1997). When water temperatures drop below 20°C,

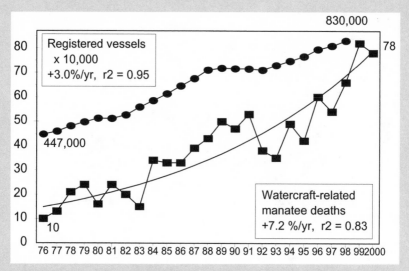

As the number of boats in Florida waters continues to increase, manatee deaths have risen also.

manatees seek out natural or artificial warm-water refuges or move to the southern tip of Florida (O'Shea 1988; USFWS 1996; Deutsch et al. 1998). Manatees are at the northern limit of their range in Florida, and water temperatures can be fatal below about 10°C (Ackerman et al. 1995). Waste heat discharged by power plants and papermills creates important congregation areas for manatees. Although the manatee has exhibited its adaptability by using natural springs and power plants as warm-water winter refugia, it can be locally overwhelmed by habitat change (O'Shea et al. 1995). Manatees use habitats ranging from remote wilderness to urban canals in one of the fastest-growing states in the United States. The primarily coastal-dwelling human population in Florida has more than doubled in 30 years, and is projected to increase at a rate of 1.8 percent a year from 15.2 million to 20.7 million in 2025 (Campbell 1996). There are almost a million watercraft in Florida (see the accompanying figure), and their number is increasing faster than the human population (+3.0 percent/year, Florida Fish and Wildlife Conservation Commission [FWC],unpublished data).

Mortality

Intensive efforts from 1974 to 1999 have recovered 3834 manatee carcasses primarily in Florida (O'Shea et al. 1985; Ackerman et al. 1995; Wright et al. 1995; FWC, unpublished data). Of the 3770 manatee carcasses recovered in Florida, 24 percent were killed by collisions with boats and ships, most probably by small, fast-moving boats (O'Shea 1995; Wright et al. 1995). Video surveillance and aerial observations show that manatees can detect approaching boats but often cannot evade them (Flamm et al. 2000). Floodgates, canal locks, entanglement in lines and monofilament, entrapment in pipes and culverts, ingesting debris, and poaching account for another 7 percent. Natural mortality including perinatal (newborn or stillborn) deaths, cold temperatures, malnutrition, disease, birth complications, natural accidents, and red tide account for 38 percent of mortality. Cause of death was undetermined in 31 percent of the cases.

The documented number of carcasses has increased at an annual rate of 5.8 percent since 1976 (O'Shea et al. 1985; Ackerman et al. 1995; FWC, unpublished data). This trend may have paralleled population increases, but deaths in 1998 and 1999 were higher than any year without red tide. The number of deaths caused by watercraft collisions has increased by 7.2 percent annually—faster than most other causes since 1976 (see the figure).

Population Trend

Despite the high annual mortality, manatee numbers in Florida appear to be stable or even increasing. Increases have been apparent in the St. Johns River (+5.7 percent/year) and northwest Florida (+7.4 percent/year) subpopulations (Eberhardt and O'Shea 1995), but no significant changes have occurred on the Atlantic coast. Other models suggest that manatee populations may have increased in all four regions since 1976 (B. Ackerman, FWC, unpublished data). Projections from models appear to be better than aerial counts for assessing population trends. Although aerial counts suggest increases or stability in all regions (Garrott et al. 1994; Ackerman 1995), there are several biases to be considered. Counts can be highly variable through time and, moreover, are strongly affected by weather conditions and changes in manatee behavior (Ackerman 1995; Lefebvre et al. 1995). A population viability analysis (PVA), by contrast, "projects a slightly negative growth rate (−0.003) and an unacceptably low probability of persistence (0.44) over 1,000 years" (Marmontel et al. 1997:467).

The Problem

Whereas manatee populations have increased from low numbers early in this century, the rapid, steady increases in human numbers and boat traffic suggest that this long-term trend is unlikely to continue. Population growth may recently have ceased in two of the populations. Certainly there is no reason to believe that manatees are any less endangered than before—or that any ongoing manatee conservation strategies should be halted. The record number of deaths in recent years remains a major impediment to the recovery of the species. Continued high rates of mortality from watercraft collisions, as well as habitat loss and environmental degradation, are serious concerns.

If the population is declining locally, it probably would not be immediately detectable by aerial surveys (Taylor and Gerrodette 1993; Marsh 1995). A slow increase in mortality would be difficult to detect by any field method apart from long-term data (Taylor and Gerrodette 1993). If managers wait for definitive trends based on surveys, management reactions may come too late to reverse a downward population spiral.

Manatee Conservation Means People Management

The FWC's manatee research and management are supported by the Save the Manatee Trust Fund, made possible by sales of a special manatee auto license plate ($2.5 million annually), boat registration fees, sales of decals, voluntary contributions, and interest income. Total funding of this program for 1998–1999 was $4.4 million. Of this amount, the FWC used $2.6 million. The rest was passed on to other research organizations, to oceanaria that rescue and rehabilitate manatees, and to educational foundations. This is one of the largest single-species wildlife programs in the United States and is strongly supported by Floridians.

Insofar as most of the manatee's aquatic habitat is already in some form of government ownership and protection, speed limits for boats are a keystone of manatee protection. The first speed zones were established in 1979 to protect aggregations of wintering manatees. In 1989, 13 counties were designated for establishment of countywide speed zones. Speed limits range from no-entry zones (all boats and swimmers prohibited) to no-motorboat zones and zones where allowable boat speeds range from "idle" to 40–56 km/hour. Statewide rules have also been used to limit the number of boat slips.

Boater education courses have been required of all boaters born since 1980, but a boat operator license is not required in Florida. The FWC sponsors boating safety classes and has a "Boat Smart" education package. Boat propeller guards are available to reduce the risk of propeller cuts to manatees and human swimmers, but they do not eliminate the larger risk of impact trauma (Wright et al. 1995).

Legal Protection

Comprehensive growth management plans are required in each Florida county and must address the effects of human population growth and environmental issues including manatee protection (FWC 1999). Boat speed zones are a primary component of these plans. The plans are intended to reduce manatee mortality from watercraft, navigation structures, and commercial fishing, address boater and swimmer safety, facilitate recreational planning, and protect aquatic habitat for manatees and other species. Whereas state and federal agencies have clear mandates, those of local governments seem less clear: only 4 of 13 county plans initiated in 1989 have been completed. Thus a statewide manatee-protection system is not yet in place.

Permit applications to state and federal regulatory agencies for dredging channels, construction of docks and marinas, expansion of port facilities, bridge construction, aquaculture, and boat races must address impacts to manatees and seagrass. Of 400 permit applications reviewed by the FWC in fiscal year 1998–1999, some 397 were approved with the adoption of design modifications to reduce manatee impacts (FWC 1999). The USFWS also reviews projects through the authority of the Endangered Species Act (Section 7) consultation process. In 1999, the USFWS performed 452 consultations (USFWS 2000). Over the years more Section 7 permit reviews and jeopardy opinions have been issued for manatees than for any other endangered species—indeed, almost more than for all other endangered species combined in the United States.

Protection of manatee habitat, in the face of Florida's ever-increasing human population and rapid coastal development, is another key component of manatee restoration. The most important habitats are seagrass, natural springs, aquatic corridors, and the artificial thermal discharges at power plants. The state owns most seagrass and submerged bottomlands. Historically, coastal development has resulted in degradation of water quality and destruction of seagrass, the manatee's primary food. Seagrass requires clear water and is damaged by boat propellers, siltation from construction (new docks, aquaculture, dredging, expansion of port facilities), turbidity from boat propellers, and eutrophication initiated by stormwater runoff, agricultural fertilizers, and septic systems (Sargent et al. 1995).

To ensure successful manatee conservation, regulatory agencies will have to develop long-range management plans that take into account the uncertainty of manatee population trends and the certainty of an increasing human population (and our recreational preferences). The key ingredients of manatee restoration will be:

- Increase suitable aquatic habitat including seagrass and warm-water refuges.

- Maintain metapopulation structure through protection of aquatic travel corridors.

- Continue intensive monitoring of mortality and population trends.
- Expand education programs for boaters and the general public.
- Improve enforcement of manatee speed zones and increase the number of protected areas.
- Limit development in feeding and migratory habitat.
- Continue permit review and encourage manatee-friendly development.
- Encourage completion of county growth management plans that promote manatee recovery.
- Improve the predictive value of population models.

Acknowledgments

Long-term field studies have been conducted since 1974 by researchers of the U.S. Geological Survey Sirenia Project and since 1985 by the FWC. Hundreds of staff members of the two organizations have participated. The FWC activities described here represent the joint efforts of about 50 current full-time FWC staff members of the manatee protection team. We are very grateful for their dedicated efforts and contributions.

Literature Cited

Ackerman, B. B. 1995. Aerial surveys of manatees: A summary and progress report. In T. J. O'Shea, B. B. Ackerman, and H. F. Percival, eds., *Population Biology of the Florida Manatee*. Information and Technology Report 1. Washington, D.C.: National Biological Service.

Ackerman, B. B., S. D. Wright, R. K. Bonde, D. K. Odell, and D. J. Banowetz. 1995. Trends and patterns in mortality of manatees in Florida, 1974–1992. In T. J. O'Shea, B. B. Ackerman, and H. F. Percival, eds., *Population Biology of the Florida Manatee*. Information and Technology Report 1. Washington, D.C.: National Biological Service.

Campbell, P. R. 1996. Population projections for states by age, sex, race, and Hispanic origin: 1995 to 2025. Report PPL-47. Washington, D.C.: U.S. Bureau of the Census, Population Division.

Deutsch, C. J., R. K. Bonde, and J. P. Reid. 1998. Radio-tracking manatees from land and space: Tag design, implementation, and lessons learned from long-term study. *Marine Technology Society Journal* 32:18–29.

Eberhardt, L. L., and T. J. O'Shea. 1995. Integration of manatee life-history data and population modeling. In T. J. O'Shea, B. B. Ackerman, and H. F. Percival, eds., *Population Biology of the Florida Manatee*. Information and Technology Report 1. Washington, D.C.: National Biological Service.

Flamm, R. O., E. C. G. Owen, C. F. W. Owen, R. S. Wells, and D. Nowacek. 2000. Aerial videogrammetry from a tethered airship to assess manatee life-stage structure. *Marine Mammal Science* 16:617–630.

FWC. 1999. *Save the Manatee Trust Fund, Fiscal Year 1998–1999 Annual Report.* Tallahassee: Florida Marine Research Institute, Florida Fish and Wildlife Conservation Commission.

———. 2000. Atlas of marine resources. CD-ROM, Version 1.3. R. O. Flamm, L. I. Ward, and M. D. White, eds. St. Petersburg: Florida Marine Research Institute, Florida Fish and Wildlife Conservation Commission.

Garrott, R. A., B. B. Ackerman, J. R. Cary, D. M. Heisey, J. E. Reynolds III, P. M. Rose, and J. R. Wilcox. 1994. Trends in counts of Florida manatees at winter aggregation sites. *Journal of Wildlife Management* 58:642–654.

Hartman, D. S. 1979. *Ecology and behavior of the manatee (*Trichechus manatus*) in Florida.* Special Publication 5. Lawrence, Kan.: American Society of Mammalogy.

Lefebvre, L. W., B. B. Ackerman, K. M. Portier, and K. H. Pollock. 1995. Aerial survey as a technique for estimating trends in manatee population size—problems and prospects. In T. J. O'Shea, B. B. Ackerman, and H. F. Percival, eds., *Population Biology of the Florida Manatee.* Information and Technology Report 1. Washington, D.C.: National Biological Service.

Marmontel, M., S. R. Humphrey, and T. J. O'Shea. 1997. Population viability analysis of the Florida manatee (*Trichechus manatus latirostris*), 1976–1991. *Conservation Biology* 11:467–481.

Marsh, H. 1995. Fixed-width aerial transects for determining dugong population sizes and distribution patterns. In T. J. O'Shea, B. B. Ackerman, and H. F. Percival, eds., *Population Biology of the Florida Manatee.* Information and Technology Report 1. Washington, D.C.: National Biological Service.

O'Shea, T. J. 1988. The past, present, and future of manatees in the southeastern United States: Realities, misunderstandings, and enigmas. In R. R. Odom, K. A. Riddleberger, and J. C. Ozier, eds,. *Proceedings 3rd Southeastern Nongame and Endangered Wildlife Symposium.* Social Circle, Ga.: Georgia Department of Natural Resources, Game and Fish Division.

O'Shea, T. J. 1995. Waterborne recreation and the Florida manatee. In R. L. Knight and K. J. Gutzwiller, eds., *Wildlife and Recreationists: Coexistence Through Management and Research.* Washington, D.C.: Island Press.

O'Shea, T. J., C. A. Beck, R. K. Bonde, H. I. Kochman, and D. K. Odell. 1985. An analysis of manatee mortality patterns in Florida, 1976–81. *Journal of Wildlife Management* 49:1–11

O'Shea, T. J., B. B. Ackerman, and H. F. Percival, eds. 1995. *Population Biology of the Florida Manatee.* Information and Technology Report 1. Washington, D.C.: National Biological Service.

Reynolds, J. E., III, and D. K. Odell. 1991. *Manatees and Dugongs.* New York: Facts on File.

Sargent, F. J., T. J. Leary, D. W. Crewz, and C. R. Kruer. 1995. Scarring of Florida's seagrasses assessment and management options. Technical Report TR-1. St. Petersburg: Florida Marine Research Institute, Florida Department of Environmental Protection.

Taylor, B. L., and T. Gerrodette. 1993. The uses of statistical power in conservation biology: The vaquita and northern spotted owl. *Conservation Biology* 7:489–500.

U.S. Fish and Wildlife Service. 1996. Florida Manatee Recovery Plan (*Trichechus manatus latirostris*), 2nd rev. Prepared by the Florida Manatee Recovery Team for the U.S. Fish and Wildlife Service, Atlanta.

———. 2000. Florida Manatee Recovery Accomplishments, 1999 Annual Report. Unpublished report. Jacksonville: U.S. Fish and Wildlife Service.

Wright, S. D., B. B. Ackerman, R. K. Bonde, C. A. Beck, and D. J. Banowetz. 1995. Analysis of watercraft-related mortality of manatees in Florida, 1979–1991. In T. J. O'Shea, B. B. Ackerman, and H. F. Percival, eds., *Population Biology of the Florida Manatee*. Information and Technology Report 1. Washington, D.C.: National Biological Service.

Chapter 16

The Biotic Province: Minimum Unit for Conserving Biodiversity

LARRY D. HARRIS, LINDA C. DUEVER,
REBECCA P. MEEGAN, THOMAS S. HOCTOR,
JAMES L. SCHORTEMEYER, AND DAVID S. MAEHR

The reintroduction of a living species (or other taxon) into the indigenous range from which it has been extirpated is justified on all ecological grounds whether or not society chooses to reject the reintroduction of lethal diseases or taxa that are objectionable for other moral reasons. But the reintroduction of a species "just to have it back in its range" is not itself sufficient for anything other than self-serving reasons (including financial motives). The only truly defensible rationale has to do with ecological and evolutionary forces as exerted via natural selection and scores of ecological processes (Hutchinson 1965). It is the role that the reintroduced species will play in the "ecological theater and evolutionary play" that constitutes the fundamental scientific basis for reintroductions.

Having said this, we now go further. For thousands of years masons have constructed awe-inspiring stone structures that pit known physical principles against the most powerful of natural forces such as gravity, flood, hurricane, and earthquake. Consider even the crudest, arched stone bridges that can span some of the largest rivers or canyons and date back thousands of years. A single wedge-shaped stone—the keystone—makes it possible for the total assembly of stones, of incredible mass in and of themselves, to interact dynamically in defiance of natural forces. The key element is not just a key-stone properly placed but the dynamic interplay between opposing forces. It is the juxtaposition and complementarity of myriad ecological forces of natural selection that have both created and conserved biodiversity as we know it. And the critical unit for monitoring our success as restoration ecologists must rest totally on ecological units of measure. In this chapter we acknowledge the benefits of adopting ecosystem and landscape approaches to restoring large carnivores and other ecological processes. However, we contend that even these scales are insufficient because they do not consider the full breadth of abiotic and evolutionary processes that explain the diversity of life on earth. The embattled Everglades makes an ideal case for applying landscape and carnivore restoration from the biotic province perspective.

Southern Florida exemplifies the serious issues regarding regional conservation and ecosystem restoration. One might be deceived by looking at a composite map of present-day public conservation lands in southern Florida. With such a large percentage of this land in public ownership, the casual observer might be led to think that southern Florida's native biological diversity is being effectively conserved (Mann 1995). The Arthur R. Marshall Loxahatchee National Wildlife Refuge (LNWR) serves as a primary example of the biodiversity conservation issues in southern Florida. LNWR was protected first for water conservation and authorized as a national wildlife refuge (NWR) later. Though probably viewed by many visitors as a "magical place" with large alligators (*Alligator mississippiensis*) and flocks of wading birds, LNWR suffers from its role as a water storage facility and in its present form it simply exacerbates the deliberate fragmentation of the southern Florida landscape. The refuge namesake's nephew John Marshall, president and CEO of the Arthur R. Marshall Foundation, recently observed: "This refuge, as it exists today, is a good example of what Art Marshall deplored about man's rearrangement of the Everglades" (Wilson 1999).

Although the 700-km^2 LNWR is the second-largest NWR in the eastern United States, it plays an insignificant role in protecting the region's unique subtropical biota, especially mammals. Abusive and incompatible uses, political disregard of the larger biogeographic context, a focus predominately on water control—all have rendered this portion of the Everglades landscape virtually useless for large mammal conservation. None of Florida's endangered or threatened mammals exist on the refuge. Nor could they survive if we reintroduced them. Over a century of conservation efforts with a myopic focus on wetlands and wading birds has left LNWR flooded and isolated. The tree islands so important as mammalian refuges during times of high water are often underwater, their formerly diverse tropical forests now thickets of exotics and other species common to more northern swamps. Dikes and canals fragment the landscape. In combination, the vertical face from top of berm to bottom of the larger canals constitutes a barrier to movement for perhaps the majority of terrestrial vertebrate species. Wet season/dry season hydrological fluctuations that drive the dynamics of annual ecological cycles have been damped. The rarely flooded upland edges of the system are buried beneath urban development. And connections with kindred populations of flora and fauna have been severed.

Art Marshall was a visionary. Long before landscape ecology became a well-known term, he understood that little landscape pieces—even big landscape pieces—cannot survive in any ecologically meaningful way unless they are managed within the context of regional "evosystems." "Balkanization" is the term he used to call attention to the foolishness of setting aside chunks of land and calling them "preserves" (Wilson 1999). LNWR may be just one piece of southern Florida, but it illustrates what has occurred throughout the region. On a map of southern Florida's conservation lands, LNWR appears as one of many green polygons (USGS 1993). Although it is impressive (Mann 1995), the seamless green landscape so easily depicted on GIS maps is not so seamless, nor so natural, on the ground. LNWR epitomizes an ecosystem management and restoration approach that does not consider the historical and regional landscape context that is necessary for effective conservation of the ecological and evolutionary legacy of southern Florida and of other regions around the globe.

The Yardstick for Restoration

This brings us to the key issue that undergirds mammal reintroduction in southern Florida: Are we managing the landscape as a whole in such a way as to restore and maintain the region's biological integrity? The relevant question here is not whether we could or should reintroduce one species or another. It is not how the "sacred flagship" Florida panther (*Puma concolor coryi*) might be persuaded to live here. These are secondary problems. The real issue is whether the regional landscape will be allowed to support even a semblance of the full complement of indigenous biota over the long term. Only by addressing this larger concern can we frame meaningful questions of reintroduction. If the key issue in mammalian species restoration is making the landscape suitable for long-term survival of the full native biota, the first thing we need to ask is: What is unique about this region or biota? How should this system function? And why is it not functioning as expected?

Phrased another way: What is the measuring stick for restoring ecological integrity in southern Florida?

The Physiographic Perspective

The southern Florida environment has been molded by a set of unique bio-physical forces. The Gulf Stream bathes most of the terrestrial areas and then brings its warm water, warm air, and tropical biota northward nearly to Cape Canaveral. Coral reefs create barriers for sheltered bays, foundations for emerging islands, and white sand for building more land, not just for attracting tourists. The labyrinthine coastline creates tremendous opportunities for large amphibious creatures ranging from manatee and porpoise to seals and sharks.

In English the prefix "pen" is added to indicate that the new word is almost like the root word. Hence, "peninsular" means almost insular—that is, almost an island. Insular ecology is clearly distinct from that of mid-continental areas (Wallace 1876; Carlquist 1965; Lack 1976; Williamson 1981). The patterns of faunal and floral distribution on peninsulas are characteristically different from those of similar-size areas embedded in larger terrestrial landscapes. The relative abundance of breeding species of land birds, for example, decreases by at least 60 percent as one proceeds down the Florida peninsula. (See the classic figure in Robertson and Kushlan 1984.) This pattern is not unique to Florida's birds or to Florida. It is characteristic of peninsulas in general. The globally consistent tendency for the number of species to dwindle as one proceeds distally from a continent or large island through a peninsula is called "the peninsula effect" (Simpson 1940). During the Pleistocene interglacial, sea levels were higher, all of the present Everglades was under water, and the Miami Ridge was simply a northern extension of the Florida Keys. What we think of as part of a major landmass was once an archipelago of coral islands and oolitic limestone deposits in a tropical sea. The upland biota (especially that of the Lake Wales Ridge) thus evolved as island flora and fauna and today retains great affinity with the Antillean biota. During the peak of the last glaciation about 18,000 years ago, sea levels were very much lower and both the Gulf coastal corridor as well as the Atlantic coastal corridor provided the major routes for colonizing mammals.

From an ecological perspective, southern Florida is still very much a subtropical biological island in that it is isolated by water on three sides and colder temperate ecosystems on the fourth. In the late nineteenth and early twentieth centuries human activity (especially dredging the Caloosahatchee and St. Lucie canals) greatly exacerbated the islandlike ecology by limiting further colonization from continental source pools while denouncing those from the Antilles as "exotics." Today the peninsula is dissected by nearly a dozen canals, major highways with strip development, extensive blocks of intensive agriculture, and urbanization (Figure 16.1). The concept of "hyper-peninsularity" must now be accepted as reality. The subtropical tip of Florida is becoming a biotically depauperate archipelago of nonforested management units at best—and simply the sawgrass-dominated "River of Grass" at worst. Tragically, the simple historical and biophysical fact that southern Florida was characterized by subtropical moist forest is inexorably being forgotten (Figure 16.2b). It is only in the context of surrounding forest that the simple English term "glade" makes any sense whatsoever.

The adjective "insular" is increasingly appropriate in almost all of these contexts because the ecology of islands, whether surrounded by water or by a distinctly different habitat type, is significantly different from the ecology of similar unconfined areas. Just as Central Park is clearly an island of green-space in the sea of Manhattan concrete, virtually all of Florida's parks and protected areas are at risk of becoming islands of habitat in a sea of dense human occupation. The increasing insularity of our parks and preserves—and of southern Florida's peninsula tip itself—is significant because the principles

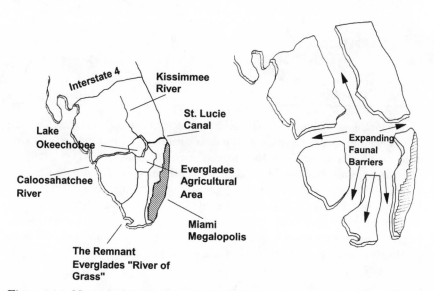

Figure 16.1. Not only does southern Florida exhibit hyperpeninsularity, but the landscape has been further fragmented by human influences that have created zones of absence and barriers to movement of native large mammals. Large mammal restoration in southern Florida will require relinking what have become virtual landscape islands.

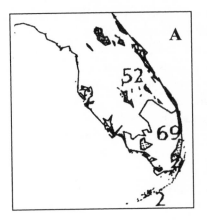

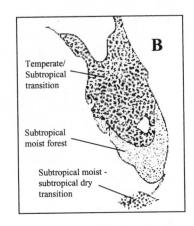

Figure 16.2. Four authorities depict the delineation of biological-ecological zonation as it pertains to southern Florida: *a*. The most coarse-grained depiction reflects loss of historical information where 69 = marsh/Everglades (from Ricketts et al. 1999). *b*. Holdridge life zones delineated with climatological data (from Dohrenwend and Harris 1975). *c*. First known phytogeographic delineation (Harshberger 1911a). *d*. Floridian faunal province as first published by Allen (1892). A principal concern is the increasing disregard for the tragic loss of North America's only subtropical moist forest. A combination of deforestation, severely reduced groundwater, and land use explains the dramatically different interpretations depicted in *(a)* and *(b)*.

of island biogeography dictate that ecological processes operate differently within such restricted landscapes. (See Harris 1984; Freemuth 1991; Forman 1995; Harris et al. 1996a, Harris et al. 1996b.) Insular ecology predicts, for example, that island forms of large-bodied vertebrates (such as the key deer [*Odocoileus virginianus clavium*]) tend to be smaller than their continental counterparts. Conversely, the smaller species of vertebrates (especially mammals) are commonly represented on islands by larger-than-usual forms. Flightlessness as well as other characteristics and ecological processes (or lack of them) operate differently than on continents. Local extirpation ("winking-out") is common on both oceanic and continental islands. Even in the presence of a source pool of colonists, biological communities of islands manifest a more dynamic nature than similar-sized communities on mainlands. Maintenance of colonization sources and processes are the essential keys to successful biodiversity conservation in insular situations.

The Biotic Province

The biotic province is a useful hierarchical unit from which to approach conservation at regional scales. Indeed, biotic provinces were delineated on the basis of biotic characteristics that reflect long-term differences among physical disturbance regimes as well as the relative strength of different ecological processes such as predation and volancy. Prior to 100 years ago explorers and naturalists had no trouble seeing distinctions in the composition of regional biota. For example, is there any biologically defined line on earth that is more significant than Wallace's Line (a narrow, deep-water passage that separates the Malay Archipelago into two distinct island biota, and that became a cornerstone of biogeography theory)? Prior to the biotic homogenization that is currently rampant, the earth's regional biota consisted of reasonably discrete assemblages of living things. For Creationists this indeed posed a major conundrum regarding God's different creations and or exactly how many Arks Noah landed (Browne 1983). It is only now that these early descriptions have become so critically important—because today we can look back at the descriptions of the eighteenth and nineteenth centuries as benchmarks for purposes of ecological restoration. These historical records enable us to base future management and restoration on the characteristic biota and coevolved ecological assemblages that distinguish biotic provinces and provide the historical and regional context to guide comprehensive conservation efforts.

Even though the earliest and most thorough of the biological collectors were botanists, malecologists, and ornithologists, mammals of the Everglades and Florida Keys did receive significant attention (Allen 1871, 1892; Schwartz 1952). Allen did not explore the Floridian province intensively, however, and Schwartz did not couch his work in the biogeographic context.

With the exception of Africa, humans and large mammals have simply not fared well in coexistence (apart from the domesticated species that furthered human welfare). In the Caribbean, for example, the arrival of humans

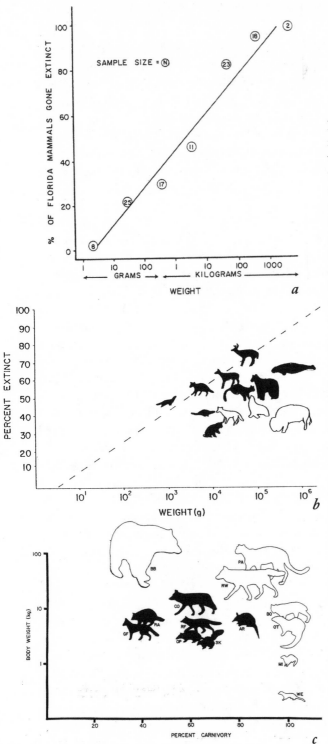

Figure 16.3. Large mammals, especially large carnivores, are particularly extinction prone: *a.* Generally, small Pleistocene mammals are all that have survived European colonization in Florida (from Harris and Eisenberg 1989). *b.* Of the 12 native Florida mammals larger than 1 kg, 11 are extinct, listed as endangered, threatened, or species of special concern, or are on the CITES international list. Silhouettes of red wolf, monk seal, and bison signify extinct in Florida. *c.* Although body and home range size are the principal determinants of vulnerability, carnivory is also important. All animals portrayed in silhouette are rare or listed whereas middle-sized (solid black) omnivore populations are often overabundant (from Harris and Gallagher 1989). Species abbreviations are as follows: BB = black bear; PA = Florida panther; RW = red wolf; BO = bobcat; OT = river otter; MI = mink; WE = Florida weasel; GF = gray fox; RA = raccoon; CO = coyote; RF = red fox; OP = Virginia opossum; SK = striped skunk; AR = nine-banded armadillo.

was concurrent with extermination of perhaps 90 percent of all mammal species existing there at the time (Morgan and Woods 1986). Harris and Eisenberg (1989) have elaborated on this matter in some detail and conclude that on Antillean islands even as large as Puerto Rico, the extermination of land mammals has been total—in other words, there are no indigenous land mammals remaining on Puerto Rico. With respect to Florida, the same phenomenon pertains: 100 percent of the extremely large indigenous mammals that occurred there when humans arrived are extinct, at least 90 percent of the very large, about 85 percent of the large, and so on down to the very small (Figure 16.3a). Although there is no evidence that any mammal the size of bats and shrews has been exterminated by humans during the last 10,000 years, the pattern is very clear. Both body size and trophic characteristics dictate the correlation between human settlement and extermination of mammals in Florida. Indeed, Figure 16.3a illustrates the tight correlation between size and vulnerability to extinction and figure 16.3c offers the simple compelling conclusion. Largeness and total fidelity to carnivory seem to explain the precarious state of mammals throughout Florida, not just in the Floridian Biotic Province. Conversely, it is exceedingly clear that midsized omnivores are overwhelming the state to the point of causing epidemics of rabies and suburban nuisances.

Faunal Composition

Animal ecologists, as a rule, recognize more associations that transcend larger spaces than do plant ecologists (Whittaker 1962). Mammals of Serengeti National Park, for example, transcend dozens, perhaps scores, of plant communities and ecosystem types. In short, wolf-sheds are larger than watersheds. Thus when we speak of a region's "fauna" (which is not equal to wildlife; Echternacht and Harris 1993), we refer to the composite assembly of indigenous animal species that characterize either a distinct geological period of time or a distinct geographical unit of space and distinguish it from adjacent temporal or spatial areas. We confess that this is becoming progressively more difficult—which is precisely why the historical record is so important to restoration ecologists.

Chapman (1908) listed the following specific characteristics that he believed distinguished the Floridian Biotic Province: "(1) occurrence of . . . West Indian species, (2) evolution of . . . geographic races or nascent species, (3) continued existence of species otherwise extinct in North America, (4) presence of . . . western species . . . not found elsewhere on the Atlantic coast, and (5) great development of those communal gatherings . . . 'rookeries.'" Allen (1930:28) simply stated that "the [moist] tropical region reaches the [continental] United States only on the tip of Florida as far north as Lake Worth and the mouth of the Caloosahatchee River."

Because of wholesale changes in land use and the invasion of numerous exotic species, once-clear distinctions between the biotas of different regions are not so obvious now. For example, land use changes have encouraged inter-specific breeding of so many species pairs (yellow-shafted and red-shafted flickers, for instance) that the American Ornithological Union (AOU) simply gave up and cleaned the ledgers of about 10,000 taxa that were previously recognized as separate species (Mayr 1982). Lumping species provides a convenient index to the degree of biotic homogenization that is occurring around the globe.

Recently, managers introduced Texas cougars (*Puma concolor stanleyana*) into southern Florida in an attempt to correct a perceived loss of genetic interchange with the Florida panther. The extant population of the Florida panther occurs near the tip of the peninsula, however, and therefore had the least genetic contact with other cougar subspecies to the north and the west. Though inbreeding depression might be a long-term issue for the Florida panther, the questionable management of the genetic introgression efforts leads to legitimate questions about the future genetic composition of a southern Florida panther/cougar population. This nascent species, originally recognized as a distinct species by Charles Cory in 1896, may now have lost the unique identity that was a product of a peninsular, subtropical environment. Still, the major faunal realms remain sufficiently intact to argue against carelessly moving animals from one region to another. Overt translocation of biota is such an overpowering human compulsion that it may eventually destroy the validity of several principles of biogeography. The planet has not yet reached that hopeless extreme of homogenization, however.

Floral Composition

Numerous plant geographers have noted the tropical nature of southern Florida's vegetation and recognized the similarities among the floras of southern Florida, the Bahamas, and the West Indies (Schwarz 1901; Shelford 1963; Gleason and Cronquist 1964; Brown et al. 1980). Harshberger (1911a, 1911b) noted that a statistical comparison of the floras confirmed significant botanical similarities with respect to how many plants these areas had in common. He mapped a Bahaman phytogeographic region consisting of "the islands and cays (keys) which form the Bahaman Archipelago, together with the southern end of the Floridian Peninsula" (Figure 16.2c).

The Floridian Biotic Province

Acknowledging the evidence that southern Florida is profoundly influenced by Bahamian and Caribbean ecology, we have oriented this chapter around the restoration and preservation of the fauna of the Floridian Biotic Province (Allen 1892) (Figure 16.2d). The fauna of this province is unique because it originated through dispersals from a diverse combination of source areas and evolutionary forces found in insular and peninsular ecosystems. Animals that moved over land bridges from the Sonoran and Carolinian provinces mingled with tropical creatures that found their way here via sweepstakes dispersal from the West Indies. The fauna of the Floridian Biotic Province is the result of land bridges, the Gulf Stream and sweepstakes dispersal, and peninsular and insular ecology resulting in significant evolution, differentiation, and distinct southern Florida subspecies such as the key deer, Florida panther, and Everglades mink (*Mustella vison mink*).

This region is special because southern Florida's peninsular form, low latitude, and proximity to the warm Gulf Stream have provided a suitably tropical environment for species ranging from flamingos to mahogany—species that occur nowhere else in the United States. The southern Florida biota is also ecotonal in that it combines fauna from several sources. Although the plants come overwhelmingly from Antillean sources, some of the most common species are of temperate origin. Remarking on this ecotonal nature, Gleason and Cronquist (1964) note that the wax myrtle (*Myrica cerifera*) so common on Everglades tree islands is a temperate species whose "abundance here serves to emphasize the transitional nature of the Everglades region."

The amphibious nature of the landscape emphasizes this neither-here-nor-there theme. The land/water interface is not just a peripheral edge. It is the dominant feature of the landscape. Species that benefit from this edge include bottlenose dolphin (*Tursiops truncatus*), manatee (*Trichechus manatus latirostris*), Caribbean monk seal (*Monachus tropicalis*), red wolf (*Canis rufus floridanus*), gray wolf (*Canis lupus*), black bear (*Ursus americanus floridanus*), Florida panther, and even sharks (*Carcharhinus leucas* and *Scoliodon terraenovae*). This is a place where the distinction between marine and terrestrial fades. How does one categorize the habitat of the monk seal, the alligator, or the American crocodile (*Crocodylus acutus*)? The amphibious black bear moves from one habitat to another to take advantage of patchily distributed food resources (Maehr 1997a). The highest black bear biomass in North America likely occurred on the coastal fringes of Florida when bears assembled to eat turtle eggs. As the season changed, these same bears moved on to eat seagrapes (*Coccoloba uvifera*), pigeon plums (*Coccoloba diversifolia*), cocoplums (*Chrysobalanus icaco*), acorns (*Quercus* spp.), saw palmetto (*Serenoa repens*), cactus (*Opuntia* spp.), and pond apple (*Annona glabra*).

The habitat features themselves change roles with the seasons in this region where droughts and fires alternate with floods and hurricanes. The alligator hole is a lifesaver to aquatic species in drought; the hammock is a lifeboat to terrestrial species in flood.

Defining Characteristics

By the standards of the typical nineteenth-century naturalist or tourist, southern Florida was exotic. But this exotic nature was in fact the distinctive characteristic of this unique biotic province. These are the characteristics that define southern Florida:

- A subtropical Caribbean climate and, to a certain extent, Antillean biology

- An extraordinarily rich biota (including large mammals) composed of numerous Sonoran and Carolinian animals interacting with a smaller Antillean faunal component and a predominantly Antillean flora

- Extreme hyperpeninsularity resulting in islandlike landscape ecology processes

- An inherently ecotonal nature accented by the amphibiousness of species and habitats

- A primeval character created by ancient species recolonizing a recently reemerged landform

Conservation at the Biotic Province Scale

From an evolutionary and ecological perspective, the biotic province is the minimum landscape area for conservation and restoration planning, monitoring, and evaluation because it is geographically distinct with well-described communities. In most instances, a range of mountains or a large river provides the basis for the floral and faunal differentiation that distinguishes biotic provinces. In the extreme, such divisions may be as globally dramatic as Wallace's Line (Wallace 1876). Unlike many politically or arbitrarily defined regional boundaries, Wallace's Line was discovered, described, and differentiated on the basis of indigenous biota. In the case of southern Florida, it was the combination of peninsula effect, source pool, and geographic location in the western Antilles that created today's biota. Classification should reflect where things came from because a species or a community's biologically defining characteristics are determined by its origin. Biotic province definitions respect this necessity.

Both the biological community and the ecosystem are inadequate units for meaningful landscape-scale restoration simply because the areas they encompass are rarely sufficiently large to manifest and maintain the characteristics that distinguish them from other units. They cannot, for example, assure visits by those rare wandering creatures that play such an important role in bringing in new genetic material for natural selection. And although such systems may be validly addressed as distinct subunits, they are seldom extensive enough to have adapted to the vast forces to which they will surely be subjected over the long term (catastrophic fires, tropical cyclones, volcanic eruptions, cyclical climatic stresses). Global disturbances such as meteorite impacts and glaciation most certainly contribute to natural selection and evolution and transcend even the biotic province in expanse. We cannot chop the landscape into smaller bits and hope the biota will survive such massive disturbances.

We need to work with landscape units big enough to allow natural extirpation and potential recolonization. A single hurricane, for example, can eliminate a vulnerable endemic species—an argument frequently made for the Florida panther. Two or more severe storms in close succession can destroy the prospective source pools for recovery of rare indigenous biota. Severe droughts and catastrophic fires can have similar effects. Arguably it is these relatively unpredictable and random events that are truly important in shaping the biota's prominent characteristics. Harmful though such events may appear, fires, floods, freezes, and hurricanes are critical to natural selection and ecosystem functions and must be accommodated within our conservation strategies.

The biota of the faunal province reflects its selection-driven ability to withstand the biophysical forces that created the region's ecological character and continue to shape it. Hence, preserving the dynamics that maintain the flora and fauna of the biotic province is the best way to assure biological resilience. Although physical restoration of key faunal elements such as Florida panther and red wolf will be needed at some point, these actions pale in importance relative to restoring key landscape processes. As we speak of flora and fauna in this context, remember that we are not just talking about "vegetation" and "wildlife." Whereas vegetation refers to whatever green stuff grows there (usually described in terms of community types), flora specifically implies the complete array of plant species that belong in the region—and native flora means the full list of plants that have evolved and colonized the area and have become fully integrated components of its communities over the long term. Similarly, "indigenous fauna" must be distinguished from "wildlife" (Echternacht and Harris 1993 and Udvardy 1969 notwithstanding).

The currently fashionable way to speak of southern Florida's ecology is through reference to the "Everglades ecosystem." But is there really any such thing? Or is the Everglades actually an intrinsic part of something bigger? The familiar concept of "Everglades ecosystem" is not very useful in addressing faunal conservation. Knowledgeable scientists and decision makers know full well that the Everglades regional system that must be restored includes a dozen or more distinctly different habitat types, some of which may be adequately protected, but most of which are not. Many conservation leaders have become so obsessed with a contrived and glorified image of the Everglades "River of Grass" (Douglas 1947) that they have become blind to the origins and needs of the region's biological riches. Southern Florida's biodiversity was not spawned in square mile after square mile of flooded sawgrass. Sawgrass marshes are in fact almost desertlike in their low diversity and sparse fauna. The wonderful creatures that characterize this special region came from the Everglades' fluctuating amphibious edges, fire-maintained pineland flood refugia, and an extensive array of tropical hammocks and tree islands. Other animals evolved as dwellers of coral islands and shallow tropical seas and must be viewed in that context. Unless we step back and respect the Bahamian heritage of the biota, we will neglect such big and important species as the manatee, the monk seal, sea turtles, crocodiles, sharks, and porpoises.

The Danger of Regional Replumbing

We fear that it is the tree islands and rocklands—remnants of our only true subtropical moist forest —that are not only in greatest jeopardy but are least considered in current restoration planning. Present-day plans to "restore the Everglades" not only ignore how diverse southern Florida's tropical forests were, and how important they were to wildlife, but also how extensive they were. (See Campbell 1926; Shellfire 1963; Craighead 1974; Richardson 1977; Austin et al. 1977; Bailey 1980; Snyder et al. 1990; Myers and Ewel 1990.) The southern shore of Lake Okeechobee once supported a berm of tropical forest sloping northward into Florida's greatest pond apple swamp (Gleason and Cronquist 1964). These forests served as land bridges aiding terrestrial animal movement throughout southern Florida. Further, tree species such as pond apple, swamp dogwood (*Cornus foemina*), and cabbage palm use omnivorous fauna such as black bear for the dispersal of their seeds (Maehr 1997a).

These unique forests are ignored by today's land-and-water planners. Thus the already imperiled fauna of southern Florida faces a new threat presented under the comforting rubric of "ecological restoration." The U.S. Army Corps of Engineers proposes to modify the water delivery system in southern Florida through the greatest composite civil engineering project in U.S. history. As currently conceptualized, this regional replumbing will essentially drive subtropical moist forest off the North American continent. As federal and state authorities move to spend billions of dollars earmarked to restore the Everglades, their disdain of large-scale processes (other than water sheet flow) and their indifference to large mammals as key restoration components will assure that the results of restoration are no less artificial than today's denatured system. The pre-European Everglades was contained within a primordial landscape where coastal aborigines harvested the bounty of maritime ecosystems and where black bear and panther drove mutualistic and predatory relations with a host of plants and animals. Such relations—products of Antillean colonization, a unique forest, and its inhabitants—will not be vital components of this "restored" Everglades.

Putting Reintroduction in Context

Reintroduction of once-indigenous species into large parks and preserves is an increasingly important method of maintaining and restoring native biological diversity. But more efficacious (though arguably artificial) methods such as gene banks, botanical gardens, game ranches, and zoological parks will do the job if the biodiversity challenge is simply preserving the taxa we happen to recognize at this time. There are scores of moral, aesthetic, and recreational reasons—and an increasing array of futuristic technological means—for the production of designer genes and pharmaceuticals and the like. But reintroducing a species without an ecologically valid rationale is self-indulgence, not ecological restoration. Only when we look at a reintroduction in the context of the evolutionary, structural, and functional roles (or work) of the indigenous biota can we find objective justification that transcends human self-gratification. If our decisions are to be rooted in ecological integrity, they must relate to the functional roles the species contributed to the region's natural landscape or an even larger unit of ecological hierarchy such as the biotic province.

The first question we must ask is whether the species is really missing from the essential flora/fauna of the ecosystem. This is not a simple matter of determining that it was reported by the early explorers but now is gone. What we are concerned with is whether it is truly representative of the region's biota. Before we can ascertain whether an animal species is truly an authentic component of the regional biota, we must determine what biotic province the fauna of the place really represents. Only native-system functions—the provision of appropriate prey for predators, the enhancement of diversity by grazing, or some other ecological engineering—can constitute an ecologically defensible rationale for restoring a depleted biological community. Everything that is a genuine component of a biological community has at least one job to do within that community. It is in its unique ability to do this work efficiently that its ultimate ecological value lies. We propose that it is the work a species performs when occurring in naturally coevolved assemblies that constitutes the critical justification for its reintroduction or the restoration of a depauperized ecosystem.

Most of the work that wild, free-ranging species do is hard for humans to comprehend—let alone identify, quantify, or appreciate. Of course, there are a few obvious functions beneficial to humans like nitrogen fixation by bacteria. But how many people are familiar with the critical work that large alligators perform in deepening Everglades waterhole drought refuges for the survival of scores of other species? How many farmers go to the bank with calculations of the monetary value of the fertilizer they did not have to buy this year because of the nitrogen-fixing legume they planted last year? Intelligent naturalists have been pointing out for decades the valuable services wild creatures perform (Craighead 1968a, 1968b; Odum 1969, 1971), and ecologists are increasingly acknowledging that certain "ecological engineers" do in fact modify the ecosystem, enhance habitat value for other species, and promote overall biodiversity (Alper 1998). As a society, however, we just do not get it. But it is precisely these elusive, yet significant, forms of work that constitute a biologically meaningful rationale for species reintroduction and ecological restoration.

Total systems function at the level of natural selection. Evolution is the principal reason why we should consider reintroducing large, possibly troublesome, and sometimes dangerous animals back into the human-inhabited environment. Top carnivores are critically important to an ecosystem. Wolves control growth rates of trees through predation on moose (*Alces alces*) (McClaren and Peterson 1994). Jaguar (*Panthera onca*) and puma affect recruitment of tropical forest trees through predation on seedeaters (Terborgh 1988). We include large reptiles in our discussions of large mammals because they are so characteristic of—and important to—the southern

Florida biota. Here, the alligator replaces a large mammal as the top carnivore. Only when viewed from the perspective of natural selection and evolution can biologists truly justify why a 12-foot-long alligator should be allowed to prey unabashedly on beautiful birds, valuable fish, the neighborhood raccoons, or even someone's dog (which should have been on a leash). Predation by a thousand raccoons cannot equal predation by ten large alligators and Florida panthers. Just as different frugivores disperse the seeds of different plants, different predators eat different prey. With very few exceptions, we cannot expect another species to take over the ecological work of one that is gone. Raccoons used to be panther and red wolf food. The unnatural overabundance of these midsize omnivores now threatens whole suites of ground-dwelling species, including beach-nesting sea turtles.

What Needs Fixing?

The history of cooccurrence of modern humans and large mammals has been catastrophic for the latter except in those rare situations where the species have coevolved. Florida's mammals have been decimated by the same human forces that have nearly exterminated other large mammals throughout the world (Harris and Eisenberg 1989).

The tragic effects of human colonization in the West Indies are illustrative. Cuba, the largest of the West Indian islands (at 114,000 km^2 it is about the same size as the Florida peninsula), has lost 81 percent of its land mammal species. As expected, Caribbean extinction rates are clearly and strongly the inverse of island size. Thus smaller Puerto Rico has lost all its native land mammal species except for a few bats. Although biologists may debate why the last few individuals of each species died, there is no question that a virtually total blockage of recolonization from the larger islands to the smaller ones has been a factor in every case.

Modern conservation strategies are obviously biased: we discriminate against big animals and their big habitats. Despite 100 years of aggressive conservation programming, large-mammal extinctions continue as they did under the influence of our Amerindian predecessors. Big carnivores and other megafauna always get in trouble around people. This is dramatically true in twenty-first-century southern Florida, where a series of mostly human-made landscape barriers prohibit movements essential to the survival of large land mammal populations (Figure 16.1).

As late as 10,000 years ago, when humans were first colonizing peninsular Florida, the biotic province still consisted of nearly 100 species of large mammals (Harris and Eisenberg 1989). Of these, about 95 percent of the very large species are extinct and nearly 80 percent of the large species (wolf and larger) have been lost. Whereas about 45 percent of mammals the size of bobcats (*Lynx rufus*) are gone, we are missing only 30 percent of the mink-size creatures. Clearly the long-term relation between body size and extinction of land mammals from Florida is very linear and very strong (Eisenberg and Harris 1989). This is not appreciably different from what has taken place on the neighboring true islands of the West Indies. Despite the peninsula's connection to North America, Florida has lost most of its large mammal species, especially at the southern tip of the peninsula. Hyperpeninsularity has turned southern Florida into a virtual island.

This trend has continued through recent times. John Bartram, the king of England's commissioned naturalists, and his son William surveyed Florida natural history about 200 years ago. At that time there were at least 12 species of native land mammals larger than about 2.5 kg. As the record stands, three of these (bison, monk seal, and red wolf) have been at least locally extirpated. Three more of the largest (manatee, Florida panther, and key deer) are federally listed as endangered and an additional three (black bear, bobcat, and river otter [*Lutra canadensis*]), are listed by the state of Florida or the Convention on International Trade in Endangered Species (CITES). Only 2 of these original 12 mammals—white-tailed deer and raccoon, which happen to be good at coexisting with humans and exploiting our habitat alterations and food subsidies—are doing well in Florida today.

Predators and their prey, plants and their pollinators, and countless other species combinations interact in ways vital to each other's survival. Self-sustaining, naturally functioning populations of these species cannot be established unless they are reintroduced in appropriate suites of interacting species. Returning just one species to its original habitat will not work if other critical evolutionary relations are absent. There is little point to the fanfare and public dollars spent on Florida panther recovery, for example, unless southern Florida's forests are also restored. It is well known that *Puma concolor* is a forest obligate in Florida—unlike its western relatives that can utilize canyons and other treeless expanses as cover (Maehr and Cox 1995; Maehr 1997b). Its future is tied to our ability to move beyond the treatment of symptoms related to a small population. Genetic restoration via the introduction of female Texas cougars is all well and good. But restoration fails to consider the long-term evolutionary conditions under which the Florida panther evolved—and will need if it is to survive outside of an artificially controlled pedigree. For instance, the panther was once connected to other cougar subspecies and other biotic provinces through uninterrupted forest.

The ebb and flow of genetic material through peninsular Florida filtered through selective forces found nowhere else on earth. Airlifted cougars from the desert do not replace the process whereby the panther became the most distinct cougar subspecies in North America (Young and Goldman 1946). Similarly, a "restored" Everglades ecosystem without the forests that integrated it into southern Florida and buffered it from the physical forces of wind and tide will not exist as a truly reintegrated regional landscape with its full complement of interacting native biota.

Conclusions

All this will be expensive. But restoration efforts such as the rechanneling of the Kissimmee River and the construction of wildlife underpasses for highway-crossing panthers suggest that the engineering savvy exists and the efforts are supported by the public.

Ecological restoration, including species reintroduction, must occur within the context of regional landscape-level planning and design. The biotic province is the key for assessing, designing, and monitoring efforts to restore and maintain ecological integrity. Large-mammal reintroductions must take into account the history, landscape context, ecological relations and interactions, and human-induced landscape transformations. The biotic province is the yardstick for assessing the significance of ecological changes in the present landscape within the context of the region's composition and distinctiveness in the recent past—and thus can provide direction for restoring ecological integrity in the near future.

So far, Everglades "restoration" is a great example of how not to proceed. To focus myopically on just one component—the "River of Grass"—of a once diverse and tropically distinct North American biotic province ignores the importance of large-scale ecological functions, tropically influenced communities, and large-bodied species to the biological diversity and ecological health of the entire southern Florida landscape. Ecological restoration in southern Florida, we contend, should embrace the distinctiveness of its physiographic and historical biogeographic setting. Such a comprehensive and integrative restoration approach would include several key elements: aggressive restoration efforts of north-south forests on both the western and eastern fringes of the Everglades and restoration of the east-west pond apple forest corridor on the southern fringe of Lake Okeechobee; hydrologic restoration that maintains or restores berm corridors and tree islands throughout the Everglades; reestablishment of healthy populations of the Florida panther and Florida black bear throughout intact or restorable

expanses in the southern Florida landscape; and consideration of the reintroduction of red wolves and monk seals.

We look forward to a future southern Florida landscape where the Florida
panther continues to evolve its tropical affinities and exert evolutionary influence on prey species including southern Florida's diminutive, peninsular-
influenced white-tailed deer. We also look forward to a future southern
Florida landscape where panthers, wolves, and alligators once again control
populations of weedy mesomammals such as the raccoon, opossum, and
armadillo, a landscape where the Florida black bear again occurs on the
beaches and coastal barrier islands to eat sea turtle eggs and disperse the seeds
of characteristic subtropical and tropical plants.

We have no quarrel with those who engage in regional or continental conservation planning. There are many reasons—most of them political—why
operating at smaller or contrived scales may be a wise and practical choice.
But we hope that those who do so will use biogeography criteria in selecting
the locations, boundaries, and characteristics to be used in assessing their
achievements and failures. Conservation plans are often based on smaller
ecological and biogeographic units: communities, ecosystems, watersheds,
and the like. These otherwise convenient landscape divisions are rarely extensive enough, however, to accommodate the full gamut of phenomena that
drive natural selection and evolution.

Acknowledgments

We thank O. L. Bass, E. D. Land, and F. W. King for constructive criticism
and thoughtful discussions of many of the ideas that were wrestled with in
this chapter.

Literature Cited

Allen, A. A. 1930. *The Book of Bird Life*. New York: Van Nostrand.

Allen, J. A. 1871. On the mammals and winter birds of East Florida, with an examination of certain assumed specific characters in birds, and a sketch of the bird-
fauna of eastern North America. *Bulletin of the Harvard Museum of Comparative
Zoology* 2:161–450.

———. 1892. The geographic distribution of North American mammals. *Bulletin of
the American Museum of Natural History* 4:199–243.

Alper, J. 1998. Ecosystem "engineers" shape habitats for other species. *Science*
280:1195–1196.

Austin, D. F., K. Coleman-Marois, and D. R. Richardson. 1977. Vegetation of southeastern Florida. *Florida Scientist* 40:331–361.

Bailey, R. G. 1980. *Description of the Ecoregions of the United States.* Miscellaneous Publications 1391. Ogden: U.S. Forest Service.

Brown, D. E., C. H. Lowe, and C. P. Pase. 1980. A digitized systematic classification for ecosystems with an illustrated summary of the natural vegetation of North America. General Technical Report RM-73. Fort Collins: U.S. Forest Service.

Browne, E. J. 1983. *The Secular Ark: Studies in the History of Biogeography.* New Haven: Yale University Press.

Campbell, D. H. 1926. *An Outline of Plant Geography.* New York: Macmillan.

Carlquist, S. 1965. *Island Life.* Garden City: Natural History Press.

Chapman, F. 1908. *Camps and Cruises of an Ornithologist.* New York: Appleton.

Cory, C. B. 1896. *Hunting and Fishing in Florida, Including a Key to the Water Birds Known to Occur in the State.* Boston: Estes and Lauriat.

Craighead, F. C., Sr. 1968a. The role of the alligator in shaping plant communities and maintaining wildlife in the southern Everglades. Pt. 1. Florida Naturalist 41:2–7.

————. 1968b. The role of the alligator in shaping plant communities and maintaining wildlife in the southern Everglades. Pt. 2. Florida Naturalist 41:69–74.

Craighead, F. C. S. 1974. Hammocks of south Florida. In P. J. Gleason, ed., *Environments of South Florida: Present and Past.* Vol. 2. Miami: Miami Geological Society.

Dohrenwend, R. E., and L. D. Harris. 1975. A climatic change impact analysis of peninsular Florida life zones. In *Impacts of Climatic Change on the Biosphere.* Vol. 2, pt. 2: Climatic Impact. Washington, D.C.: Climatic Impact Assessment Program, U.S. Department of Transportation.

Douglas, M. S. 1947. *The Everglades: River of Grass.* 3rd ed. Miami: Hurricane House.

Echternacht, A. C., and L. D. Harris. 1993. The fauna and wildlife of the southeastern United States. In W. H. Martin, S. G. Boyce, and A. C. Echternacht, eds., *Biodiversity of the Southeastern United States: Lowland Terrestrial Communities.* New York: Wiley.

Eisenberg, J. F., and L. D. Harris. 1989. Conservation: A consideration of evolution, population, and life history. In D. Western, and M. Pearl, eds., *Conservation for the Twenty-First Century.* New York: Oxford University Press.

Forman, R. T. T. 1995. *Land Mosaics: The Ecology of Landscapes and Regions.* Cambridge: Cambridge University Press.

Freemuth, J. C. 1991. *Islands under Siege: National Parks and the Politics of External Threats.* Lawrence: University Press of Kansas.

Gleason, H. A., and A. Cronquist. 1964. *The Natural Geography of Plants.* New York: Columbia University Press.

Harris, L. D. 1984. *The Fragmented Forest.* Chicago: University of Chicago Press.

Harris, L. D., and J. F. Eisenberg. 1989. Enhanced linkages: Necessary steps for success in conservation of faunal diversity. In D. Western, and M. Pearl, eds., *Conservation for the Twenty-First Century.* New York: Oxford University Press.

Harris, L. D., and P. B. Gallagher. 1989. New initiatives for wildlife conservation: The need for movement corridors. In G. Mackintosh, ed., *In Defense of Wildlife: Preserving Communities and Corridors.* Washington, D.C.: Defenders of Wildlife.

Harris, L. D., T. S. Hoctor, and S. E. Gergel. 1996a. Landscape processes and their significance to biodiversity conservation. In O. E. Rhodes Jr., K. Chesser, and M. H. Smith, eds., *Population Dynamics in Ecological Space and Time.* Chicago: University of Chicago Press.

Harris, L. D., T. Hoctor, D. Maehr, and J. Sanderson. 1996b. The role of networks and corridors in enhancing the value and protection of parks and equivalent areas. In R. G. Wright, ed., *National Parks and Protected Areas: Their Role in Environmental Areas.* Cambridge, Mass.: Blackwell Science.

Harshberger, J. W. 1911a. *Map Showing the Distribution of Plants and the Phytogeographic Areas, Districts, and Regions of North America.* Leipzig: Wilhelm Engelmann.

————. 1911b. *Vegetation der Erde: Phytogeographic Survey of North America.* Vol. 13. New York: G. E. Stechert.

Hoctor, T. S., M. H. Carr, and P. D. Zwick. 2000. Identifying a linked reserve system using a regional landscape approach: The Florida Ecological Network. *Conservation Biology* 14:984–1000.

Hutchinson, G. E. 1965. *The Ecological Theater and the Evolutionary Play.* New Haven: Yale University Press.

Lack, D. 1976. *Island Biology.* Berkeley: University of California Press.

Maehr, D. S. 1997a. Comparative ecology of bobcat, black bear, and Florida panther in South Florida. *Bulletin of the Florida Museum of Natural History* 40:1–176.

————. 1997b. *The Florida Panther: Life and Death of a Vanishing Carnivore.* Washington, D.C.: Island Press.

Maehr, D. S., and J. A. Cox. 1995. Landscape features and panthers in Florida. *Conservation Biology* 9:1008–1019.

Mann, C. C. 1995. Filling in Florida's gaps: Species protection done right? *Science* 269:318–320.

Mayr, E. 1982. *The Growth of Biological Thought.* Cambridge: Harvard University Press.

McLaren, B. E., and R. O. Peterson. 1994. Wolves, moose, and tree rings on Isle Royale. *Science* 266:1555–1558.

Morgan, G. S., and C. A. Woods. 1986. Extinctions and the zoogeography of West Indian land mammals. *Biological Journal of the Linnean Society* 28:167–203.

Myers, R. L. and J. J. Ewel, eds. 1990. *Ecosystems of Florida.* Orlando: University of Central Florida Press.

Odum, E. P. 1969. The strategy of ecosystem development. *Science* 164:262–270.

———. 1971. *Fundamentals of Ecology.* 3rd ed. Philadelphia: Saunders.

Richardson, D. R. 1977. Vegetation of the Atlantic coastal ridge of Palm Beach County, FL. *Florida Scientist* 40:281–330.

Ricketts, T. H, E. Dinerstein, D. M. Olson, C. J. Loucks, W. Eichbaum, D. Del-laSala, K. Kavanagh, P. Hedao, P. T. Hurley, K. M. Carney, R. Abell, and S. Walters. 1999. *Terrestrial Ecoregions of North America: A Conservation Assessment.* Washington, D.C.: Island Press.

Robertson, W. B., and J. A. Kushlan. 1984. The southern Florida avifauna. In P. J. Gleason, ed., *Environments of South Florida Present and Past II.* Coral Gables, Fla.: Miami Geological Society.

Schwarz, G. F. 1901. *Forest Trees and Forest Scenery.* New York: Grafton Press.

Schwartz, A. 1952. The land mammals of southern Florida and the upper Florida Keys. Ph.D. dissertation. University of Michigan, Ann Arbor.

Shelford, V. E. 1963. *The Ecology of North America.* Urbana: University of Illinois Press.

Simpson, G. G. 1940. Mammals and land bridges. *Washington Academy of Science* 30:137–163.

Snyder, J. R., A. Herndon, and W. B. Robertson Jr. 1990. South Florida rockland. In R. L. Myers and J. J. Ewel, eds., *Ecosystems of Florida.* Orlando: University of Central Florida Press.

Terborgh, J. 1988. The big things that run the world—a sequel to E. O. Wilson. *Conservation Biology* 2:402–403.

Udvardy, M. D. F. 1969. *Dynamic Zoogeography with Special Reference to Land Animals.* New York: Van Nostrand Reinhold.

U.S. Geological Survey. 1995. *South Florida 1993.* Reston, Va.: U.S. Geological Survey.

Wallace, A. R. 1876. *The Geographical Distribution of Animals with a Study of the Relations of Living and Extinct Faunas as Elucidating the Past Changes of the Earth's Surface.* New York: Harper & Brothers.

Whittaker, R. H. 1962. Classification of natural communities. *Botanical Review* 28:1–239.

Williamson, M. 1981. *Island Populations.* Oxford: Oxford University Press.

Wilson, S. 1999. What would Art Marshall say today? Unpublished speech presented at annual Marshall-Loxahatchee National Wildlife Refuge Day. Palm Beach, Fla.: Arthur R. Marshall Foundation.

Young, S. P., and E. A. Goldman. 1946. *The Puma: Mysterious American Cat.* Washington, D.C.: American Wildlife Institute.

Large Mammal Restoration: Too Real to Be Possible?

DAVID S. MAEHR

Not since the scientific revelations of Alfred Wallace (1889) and Charles Darwin (1859) has there been such an upheaval in biological thinking. Certainly there have been great strides in understanding our living world, but the species concept, biogeography, ecology, and even the discovery of DNA are all products of the inspirational observations and thinking of these two evolutionary philosophers. Through much of the twentieth century, Ernst Mayr (1976) has tirelessly refined the Darwinian view of life on Planet Earth and has been a driving force in the growth of evolutionary biology. Aldo Leopold (1949), in writings that would become the underpinning of both wildlife management and conservation biology, repeatedly stressed the importance of evolution to the development of a land ethic—a concept that would also help define the discipline of landscape ecology (Forman 1995). He also presaged the creation of the keystone concept when he recognized the top-down role of large carnivores by noting that when "the larger predators are lopped off the apex of the pyramid, food chains . . . become shorter rather than longer." Such notions led to his "thinking like a mountain" and his observations of the destructiveness of deer herds in the absence of predation. Not since the founding of The Wildlife Society nearly six decades ago have the products of such thinking been the common vernacular of the field of wildlife management.

A quick glance in any text on wildlife management and conservation biology reveals little if any mention of evolution. Although the classic text, *Conservation and Evolution* (Frankel and Soulé 1981), is devoted entirely to the maintenance of perhaps the most important biological process of all, most wildlife managers seem to have lost sight of this concept among a forest of

345

single-species and habitat-related studies and programs. Even ecosystem management and the park paradigm (Harris et al. 1996; Maehr et al. 1999a) fall light-years short of an approach that seeks to conserve the process that gave rise to the diversity of life as we know it. Perhaps the challenge of conserving evolution is too tall an order for a discipline that has traditionally taken a reductionist approach to natural resources. Evolution requires not only time, but space—and plenty of both. As Eisenberg and Harris (1989:108) have noted: "In order to preserve larger, unique forms of distinct evolutionary lineages, large areas must be set aside." Perhaps wildlife managers with goals that are practical and within the realm of tradition (the Get Real school) will avoid disappointment as species continue to disappear and biodiversity recedes. Yet there are others who risk being labeled hopeless romantics in conjuring a vision of the future that includes evolution and the persistence of common descent by modification.

Most of the chapters in this book emerged from single-species research and management, yet they do not forsake the power of evolution. Indeed, it is the hopeless idealism of the authors that makes their message so relevant to modern biodiversity conservation and the future of wildlife management. They have made it clear that the efforts of individuals, organizations, agencies, and the species themselves—all in combination—are necessary in restoring evolutionary processes to the land. Whether or not each study was designed to measure the degree to which community processes were lost or returned, all give us insightful glimpses into the challenges, benefits, and pitfalls of large mammal restoration. Evolutionary overtones and community-shaping influences were brought to light in the chapters on wolves (*Canis lupus*), elk (*Cervus elaphus*), and big cats. The work that all of these species perform should be sufficient to justify ecological restoration efforts worldwide, despite the formidable challenges that exist in the Tropics.

Leopold (1949:224) noted: "A thing is right when it tends to preserve the integrity, stability, and beauty of the biotic community. It is wrong when it tends otherwise." Certainly it can be argued that this quote encapsulates what wildlife management is all about. Others might point out that beauty is in the eye of the beholder, and that "stability" is the opposite of evolution. An update of Leopold's idea—one that maintains its basic message—might be: "A thing is right when it tends to preserve the integrity of evolution. It is wrong when it tends otherwise." This notion would permit the inclusion of Forman's (1995:451) missing element in land planning—the inclusion of the "ethics of isolation" whereby "it is unethical to evaluate an area in isolation from its surroundings, or from its development over time." According to Levin (1999:6), "The essential constant is change: the balance of Nature describes a system far from equilibrium, alternating between periods of relative stasis and dramatic change." From a species perspective, such a view would compel managers and agencies to go beyond the maintenance of populations that can survive 100 or 200 years (Shaffer 1981). Indeed, it could lead to the creation of plans that would encourage ecologically acceptable development and drive the restoration of species and their landscapes. The work of evolution, after all, is a gradual process that is influenced by centuries only insofar as the centuries act as minimal units in a millennia-spanning process (Darwin 1859; Leopold 1949).

Even today, evolution is not universally accepted (or understood) by the public and its relevance to management is not yet an important issue among wildlife biologists. Yet our ability to manage a game species successfully, to understand the vulnerabilities of a pest species, and to husband domestic species properly depends largely on a detailed understanding of interspecies relations—one of the fuels of evolution (Thompson 1994). Although we may not fully grasp the great complexity of life that Darwin (1859) termed the "entangled bank," it is increasingly apparent that as it becomes less entangled, human life becomes less rich and probably less sustainable. In this sense, many large mammals are effective ambassadors for biodiversity conservation because they can serve as management umbrellas and their charisma generates popular support. As flagships, large mammals can serve as effective management targets if their habitat requirements are sufficiently broad, their spatial needs are extensive, and there is adequate funding to impel effective action.

Twenty years ago Jim Schortemeyer was southern Florida's guru on deer management in the Everglades. Nearly a century of water diversions, canals, dikes, and berms associated with agriculture and U.S. Army Corps of Engineers "reclamation" of useless wetlands had created a system that was very unfriendly to the region's most abundant large mammal: the white-tailed deer (*Odocoileus virginianus*). Water drainage, restriction of natural sheet flows, and exaggerated wet and dry cycles encouraged unseasonable flooding and destructive fires, facilitated damaging freezes, and gradually whittled away at the tree islands that were characteristic of the region. Today a typical summer tropical storm is likely to strand deer and other terrestrial wildlife on increasingly small and wet refugia. As many as 80 percent of the herd might be lost in a given summer. Even worse, the herd had declined by an order of magnitude from well over 5,000 to just over 100 from the early 1950s to the late 1990s (Loveless 1958; Florida Fish and Wildlife Conservation Commission, unpublished data). This was not just bad for the hunting public who had developed a rich tradition in this region. It threatened the very existence of a regional deer herd. Jim was a game manager by training and practice, yet he had retained an innovative eye toward wildlife problem solving. Over the span of a very few years, Jim and his Florida Game and Fresh Water Fish Commission colleagues planted bare-root live oak (*Quercus virginianus*), laurel oak (*Q. laurefolia*), cabbage palm (*Sabal palmetto*), mahogany (*Swietenia mahogani*), red maple (*Acer rubrum*), cypress (*Taxodium distichum*), and other native trees along and atop the very spoil banks that had been heaved up to hurry the Everglades water to the coast. The idea was to create a stopgap replacement for the incredible losses of forest caused by water management and to offer a life raft for deer in times of flood, fire, and freeze. In the meantime, Miami, Palm Beach, Fort Lauderdale, and the east coast megalopolis had consumed the Atlantic Coastal Ridge forest—once home to the bears, wolves, and panthers that ventured from this subtropical forest and into the marsh-dominated Everglades to hunt pond apple fruit and deer. Most of the region's forests were gone, and those remaining were on the decline.

A visit by airboat along the Miami Canal today would reveal not only the continued shrinkage and disappearance of natural tree islands but also the extent to which the bare-root seedlings have prospered. They are the dominant life forms on the islands that bear little resemblance to the spoil that was draglined out of the marsh more than half a century ago. Wetland birds of all description (including several listed species) utilize 20-m-tall live oak, cypress, and red maple to roost, nest, and consume vertebrate, arthropod, and mollusk prey. Deer seek shelter in the cover and consume the browse that is all but absent on eroding tree islands. Florida panthers use this linear forest during dispersal from the nearby Big Cypress Swamp and certainly consume deer that have sought its cover for other reasons. This new replacement forest is even impressive from outer space: a distinct, and remarkably wide, corridorlike swath on satellite-generated photos (USGS 1993).

Though it was only a modest effort to stem the tide of a landscape-scale problem, the Schortemeyer forest is now the topographic and biodiversity center of a region that will soon experience a multi-billion-dollar restoration. The new forest was created at the cost of a few thousand seedlings, airboat fuel, and the time of game managers who had no idea whether or not their efforts would benefit just deer. While they did not fully consider all the future benefits of their work, this ersatz forest is now a biodiversity magnet. Just as wolves are restoring ecological processes to Yellowstone via translocation and to Alberta through the restoration of landscape corridors, Jim Schortemeyer's forest restoration work has maintained species and ecological processes that might otherwise have disappeared. Whether Everglades restoration managers recognize the value of the Miami Canal forest is anyone's guess. At the very least, it is another example on a growing list of projects that has demonstrated the feasibility of large-scale ecological restoration—as well as the fact that agencies need not succumb to the "get real!" syndrome. Had Jim suggested 20 years ago that his plantings were designed to stem the loss of regional biodiversity, he would have been declared infirm. It would have been even greater folly to suggest future panther use.

It is relatively easy to make ecologically based arguments in support of large mammal restoration. As I noted elsewhere (Maehr 1999:1121):

> Wildlife restoration in North America has a 60-year history, with billions of dollars generated by the Pittman-Robertson Act and spent by the states on behalf of popular game species and their habitats (Kallman 1987). It is strictly a recent phenomenon that large predators have become the target of restoration. In Florida, the American alligator (*Alligator mississippiensis*) is once again near its pre-settlement abundance, a result not of its intrinsic ecological value but of a lucrative market for alligator products. Large carnivore restoration has grown beyond short-term economic rewards, as demonstrated by the amazing story that is unfolding in Yellowstone National Park. The gray wolf (*Canis lupus*) is back. Despite the controversy surrounding its return, the wolf's short tenure in this wilderness preserve has provided the evidence of the powerful top-down effects of predation. The coyote (*C. latrans*) has been reduced in number through direct interference competition, and large-ungulate carcasses are now available to scavenging common ravens (*Corvus corax*), bald eagles (*Haliaeetus leucocephalus*), and grizzly bears (*Ursus arctos*) fresh out of hibernation. The mountain lion (*Puma concolor*), coevolved scourge of North American deer, exerts constant tension on prey populations where both *Puma* and *Odocoileus* reside. In the absence of large carnivores, deer numbers reach plague-like proportions and mesomammals such as raccoons (*Procyon lotor*), gray fox (*Urocyon cinereoargenteus*), and Virginia opossum (*Didelphis virginiana*) become superabundant reservoirs of disease and songbird nest predation. The loss of such predatory work is the sound of evolution coming to a screeching halt.

The return of large mammals, by contrast, is the sound of life returning to artificially simplified landscapes.

Restoring large mammals, especially carnivores, is an entirely different matter when viewed from the perspective of human social values. A nongovernmental organization may wish to return a preserve to some vestige of a past era. Let's say the maintenance of such a system depends on some long-gone herbivore. But how will the neighbors of the preserve tolerate elk and bison in their gardens? A wildlife agency may see economic and political benefits to reestablishing huntable big game populations. But what will they do when deer and elk overrun the countryside and there are no states left to accept the surplus? What happens when the offspring of a reintroduced cougar kills a jogger? It is no wonder that many wildlife agencies sidestep large mammal restoration in favor of quality deer management and other paradigms that address simplified, usually predatorless, communities.

While some of us may fantasize that large mammal restoration is necessary to return evolutionary process to the landscape, such a notion is difficult to sell. The marketable reasons behind restoration are, basically, utilitarian (that is, for eventual harvest) and aesthetic (that is, wildlife viewing and vicarious enjoyment). Regardless of the motives for restoration, there are several hurdles to overcome. Yalden (1993) suggests that five criteria must be met before a reintroduction attempt can be considered:

- Conditions have improved since the species first was extirpated.
- Suitable reintroduction stock must be available.
- Sufficient numbers are available from the donor population.
- The proper age and gender classes are available for reintroduction.
- There are sufficient monetary and human resources to support the effort (including adequate law enforcement and postrelease monitoring).

To this list I would add a sixth critical element that was a common theme throughout nearly every chapter dealing with the restoration of a large mammal species:

- The public is sufficiently supportive to avoid direct conflict with reintroduced animals.

Carnivores are generally recognized as the most difficult species to restore because of the difficulties in reversing the human social landscape that caused extinction in the first place (Griffith et al. 1989; Stanley Price 1989; Yalden 1993). The cultural antipathy toward large, free-ranging, meat-eating animals is very apparent when there is the perception that they will hide behind houses and steal away children. Large ungulates such as elk and bison are often feared as highway hazards, disease reservoirs, and crop-eaters. This sort of resistance is not unusual. It has happened with wolves in Michigan (Yalden 1993), with panthers in Florida (Belden and Hagedorn 1993), and with elk in Kentucky (Maehr et al. 1999b). In these situations, vocal minorities opposed reintroduction and succeeded in constraining restoration efforts, especially for the two carnivores. But restoration attempts have proceeded despite pockets of negativism. Gray wolves in the American West (chapters 6 and 8 in this volume) and an exhaustive effort to return the grizzly bear to a portion of the northern Rocky Mountains (chapter 10) demonstrate that the proper landscape and the dogged efforts of individuals are keys to opening the crate and releasing a flash of fur into the forest.

In some areas, successful reintroduction may have more to do with release method than with ownership patterns and potential human conflicts. When a significant aspect of restoration decision making comes down to the choice of hard versus soft release, the planned reintroduction likely has a much higher probability of success than one that is made all but impossible by the NIMBY mentality, property rights issues, agency liability anxiety, and human overpopulation. Some landscapes are currently so denatured that the return of large mammals is simply fanciful. In other places, the cultural resistance is sufficient to prevent wildlife resource agencies from considering the possibility even if sufficient habitat exists. Several of these problems can be overcome if a species can be compelled to restore itself. These situations are probably the most effective antidote to the "get realists" who pervade the decision-making ranks of government agencies. If suitable habitat exists within striking distance of a nearby population, black bears (chapter 12) and Florida panthers (chapter 15) will get there on their own—so long as they don't get shot or run over in the process. If wolves are provided a habitat corridor that is reasonably free of human activity and artificial obstacles, entire segments of the landscape reinherit biotic processes and evolutionary relations in a matter of months (chapter 13). The advantage to encouraging natural colonization and range expansion—a process that is slow compared to translocation—is that agencies do not appear heavy-handed in their advocacy, they are absolved of some of the more obvious liability issues, and the local inhabitants (human and otherwise) have time to adjust to the returning neighbors. In some areas, particularly in the South, local human populations are already prepared for the return of certain creatures such as mountain lions because they believe they are already there (McBride et al. 1993). Indeed, the popular belief in the existence of entire populations of phantom predators is exemplified in Great Britain, where the queen has called out the Royal Air Force to hunt down the Beast of Bodmin Moor and the Surrey Puma is said to lurk in the pastoral hedgerow landscape of southwestern England. These animals, believed to be descendants of escaped and liberated pet cougars and leopards (*Panthera pardus*), are thought to number in the hundreds despite a complete lack of evidence (Maehr 2001). More important, perhaps this is evidence that the human species has subconsciously retained an evolutionary link to the predators that helped shape who we are today—the ghosts of predators past in the sense of Connell (1980).

If there is one overriding shortcoming of this book, it is our failure to find the cure for "get realism"—the widespread malaise that infects natural resource agencies around the world and stifles grassroots efforts to correct environmental mistakes. The examples in this book, however, hint at the possibilities. "Continental conservation" is a new buzzword that was the focus of a recent book (Soulé and Terborgh 1999) and a product of The Wildlands Project—an effort to reconnect and repredatorize natural areas across North America. Many of our agency stewards believe this is too tall a task in a world that is overpopulated with people who are sprawling across a fragmented and denatured landscape. But with the exception of the tremendous obstacles to tiger restoration in Sumatra (chapter 14) and a tangled web of genetic, disease, and social issues that affect wood bison (chapters 7 and 9), the other chapters present strong statements of optimism and success. Yet even the return of wolves in Yellowstone may be insufficient to convince skeptics that large mammal restoration is feasible, let alone needed. Perhaps it is viewed as too expensive or a diversion of scarce public funds that might be better spent on other social issues. And while many private landowners cry foul over the eco-zealotry of natural resource agencies that attempt to restrict development on their land, others have forged ahead with their own private initiatives to restore large-mammal communities. The bighorn (*Ovis canadensis*) is now reclaiming portions of its historic range through the privately funded efforts of the Turner Endangered Species Fund (chapter 11). Private/public partnerships such as this may be the future for many species that have experienced rangewide population constrictions.

Soulé and Terborgh (1999:200) observe:

> Humans and nature can coexist, but peaceful coexistence cannot come about under present conditions. The revival and survival of nature across North America will require the establishment of a network of large nature reserves. Large areas managed for biodiversity are needed to ward off a host of ecological pathologies. Through conservation-oriented management of extensive core and multiple-use areas, the vital abiotic and biotic processes that sustain biodiversity can be perpetuated.

Such an idea is hardly foreign to traditional wildlifers. The term "conservation biology" originated in the first sentence of the first article of the first issue of the *Journal of Wildlife Management* (Errington and Hamerstrom

1937), and The Wildlife Society's original statement of policy supported the continental conservation concept. Bennitt et al. (1937:2) noted that "wildlife management along sound biological lines is also part of the greater movement for conservation of our entire fauna and flora." Thus a new paradigm of regional reserve networks, replete with their complement of native large mammals, is simply a modern extension of our professional roots. It demands that we recognize the dangers of growing consumerism and continued environmental exploitation. It demands that restoration efforts go beyond individual game species and isolated preserves. And it demands that our profession move in the direction that Aldo Leopold and the founders of The Wildlife Society so eloquently encouraged us to take a half-century ago. We may laud the state of Kentucky for its success in returning two of its largest land mammals to huntable herds, but is that a sufficient measure of success? There are many who believe that such restoration is incomplete without the very real force of carnivore predation wielding its evolutionary club on deer and elk. The stories told throughout this book are blueprints for restoration and compelling lessons that will help us all "get real."

Literature Cited

Belden, R. C., and B. W. Hagedorn. 1993. Feasibility of translocating panthers into northern Florida. *Journal of Wildlife Management* 57:388–397.

Connell, J. H. 1980. Diversity and the coevolution of competitors, or the ghost of competition past. *Oikos* 35:131–138.

Darwin, C. 1859. *On the Origin of Species by Means of Natural Selection, or the Preservation of Favoured Races in the Struggle for Life.* London: J. Murray.

Eisenberg, J. F., and L. D. Harris. 1989. Conservation: A consideration of evolution, population, and life history. In D. Western and M. Pearl, eds., *Conservation for the Twenty-First Century.* New York: Oxford University Press.

Forman, R. T. T. 1995. *Land Mosaics: The Ecology of Landscapes and Regions.* Cambridge: Cambridge University Press.

Frankel, O. H., and M. E. Soulé. 1981. *Conservation and Evolution.* Cambridge: Cambridge University Press

Griffith, B., J. M. Scott, J. W. Carpenter, and C. Reed. 1989. Translocation as a species management tool: Status and strategy. *Science* 245:477–480.

Harris, L. D., T. S. Hoctor, and S. E. Gergel. 1996. Landscape processes and their significance to biodiversity conservation. In O. E. Rhodes Jr., R. K. Chesser, and M. H. Smith, eds., *Population Dynamics in Ecological Space and Time.* Chicago: University of Chicago Press.

Kallman, H., ed. 1987. *Restoring America's Wildlife 1937–1987: The First 50 Years of* the Federal Aid in Wildlife Restoration (Pittman-Robertson) Act. Washington, D.C.: U.S. Department of Interior, U.S. Fish and Wildlife Service.

Leopold, A. 1949. A Sand County Almanac. New York: Oxford University Press.

Levin, S. A. 1999. Toward a science of ecosystem management. Conservation Ecology 3(2):6. [online] URL: http://www.consecol.org/vol3/iss2/art6.

Loveless, C. M. 1958. The Everglades Deer Herd Life History and Management. Technical Bulletin 6. Tallahassee: Florida Game and Fresh Water Fish Commission.

Maehr, D. S. 1999. Book review of Continental Conservation: Scientific Foundations of Regional Reserve Networks. Wildlife Society Bulletin 27:1121–1124.

———. 2001. Restoring the large mammal fauna in the East: What follows the elk? Wild Earth 11(1): 50–53.

Maehr, D. S., L. D. Harris, and T. S. Hoctor. 1999a. Remedies for Denatured Biota: Using Carnivores to Restore Native Landscapes. Godollo, Hungary: Second International Congress on Wildlife Management.

Maehr, D. S., R. Grimes, and J. L. Larkin. 1999b. Elk restoration in Kentucky: Ecological and sociological perspectives. Proceedings of the Annual Conference of Southeastern Fish and Wildlife Agencies 53.

Mayr, E. 1976. Evolution and the Diversity of Life: Selected Essays. Cambridge, Mass.: Belknap Press.

McBride, R. T., R. M. McBride, J. L. Cashman, and D. S. Maehr. 1993. Do mountain lions exist in Arkansas? Proceedings of the Annual Conference of the Southeastern Association of Fish and Wildlife Agencies 47:394–402.

Shaffer, M. L. 1981. Minimum population sizes for species conservation. BioScience 31:131–134.

Soulé, M. E., and J. Terborgh. 1999. Continental Conservation: Scientific Foundations of Regional Reserve Networks. Washington, D.C.: Island Press.

Stanley Price, M. R. 1989. Reconstructing ecosystems. In D. Western and M. Pearl, eds., Conservation for the Twenty-First Century. New York: Oxford University Press.

Thompson, J. N. 1994. The Coevolutionary Process. Chicago: University of Chicago Press.

U.S. Geological Survey (USGS). 1993. South Florida. Denver: U.S. Geological Survey.

Wallace, A. R. 1889. Darwinism: An Exposition of the Theory of Natural Selection. London, Macmillan.

Yalden, D. W. 1993. The problems of reintroducing carnivores. In N. Dunstone and M. L. Gorman, eds., Mammals as Predators. Zoological Society of London Symposia 65. New York: Oxford University Press.

Contributors

BRUCE B. ACKERMAN
Florida Fish and Wildlife Conservation Comm.
100 8th Avenue SE
St. Petersburg, FL 33701

BASTONI
Sumatran Tiger Project
c/o Minnesota Zoo
1300 Zoo Boulevard
Apple Valley, MN 55124

EDWARD E. BANGS
U.S. Fish and Wildlife Service
100 North Park, Suite 320
Helena, MT 59601

PETER BANGS
Wildlife and Fisheries Program
School of Renewable Natural Resources
University of Arizona
Tucson, AZ 85721

WAYNE G. BREWSTER
U.S. National Park Service
Center for Resources
Box 168
Yellowstone National Park, WY 82190

WENDY M. BROWN
U.S. Fish and Wildlife Service
P.O. Box 1306
Albuquerque, NM 87103

CAROLYN CALLAGHAN
Central Rockies Wolf Project
Zoology Dept.
University of Guelph
Guelph, ON N1G 2W1

LUDWIG N. CARBYN
Canadian Wildlife Service
4999-98 Avenue
Edmonton, AB T6B 2X3

CARLOS CARROLL
Klamath Center for Conservation Research
P.O. Box 104
Orleans, CA 95556

JOSEPH L. CORN
Southeastern Wildlife Disease Study
College of Veterinary Medicine
University of Georgia
Athens, GA 30602

LOUIS CORNICELLI
Utah Dept. of Natural Resources
Division of Wildlife Resources
515 East 5300 South
Ogden, UT 84405

JOHN J. COX
Dept. of Forestry
University of Kentucky
Lexington, KY 40546

LINDA C. DUEVER
Conway Conservation, Inc.

P.O. Box 949
10952 County Route 320
Micanopy, FL 32667

BARBARA DUGELBY
Wildlands Project
HC 4 Box 144
Blanco, TX 78606

DANAH L. DUKE
Central Rockies Wolf Project
Biological Sciences
University of Alberta
Edmonton, AB T6G 1A6

JOSEPH A. FONTAINE
U.S. Fish and Wildlife Service
100 North Park, Suite 320
Helena, MT 59601

DAVE FOREMAN
The Wildlands Project
P.O. Box 13768
Albuquerque, NM 87192

NEIL FRANKLIN
Sumatran Tiger Project
c/o Minnesota Zoo
1300 Zoo Boulevard
Apple Valley, MN 55124

STEVEN H. FRITTS
U.S. Fish and Wildlife Service
P.O. Box 25486
Denver, CO 80225

JONATHAN W. GASSETT
Kentucky Dept. of Fish and Wildlife Resources
1 Game Farm Road
Frankfort, KY 40601

JOSEPH K. GAYDOS
Southeastern Cooperative Wildlife Disease Study
College of Veterinary Medicine
University of Georgia
Athens, GA 30602

ROY A. GRIMES
Kentucky Dept. of Fish and Wildlife Resources
1 Game Farm Road
Frankfort, KY 40601

LARRY D. HARRIS
Dept. of Wildlife Ecology and Conservation
University of Florida
Gainesville, FL 32611

M. HEBBLEWHITE
Central Rockies Wolf Project
Wildlife Biology Program
School of Forestry
University of Montana
Missoula, MT 59812

ERIC C. HELLGREN
Dept. of Zoology
Oklahoma State University
Stillwater, OK 74078

THOMAS S. HOCTOR
Dept. of Wildlife Ecology and Conservation
University of Florida
Gainesville, FL 32611

ROBERT HOWARD
14 Reno Place
Santa Fe, NM 87505

JACK HUMPHREY
237 Cartwright Drive
Richmond, IN 47374

MICHAEL D. JIMENEZ
U.S. Fish and Wildlife Service
100 North Park, Suite 320
Helena, MT 59601

MARK R. JOHNSON
Wildlife Veterinary Resources, Inc.
P.O. Box 10248
Bozeman, MT 59719

TIMMOTHY J. KAMINSKI
Central Rockies Wolf Project
Box 7973
Missoula, MT 59807

WAYNE KASWORM
U.S. Fish and Wildlife Service
475 Fish Hatchery Road
Libby, MT 59923

PAUL R. KRAUSMAN
Wildlife and Fisheries Program
School of Renewable Natural Resources
University of Arizona
Tucson, AZ 85721

KYRAN KUNKEL
Turner Endangered Species Fund
1123 Research Road
Bozeman, MT 59718

JEFFERY L. LARKIN
Dept. of Forestry
University of Kentucky
Lexington, KY 40546

RURIK LIST
The Wildlands Project
AP 98 Metetec 3
Mexico City, Mexico 52176

CURTIS M. MACK
Nez Perce Tribe
P.O. Box 1922
McCall, ID 83638

DAVID S. MAEHR
Dept. of Forestry
University of Kentucky
Lexington, KY 40546

JULIE A. MCCLAFFERTY
Conservation Management Institute
Virginia Polytechnic Institute and State University
203 West Roanoke Street
Blacksburg, VA 24061

REBECCA P. MEEGAN
Dept. of Forestry
University of Kentucky
Lexington, KY 40546

BRIAN MILLER
Denver Zoological Society
2900 East 23rd Avenue
Denver, CO 80205

KERRY M. MURPHY
U.S. National Park Service
Center for Resources
Box 168
Yellowstone National Park, WY 82190

CARTER C. NIEMEYER
U.S. Fish and Wildlife Service
1387 South Vinnell Way
Boise, ID 83709

REED F. NOSS
Conservation Science, Inc.
7310 NW Acorn Ridge
Corvallis, OR 97330

PHILIP NYHUS
Environmental Studies Program
Franklin and Marshall College
Lancaster, PA 17604

DAVID P. ONORATO
Oklahoma Cooperative Fish and Wildlife Research Unit and
Dept. of Zoology
Oklahoma State University
Stillwater, OK 74078

PAUL C. PAQUET
Central Rockies Wolf Project
Box 150
Mecham, SK S0K 2V0

JAMES A. PARKHURST
Dept. of Fisheries and Wildlife Sciences
Virginia Polytechnic Institute and State University
Blacksburg, VA 24061

DAVID R. PARSONS
Parsons Biological Consulting
8613 Horatio Place NE
Albuquerque, NM 87111

ZACK PARSONS
Turner Endangered Species Fund
HC 32, Box 191
Truth or Consequences, NM 87131

M. PERCY
Central Rockies Wolf Project
Biological Sciences
University of Alberta
Edmonton, AB T6G 1A6

MICHAEL K. PHILLIPS
Turner Endangered Species Fund
1123 Research Road
Bozeman, MT 59718

JAMES A. POWELL
Florida Fish and Wildlife Conservation Comm.
100 8th Avenue SE
St. Petersburg, FL 33701

ERIC ROMINGER
Dept. of Biology
University of New Mexico
Albuquerque, NM 87131

JOHNNA ROY
U.S. Fish and Wildlife Service
1387 South Vinnell Way
Boise, ID 83709

NATHAN H. SCHUMAKER
U.S. Environmental Protection Agency
Western Ecology Division
Corvallis, OR 97333

CHARLES C. SCHWARTZ
Interagency Grizzly Bear Study Team
U.S. Geological Survey
Northern Rocky Mountain Center
Bozeman, MT 59717

MIKE SEIDMAN
6236 South 10th Street
Phoenix, AZ 85040

CHRISTOPHER SERVHEEN
U.S. Fish and Wildlife Service
309 Main Hall
University of Montana
Missoula, MT 59812

JAMES L. SCHORTEMEYER
Florida Fish and Wildlife Conservation Comm.
566 Commercial Boulevard
Naples, FL 34104

Douglas W. Smith
U.S. National Park Service
Center for Resources
Box 168
Yellowstone National Park, WY 82190

Sriyanto
Sumatran Tiger Project
c/o Minnesota Zoo
1300 Zoo Boulevard
Apple Valley, MN 55124

Nancy L. Staus
Conservation Biology Institute
800 NW Starker Avenue
Corvallis, OR 97330

James R. Strittholt
Conservation Biology Institute
800 NW Starker Avenue
Corvallis, OR 97330

Sumianto
Sumatran Tiger Project
c/o Minnesota Zoo
1300 Zoo Boulevard
Apple Valley, MN 55124

Ronald Tilson
Sumatran Tiger Project
c/o Minnesota Zoo
1300 Zoo Boulevard
Apple Valley, MN 55124

John Waller
U.S. Fish and Wildlife Service
309 Main Hall
University of Montana
Missoula, MT 59812

DAVID WATSON
Canadian Wildlife Service
5320-122 Street
Edmonton, AB T6H 3S5

MOHAMMAD YUNUS
Sumatran Tiger Project
c/o Minnesota Zoo
1300 Zoo Boulevard
Apple Valley, MN 55124

Index